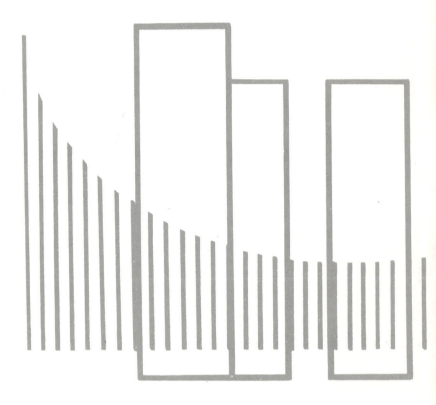

GUILLERMO OWEN

Associate Professor of Mathematical Sciences, Rice University

W. B. SAUNDERS COMPANY

Philadelphia • London • Toronto

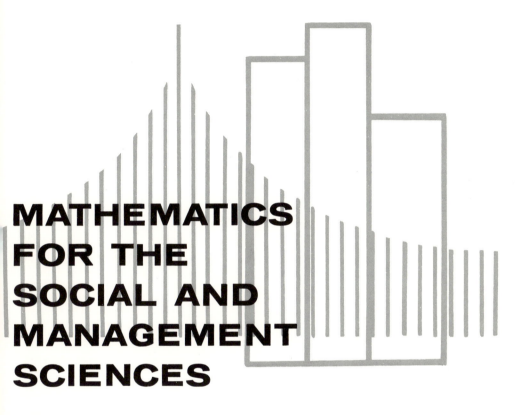

MATHEMATICS FOR THE SOCIAL AND MANAGEMENT SCIENCES

Finite Mathematics

W. B. Saunders Company: West Washington Square
Philadelphia, Pa. 19105

12 Dyott Street
London, WC1A 1DB

1835 Yonge Street
Toronto 7, Ontario

Mathematics for the Social and Management Sciences: *Finite Mathematics*   SBN 0-7216-7033-4

Print No.:   9   8   7   6   5   4   3   2

TO MY WIFE

PREFACE

This book was written to fill a need that has been evident to me for some time. There are, of course, many elementary books on finite mathematics available, some of them written specifically for the social and management sciences; there is none, however, that includes all the subjects in this book. These are subjects which should, in my opinion, be included. There are, in effect, certain topics here that are not found in other books at the level. Now, none of these requires complicated mathematics. Rather, they seem to belong to the "common-sense" school: they are applications to large problems of methods that are automatically used for small problems, and this because they are so obvious.

In general, I have tried to introduce the problem first, and have, with this motivation, developed the mathematics. It is my hope that the student will learn more easily in this manner. In some cases I have discussed alternative, but impractical, methods (e.g., enumeration of all extreme points) to show that common sense is not, in itself, sufficient: some practical experience is usually necessary.

Let us look at the book in some slight detail. Chapter I covers topics (systems of linear equations) that have almost certainly been seen before; it takes advantage of them to introduce the more complicated inequalities. Chapter II takes the topics of Chapter I and puts them in the more formal setting of linear algebra. Chapter III uses the results of the first two chapters, applying them to more practical problems.

Chapter IV again deals with subjects that have probably been seen before (sets and logic). These are then used to help in the formalism of Chapter V, which is almost certainly the most difficult one in the book.

Chapter VI, which depends on both III and V, introduces the reader to the important field of game theory. Chapter VII, which is almost independent of the others, develops some new but important methods. Finally, Chapter VIII is of importance in that it introduces the reader to graphical methods. The Appendix includes, for reference, some subjects that have probably been seen by the reader.

The logical connection among the chapters can be seen in the following table:

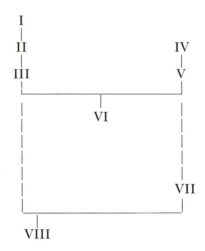

in which a solid line means that the lower chapter requires reading of the upper chapter, but a broken line means merely that knowledge of the upper chapter is helpful.

It is a pleasure to express my gratitude to those who have helped me in the writing of this book: to my wife, who has been most patient and has encouraged me throughout; to Fordham University, which contributed funds for the preparation of the manuscript; and finally, to Mrs. Adrienne DiFranco, who did some excellent typing and otherwise helped to prepare this manuscript.

G. O.

CONTENTS

CHAPTER IV

CHAPTER V

CHAPTER VI

CHAPTER I

ANALYTIC GEOMETRY

1. THE CARTESIAN PLANE

Among the intellectual achievements of the Greeks, their mathematics, and especially their geometry, must be given a place of honor. Thales, Pythagoras, Apollonius, Archimedes—to name but a few—have lent their names to important geometric ideas, while Euclid's *Elements* remains one of the fundamental works of classical antiquity.

Yet for all their excellence in geometry, the Greeks often found themselves baffled by problems that could be solved today by a schoolboy. The trouble lay, not in a lack of ingenuity, but rather in the fact that the ancients lacked one of the fundamental tools of modern mathematics.

It remained to René Descartes (1596-1650) to discover this tool. Descartes is primarily known for his philosophic works—the *Discourse on Method* and the *Meditations on Prime Philosophy*. While no one can deny the extent of his influence on modern philosophy, we would yet venture to say that his mathematical work will prove the more important and enduring.

Descartes' great discovery can best be appreciated if we consider the differences between the two mathematical sciences of geometry and algebra. Geometry deals with the relations between points and lines, while algebra deals with numbers. It is generally easy, when dealing with numbers, to decide what should be done with them; because this is not true of points and lines, algebraic problems have always been easier to solve than geometric problems. Descartes' discovery was, precisely, that it is generally possible to solve geometric problems by algebraic means. He showed that the set of all points on a line has a structure identical to that of the set of real numbers.

1

2

FINITE MATHEMATICS

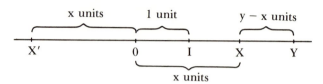

FIGURE I.1.1 The Cartesian line.

Consider, for example, a line (Figure I.1.1). On this line, two points may be taken arbitrarily (the only condition being that they be distinct points), and labeled O and I, respectively. If X is any other point on the line, we associate with X the number

1.1.1
$$x = \frac{\overline{OX}}{\overline{OI}}$$

which is the ratio of the lengths of the two line-segments, $\overline{OX}$ and $\overline{OI}$, with the stipulation that this ratio will be positive if X lies on the same side of O as I (so that $\overline{OX}$ and $\overline{OI}$ have the same direction), while it will be negative if X and I lie on opposite sides of O (so that $\overline{OX}$ and $\overline{OI}$ have opposite directions). The number x is called the *coordinate* of the point X. It is easy to see that the coordinate, x, is merely the distance $\overline{OX}$, measured in units of size $\overline{OI}$, so that the point O, called the *origin*, will have coordinate 0, while the point I will have coordinate 1.

We see, then, that each point on the line can be assigned a number, its coordinate. Conversely, given any positive number x, there are two points X and X' whose distance from the origin is equal to x units. One of these is on the same side of the origin as I and corresponds to x; the other is on the other side of the origin and will correspond to the number $-x$. In this way, each number corresponds to a unique point on the line. We have thus established a one-to-one correspondence between points on the line and real numbers. What is more, the structures of the two systems, in terms of operations that can be performed, are similar. To give an example, the distance $\overline{XY}$ between two points can be expressed in terms of their coordinates by

1.1.2
$$\overline{XY} = y - x$$

so that the geometric relation of distance reduces to the arithmetic operation of subtraction.

Dealing with the geometry of the plane, we find, however, requires a slightly more complicated procedure. It is, in fact, possible to give a one-to-one correspondence between the set of points in the plane and the set of real numbers, but this correspondence is not natural and does not preserve the structure of the system. Instead of assigning a number to each point, then, we assign a pair of numbers.

Consider, in the plane, two straight lines intersecting at a point O. The angle of intersection is not important, but for the sake of convenience, it is best to assume that they intersect at right angles (Figure I.1.2).

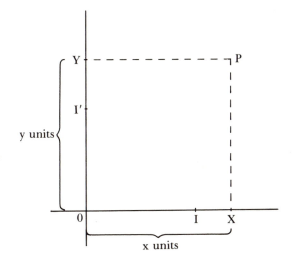

FIGURE I.1.2 The Cartesian plane.

On each of these lines, points I and I' are taken. Although the only need is that they be distinct from O, for convenience they are usually taken to be equidistant from O, and I' is usually obtained from I by a counterclockwise rotation through one right angle.

Consider, now, a point, P, in this plane. Through P, we may draw lines PX and PY, parallel to OI' and OI, respectively. The line PX meets OI at the point X, while PY meets OI' at Y. As in the one-dimensional case, just mentioned, we write

1.1.3
$$x = \frac{\overline{OX}}{\overline{OI}}$$

and

1.1.4
$$y = \frac{\overline{OY}}{\overline{OI'}}$$

where, once again, we agree to let the ratio of two line-segments be positive if they have the same direction, and negative if they have opposite directions.

The two numbers, x and y, given by (1.1.3) and (1.1.4) respectively, are called the *coordinates* of the point P. The number x is generally called the *abscissa*, and y the *ordinate*, of P. The lines OI

and OI' are called the coordinate *axes*; OI is the x-axis, and OI' is the y-axis.

We have thus assigned, to each point in space, a pair of real numbers (x,y), its coordinates. It may be pointed out that, since $OXPY$ is a rectangle, we must have $\overline{OX} = \overline{YP}$, while $\overline{OY} = \overline{XP}$. Thus x and y will be the perpendicular distances of the point P from the y-axis and x-axis, respectively, in terms of the common unit $\overline{OI}$ or $\overline{OI'}$.

2. GRAPHS AND EQUATIONS

We proceed now to study some of the advantages derived from the analytic treatment of geometry. Let us suppose that we are given a relation between two unknowns (or variables), x and y. This relation might be in the form of an equation, say,

1.2.1 $$y = x^2 + 3x$$

or of a word problem,

1.2.2 "x is not smaller than y, but not larger than twice y"

or in many other possible forms. There are necessarily certain pairs of values (x,y) for which the given relation (1.2.1) or (1.2.2) will be true and others for which it will not be true. If we consider the pairs of numbers for which the relation is true, we may plot the position of the corresponding points (i.e., the points having these pairs as their coordinates) on a coordinate plane. The set of these points is called the *graph* of the relation. Figures I.2.1 and I.2.2 show, respectively,

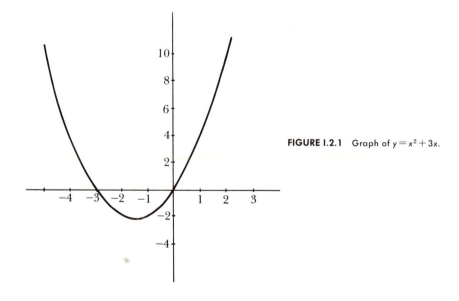

FIGURE I.2.1 Graph of $y = x^2 + 3x$.

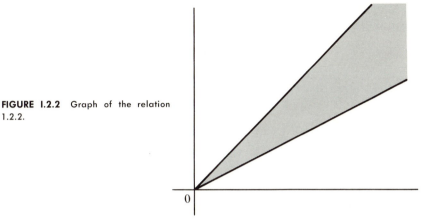

FIGURE I.2.2 Graph of the relation 1.2.2.

the graphs of the relations (1.2.1) and (1.2.2). For the equation (1.2.1), the graph is a curve (in this case, a *parabola*). As for the relation (1.2.2), we find that its graph consists of all the points inside a wedge with its angle at the origin.

Conversely, if we are given a curve in the plane, it is often possible to find a numerical relation that is satisfied by the coordinates of the points on the curve and by no others. If an equation, this relation is said to be the equation of the curve; (1.2.1) is the equation of the parabola shown in Figure I.2.1.

Relations between points can also be expressed analytically by this means. Consider, for example, two points, P and Q (Figure I.2.3), with coordinates (x_1,y_1) and (x_2,y_2) respectively. If we let R have the same ordinate as P, and the same abscissa as Q, we see that the lines

FIGURE I.2.3

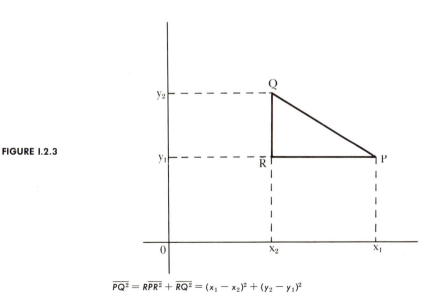

$$\overline{PQ}^2 = \overline{RPR}^2 + \overline{RQ}^2 = (x_1 - x_2)^2 + (y_2 - y_1)^2$$

PR and QR, being parallel to the two coordinate axes, must intersect at right angles, and thus, by Pythagoras' Theorem,

1.2.3
$$\overline{PQ}^2 = \overline{PR}^2 + \overline{QR}^2$$

We note, now, that $\overline{PR} = \overline{X_1 X_2}$ and $\overline{RQ} = \overline{Y_1 Y_2}$. Replacing these values in (1.2.3), introducing the coordinates, and taking square roots, we obtain

1.2.4
$$\overline{PQ} = \sqrt{(x_2 - x_1)^2 + (y_2 - y_1)^2}$$

the formula for *distance* between two points in the plane.

Consider, finally, the line PQ. It makes an angle, θ, with the x-axis. To find this angle, we note that it is the same as the angle between PQ and PR (since PR is parallel to the x-axis). From elementary trigonometry, we know that

$$\tan \theta = \frac{\overline{RQ}}{\overline{PR}}$$

Now, the tangent of the angle is called the *slope* of PQ (see Figure I.2.4). If we let m represent this slope, and introduce coordinates, we have

1.2.5
$$m = \frac{y_2 - y_1}{x_2 - x_1}$$

as the formula for slope. Although the slope is given in terms of the coordinates of P and Q, it is a property of the line PQ, and the formula (1.2.5) will give the same value if we substitute the coordinates of any two points on PQ. (Sometimes the slope is defined as a property of the two points, P and Q; it is then necessary—though easy—to prove that the slope is constant along a line.)

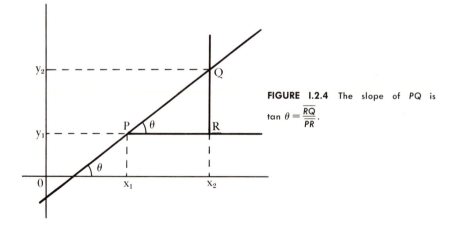

FIGURE I.2.4 The slope of PQ is $\tan \theta = \dfrac{\overline{RQ}}{\overline{PR}}$.

We point out that if a line is horizontal (parallel to the x-axis) its slope must be 0. In fact, points on such a line must have the same ordinate—and the numerator in (1.2.5) vanishes. On the other hand, if two points lie on a vertical line (i.e., parallel to the y-axis), then the denominator in (1.2.5) will vanish. In this case the slope does not exist, although it is sometimes said that the line has infinite slope.

We are now in a position to prove a few theorems concerning lines and their slopes.

I.2.1 Theorem. Two lines are parallel if and only if both have the same slope (or no slope at all).

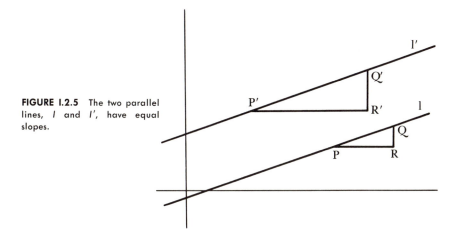

FIGURE I.2.5 The two parallel lines, *l* and *l'*, have equal slopes.

Proof (see Figure I.2.5). On the lines *l* and *l'*, take two pairs of points *P,Q* on *l* and *P'*, *Q'* on *l'*. If *l* and *l'* are parallel, the two triangles *PQR* and *P'Q'R'* are similar (having corresponding sides parallel) and so

1.2.6
$$\frac{\overline{RQ}}{\overline{PR}} = \frac{\overline{R'Q'}}{\overline{P'R'}}$$

Conversely, if (1.2.6) holds, the two triangles *PQR* and *P'Q'R'* are similar. Since *PR* and *P'R'* are parallel, this means that *PQ* and *P'Q'* are also parallel. This covers the case in which *l* and *l'* both have slopes. If the slopes do not exist, then both lines are parallel to the y-axis, and hence to each other.

I.2.2 Corollary. Through a given point *P* there passes one and only one line with a given slope *m*.

Proof. This follows from Theorem I.2.1 and the well-known Euclidean fact that through *P* there passes exactly one line parallel to the line with slope *m*.

I.2.3 Theorem. Let l and l' be lines with slopes m and m', respectively. Then l and l' are perpendicular if and only if $mm' = -1$.

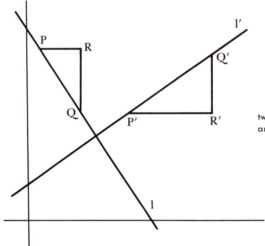

FIGURE I.2.6 The slopes of the two mutually perpendicular lines, l and l', are negative reciprocals.

Proof (see Figure I.2.6). Suppose l and l' are perpendicular. Let P, Q lie on l, while P', Q' lie on l'. Then PR and $R'Q'$ are perpendicular, RQ and $P'R'$ are perpendicular, and the triangles PQR and $Q'P'R'$ have equal angles and are therefore similar. This means that the ratios of corresponding sides are equal, if we disregard the difference in directions. It is not difficult to see, however, that the two slopes must have opposite signs. Thus we have

$$m = \frac{\overline{RQ}}{\overline{PR}} = -\frac{\overline{P'R'}}{\overline{R'Q'}} = -\frac{1}{m'}$$

and so $mm' = -1$. Conversely, if $mm' = -1$, then l' must be perpendicular to l, since there must be a perpendicular to l through P', and this perpendicular must have slope m'; l' is the only line through P' with slope m'.

3. LINEAR EQUATIONS

We now consider the equation of a line. The fundamental property of a straight line, from the point of view of analytic geometry, is that its slope is constant; we will use this property to develop the equation.

Consider, then, a line l passing through a point $P = (x_1, y_1)$ with slope m. (Corollary I.2.2 tells us there is exactly one such line.)

If $Q = (x,y)$ is any other point on this line, we know we must have

1.3.1
$$\frac{y - y_1}{x - x_1} = m$$

or, equivalently,

$$y - y_1 = m(x - x_1)$$

1.3.2
$$y = mx + y_1 - mx_1$$

as the equation of the line l.

Sometimes we are given two of the points on the line, rather than a point and the slope. In such cases we must first find the slope m, by using equation (1.2.5), and then substitute this value in (1.3.1) or (1.3.2).

Suppose that we are given two points, $P_1 = (x_1,y_1)$ and $P_2 = (x_2,y_2)$. The equation for the line passing through these two points, from (1.2.5) and (1.3.1), is seen to be

1.3.3
$$\frac{y - y_1}{x - x_1} = \frac{y_2 - y_1}{x_2 - x_1}$$

This equation is sometimes called the *two-point formula*, while (1.3.1) is the *point-slope formula*.

I.3.1 Example. Find the equation of the line passing through $(5,-1)$ with the slope $m = 2$.

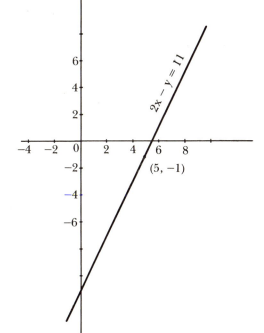

FIGURE I.3.1   Example I.3.1.

In this case, $x_1 = 5$, and $y_1 = -1$. Therefore, from (1.3.1)

$$\frac{y - (-1)}{x - 5} = 2$$

Multiplying through by the denominator, we obtain

$$y + 1 = 2x - 10$$

and, finally

$$y = 2x - 11$$

or, equivalently

$$2x - y = 11$$

1.3.2 Example. Find the equation of the line passing through (2,3) and $(-1,4)$.

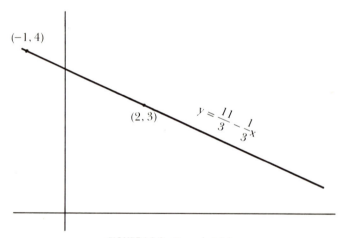

FIGURE I.3.2 Example I.3.2.

In this case, we must use the two-point formula (1.3.3). We have $x_1 = 2$, $y_1 = 3$, $x_2 = -1$, $y_2 = 4$, and so

$$\frac{y - 3}{x - 2} = \frac{4 - 3}{-1 - 2}$$

which, after simplification of the right-hand side, gives us

$$\frac{y - 3}{x - 2} = -\frac{1}{3}$$

or, proceeding as before,

$$y - 3 = \frac{2 - x}{3}$$

$$y = \frac{11}{3} - \frac{1}{3}x$$

which is the desired equation.

The formulas (1.3.1) and (1.3.3) fail when the desired line is parallel to the y-axis. In fact, we mentioned earlier that such a line does not have a slope (i.e., the slope does not exist). Equation (1.3.1) cannot be used since, indeed, it involves the slope. Moreover, two points on such a line are at an equal distance from the y-axis and thus have the same abscissa. This means that the denominator in the right-hand side of (1.3.3) vanishes; this formula, therefore, cannot be used.

Still we would like to have an equation for this sort of line. We can obtain it if we remember that on such a line all points have the same abscissa; if we let a be the common abscissa of all these points, the equation of this line will be

1.3.4 $x = a$

1.3.3 Example. Find the equation of the line passing through (3,5) and (3,−2).

We see that these two points have the same abscissa, so that the line is parallel to the y-axis. Thus by (1.3.4), the equation of the line must be

$$x = 3$$

Looking at the solutions of 1.3.1 to 1.3.3, we see that in each case the equations of the straight lines may be rearranged to obtain an equation of the form

1.3.5 $ax + by = c$

in which a, b, and c are constants. In fact, we may see from (1.3.2) and (1.3.4) that every line can be given an equation of the form (1.3.5). Such an equation, in which the variables appear only as first-degree terms (i.e., no higher powers or radicals appear, and no variables ever appear in a denominator) is called a *first-degree equation.*

We show, now, that every equation of the form (1.3.5), subject only to the condition that a and b cannot both be zero, is the equation of a straight line. Suppose we have an equation (1.3.5)

$$ax + by = c$$

If we assume that b is not zero, we can subtract the term ax from both sides and divide by b, thus "solving" for y:

1.3.6
$$y = \frac{c}{b} - \frac{a}{b}x$$

Now, assume $P = (x_1, y_1)$ and $Q = (x_2, y_2)$ are distinct points satisfying the equation (1.3.5). We see from (1.3.6) that x_1 and x_2 cannot be equal, since this would make y_1 and y_2 equal as well, meaning that the two points were not distinct. Since x_1 and x_2 are distinct, we can calculate the slope of PQ; it is, by (1.2.5) and (1.3.6),

$$\frac{y_2 - y_1}{x_2 - x_1} = \frac{\frac{c}{b} - \frac{a}{b}x_2 - \frac{c}{b} + \frac{a}{b}x_1}{x_2 - x_1}$$

which, on simplification, gives us

1.3.7
$$m = -\frac{a}{b}$$

The slope, m, of PQ, therefore, is constant for any two points satisfying (1.3.5). This is a characteristic property of straight lines: the graph of the equation must be a straight line; its slope, incidentally, is given by (1.3.7).

Suppose, now, that $b = 0$. Since we have assumed a and b cannot both be zero, it follows that a is different from zero. The equation (1.3.5) will now have the form

$$ax = c$$

and division by a gives us

$$x = \frac{c}{a}$$

so that the equation is satisfied by all points with the abscissa c/a; its graph is the line parallel to the y-axis at a distance c/a from it.

It is apparent that, in any case, the graph of (1.3.5) will be a line. For this reason, first-degree equations are also called *linear equations*.

1.3.4 Example. Find the equation of the line passing through $(1, -3)$ parallel to the line $x + 2y = 7$.

By Theorem I.2.1 the desired line must have the same slope as $x + 2y = 7$. Formula (1.3.7) gives us the slope; since $a = 1$ and $b = 2$,

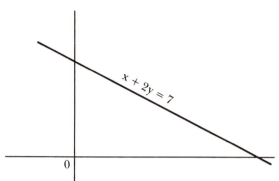
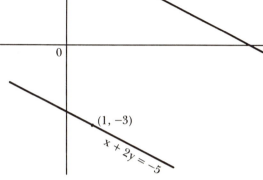

FIGURE I.3.3 Example I.3.4.

we have

$$m = -\frac{1}{2}$$

and the desired equation is

$$\frac{y - (-3)}{x - 1} = -\frac{1}{2}$$

which reduces to

$$y = \frac{-1}{2}x - \frac{5}{2}$$

or, equivalently,

$$x + 2y = -5$$

We note that, for these two lines, we have obtained equations in which the constants a and b (called the *coefficients* of x and y, respectively) are equal to 1 and 2, respectively. This is because the lines are parallel: (1.3.7) gives us the slope in terms of these coefficients only; thus, if the coefficients of x and y are the same in two linear equations, the two lines represented must have equal slopes and be parallel (unless they are the same). The converse of this is not quite true: if two lines are parallel, the coefficients in their equations need not be equal, although it is always possible to rewrite their equations in such a way that the coefficients will be the same (because one line may be represented by several equations).

I.3.5 Example. Find the line passing through (3,1) parallel to the line $2x + 3y = 7$.

We saw earlier that any line with equation $2x + 3y = c$ will be parallel to $2x + 3y = 7$ (unless $c = 7$, in which case the two lines are identical). We need only find a line with such an equation passing through $(3,1)$. It is easy to see that this happens if we set $c = 2 \cdot 3 + 3 \cdot 1 = 9$; the desired line has the equation

$$2x + 3y = 9$$

I.3.6 Example. Find the equation of the line passing through $(1,2)$ perpendicular to $x - 2y = 5$.

By (1.3.7), the line $x - 2y = 5$ has slope $m = 1/2$. According to Theorem I.2.3, the desired line must have slope $m' = -2$. Introducing this slope and the coordinates of the given point into (1.3.1), we have

$$\frac{y - 2}{x - 1} = -2$$

which simplifies to

$$2x + y = 4$$

I.3.7 Example. Find the equation of the line through $(-1,3)$ perpendicular to the line $y = 6$.

The line $y = 6$ can be seen to have slope $m = 0$. Hence the condition $mm' = -1$ cannot be satisfied, and it follows that the slope of the desired line does not exist, i.e., it must be parallel to the y-axis. (This can also be seen from the fact that $y = 6$ is parallel to the x-axis.) Hence the equation we want must be of the form $x = c$. Since the line passes through $(-1,3)$, the equation must be

$$x = -1$$

I.3.8 Example. Find the equation of the line through $(2,-2)$ perpendicular to the line $x = 2$.

In this case, since the line we are given does not have a slope, Theorem I.2.3 does not apply. We see, however, that the line is parallel to the y-axis; its perpendicular must be parallel to the x-axis and thus have slope 0. We therefore have

$$\frac{y - (-2)}{x - 2} = 0$$

This reduces to

$$y = -2$$

which is the desired equation.

In drawing the graph of a linear equation, it is sufficient to know two of the points. Since the graph is a straight line, this determines the line that can then be drawn by using a straight-edge. Drawing the graph is quite simple if two of the points are easy to obtain. Now, a linear equation such as (1.3.5) contains three arbitrary constants (parameters) that determine whether a given point lies on the line. We must generally take these three parameters, a, b, c, into consideration when plotting a point on the line.

On the other hand, there are cases in which only two of them need be taken into consideration. Suppose that we are looking for a point on the line (1.3.5) with ordinate equal to zero. Putting the value $y = 0$ into the equation, we obtain

$$ax = c$$

which, assuming a is not zero, reduces to

1.3.8
$$x_0 = \frac{c}{a}$$

which is the abscissa of the point of intersection of (1.3.5) with the x-axis, and is called the *x-intercept* of the equation.

Similarly, the ordinate of the point of intersection of (1.3.5) with the y-axis will be given by

1.3.9
$$y_0 = \frac{c}{b}$$

if b is not equal to zero. The value (1.3.9) is called the *y-intercept*.

We obtain, in this manner, two points $(x_0, 0)$ and $(0, y_0)$ on the line. Each of these depends only on two of the three parameters a, b, and c, and is, therefore, comparatively easy to find.

Two things might conceivably go wrong. One is that the intercepts need not both exist. In fact, if a line is parallel to one of the coordinate axes, it cannot intersect that axis. A line parallel to the y-axis will not have a y-intercept; one parallel to the x-axis will not have an x-intercept. Another possibility is that the two intercepts will both exist, but give us only one point. This happens when the line passes through the origin; in this instance, both intercepts are zero, and so the line intersects both axes at a single point. Since a single point does not determine a line, it follows that we cannot draw a line from its intercepts in these cases. This is really not a serious problem, but it does lessen the possibility of giving a systematic procedure for plotting lines.

We may be given a line in terms of its two intercepts. Each of these intercepts gives us a point so that, in the usual case (when the line does not pass through the origin), the two intercepts determine a line. If the two intercepts are x_0 and y_0, we know that the line passes through $(x_0,0)$ and $(0,y_0)$ and application of the two-point formula yields

$$\frac{y}{x-x_0} = \frac{y_0}{-x_0}$$

which reduces to

1.3.10 $$y_0 x + x_0 y = x_0 y_0$$

Formula (1.3.10) is known as the two-intercept formula. Note that if $x_0 = y_0 = 0$, (1.3.10) reduces to the identity $0 = 0$.

Another manner in which a line may be given is by one of the intercepts (generally the y-intercept) and the slope. In fact, if the slope is m and the y-intercept is y_0, we have, from (1.3.1),

$$\frac{y-y_0}{x} = m$$

which reduces to

1.3.11 $$y = mx + y_0$$

This is known as the slope-intercept formula.

1.3.9 Example. Find the equation of the line with x-intercept 3 and y-intercept -1.

Using (1.3.10), we obtain immediately

$$-x + 3y = -3$$

as the desired equation.

1.3.10 Example. Find the intercepts of the line $3x + 4y = 12$.

In this case we apply (1.3.8) and (1.3.9), obtaining $x_0 = 4$ and $y_0 = 3$ as the two intercepts.

1.3.11 Example. Find the intercepts of the line $3x = 6$.

Here the x-intercept is easily found to be $x_0 = 2$. The y-intercept, however, does not exist. (Some people would say that the y-intercept is infinite.)

1.3.12 Example. Find the equation of the line with y-intercept $y_0 = -4$ and slope $m = 2$.

Applying (1.3.11), we obtain $y = 2x - 4$ as the desired equation.

I.3.13 *Example.* Find the intercepts of $3x - 2y = 0$.

For this line, we find that both intercepts are equal to zero, i.e., the line passes through the origin. If we wish to plot the line, we need another point; this may be obtained by solving for y, which gives us

$$y = \frac{3}{2}x$$

Then $(1,3/2)$ or $(2,3)$ will be points on the line.

PROBLEMS ON LINEAR EQUATIONS

1. Find the equations of the lines described as follows:

(a) Passing through $(0,1)$ and $(2,5)$.

(b) Passing through $(1,3)$ and $(2,4)$.

(c) Passing through $(1,-2)$ and $(0,0)$.

(d) Passing through $(-1,-2)$ and $(-1,5)$.

(e) Passing through $(1,4)$ and $(6,4)$.

(f) Passing through $(2,5)$ with slope $1/2$.

(g) Passing through $(3,6)$ with slope 2.

(h) Passing through $(5,1)$ with slope 0.

(i) Passing through $(2,-3)$ with slope 1.

(j) Passing through $(5,1)$ with slope $1/4$.

(k) Passing through $(3,2)$ and parallel to $2x + 4y = 6$.

(l) Passing through $(0,0)$ and parallel to $x = 5$.

(m) Passing through $(1,4)$ and parallel to $x - 2y = 7$.

(n) Passing through $(2,-2)$ and parallel to $y = x + 2$.

(o) Passing through $(3,0)$ and perpendicular to $x + y = 5$.

(p) Passing through $(1,4)$ and perpendicular to $3x - y = 2$.

(q) Passing through $(6,2)$ and perpendicular to $x + 2y = 4$.

(r) Passing through $(2,1)$ and perpendicular to $2x - y = 10$.

(s) With x-intercept 3 and y-intercept -2.

(t) With x-intercept -1 and y-intercept -4.

(u) With y-intercept 6 and slope 3.

(v) With y-intercept 5 and slope -1.

(w) With y-intercept 2 and slope 0.

(x) With x-intercept -3 and parallel to $2x + y = 3$.

(y) With y-intercept 4 and perpendicular to $x - 2y = 2$.

ANALYTIC GEOMETRY

FINITE MATHEMATICS

2. Find the slopes, x-intercepts, and y-intercepts of the following lines; draw the lines on graph paper.

(a) $x + 3y = 6$

(b) $3x - 2y = 5$

(c) $x + 4y = 12$

(d) $-x + 2y = -3$

(e) $x - 4y = 16$

3. Find the distances between the following pairs of points:

(a) (5,1) and (6,1).

(b) (3,1) and (2,4).

(c) (1,4) and (3,1).

(d) (2,6) and (1,1).

(e) (1,4) and (−2,−4).

4. The three points (5,1), (2,4), and (1,0) determine a triangle. Find:

(a) The mid-points of the three sides of this triangle.

(b) The equations of the three medians of this triangle (a median is the line joining one vertex to the mid-point of the opposite side).

(c) The equations of the three altitudes of the triangle (an altitude is the line through one vertex and perpendicular to the opposite side).

4. INEQUALITIES

Up to this point, we have dealt throughout with equations; we shall deal now with the somewhat more general concept of inequalities.

The characteristic symbol of an equation is, of course, the equals sign =. We know its meaning, of course: if two expressions are connected by this sign, they are equal. We propose to deal now, with certain other signs; these are

$$\neq, <, \leq, >, \geq$$

The first sign, $\neq$, is simply an equals sign crossed out, and represents, therefore, the negation of equality. That is, the statement "$a \neq b$" is to be read as "a is different from b." Thus, "$3 \neq 2$" is a true statement, while "$5 \neq 5$" is a false statement.

The second sign, $<$, is to be read as "is less than." Thus, "$a < b$" is read "a is less than b." Since this is defined to mean that the difference $b - a$ is a positive number, the statements

$$2 < 5, -5 < 1, 0 < 3$$

are all true (note for instance that $1 - (-5) = 6$, which is positive), while the statements

$$3 < -7, 2 < 2$$

are both false.

The sign $\leq$ is to be read as "is less than or equal to." We define it by saying that the statement "$a \leq b$" is true whenever either of the two statements "$a = b$" and "$a < b$" is true. Thus the statements

$$0 \leq 3$$
$$2 \leq 2$$
$$-5 \leq 1$$

are all true, while the statement

$$3 \leq -7$$

is false.

The sign $>$ is read as "is greater than." It can best be defined by saying that the two statements "$a > b$" and "$b < a$" are equivalent, and therefore

$$5 > 1$$
$$-1 > -8$$
$$2 > 0$$

are all true, while

$$6 > 6$$
$$5 > 10$$

are both false.

Finally, the sign $\geq$ should be read as "is greater than or equal to." We define it by saying that "$a \geq b$" and "$b \leq a$" are equivalent statements. Examples of true statements are

$$6 \geq 2$$
$$-1 \geq -1$$
$$3 \geq 0$$

while the statements

$$-5 \geq 1$$
$$3 \geq 6$$

are both false.

In high school algebra, we are taught that there are many things that may be done with equations. One may add (or subtract) the same number to both sides of an equation; one may multiply both sides of an equation by the same number, or divide both sides by the same number (so long as this number is not equal to zero). Equations may also be added together, giving rise to new equations.

Inequalities are not equations, so we must be careful about treating them in the same manner. Nevertheless, with care, some of these same things may be done with inequalities.

We have seen that the inequalities are defined in terms of positive numbers. For instance,

1.4.1 $a < b$ means $b - a$ is positive.

1.4.2 $a \leq b$ means $b - a$ is positive or zero.

1.4.3 $a > b$ means $a - b$ is positive.

1.4.4 $a \geq b$ means $a - b$ is positive or zero.

It follows that, if we want to know which operations may legitimately be performed on an inequality, we must study the properties of positive numbers (as contrasted with negative numbers). These properties are simply as follows:

1.4.5   The sum of two positive numbers is positive.

1.4.6   The product of two positive numbers is positive.

1.4.7   The product of a positive and a negative number is negative.

With these properties established, we can look once again at the inequalities. First, we see that the same number can be added or subtracted to both sides of an inequality. Suppose that we know $a > b$. This means that $a - b$ is a positive number. For any number c, now, we have

$$(a + c) - (b + c) = a - b$$

so that $(a + c) - (b + c)$ is positive, and so

$$a + c > b + c$$

Thus, the inequality is preserved by the addition (or subtraction) of any number to both sides of the inequality. This would still be true if the inequality were of a different type.

Suppose, next, that we have two inequalities of the same type, say

$$a > b$$
$$c > d$$

This tells us that both $a - b$ and $c - d$ are positive. It means that $(a + c) - (b + d)$, as the sum of the two positive numbers $a - b$ and $c - d$, is also positive, and thus

$$a + c > b + d$$

It is apparent that two inequalities, if they are of the same type, can be added together. Note, on the other hand, that if two inequalities have opposite sense, such as $a > b$ and $c < d$, it is not possible to add them. Nor is it possible to subtract inequalities such as $a > b$ and $c > d$. In fact, we have no guarantee that the difference of two positive numbers will be positive.

Finally, suppose we are given the inequality $a > b$, which, we repeat, means that $a - b$ is positive. If c is a positive number, then $ca - cb = c(a - b)$ is the product of two positive numbers, and so is positive. Hence we have

$$ca > cb$$

and the inequality is preserved. If, on the other hand, c is negative, then $ca - cb$ is negative, and $cb - ca$ is positive, so

$$ca < cb$$

and the sense of the inequality is reversed. If $c = 0$, we find that $ca = cb$, and neither of the "strict" inequalities, $ca > cb$, or $ca < cb$, will hold; however, both the "loose" inequalities, $ca \geq cb$ and $ca \leq cb$, will hold.

We have only considered, here, inequalities of the type $a > b$. It may be seen, however, that similar considerations hold also for inequalities of the types $a < b$, $a \geq b$, and $a \leq b$. We thus obtain the following rules for the treatment of inequalities:

1.4.8   Addition of the same number to both sides of the inequality preserves the inequality.

1.4.9   Two inequalities may be added if they are of the same type, giving an inequality of the same type.

1.4.10   Multiplication of both sides of an inequality by the same positive number preserves the inequality.

1.4.11 Multiplication of both sides of an inequality by the same negative number *reverses* the inequality.

1.4.12 In rules (1.4.10) and (1.4.11), if the inequality is "loose" (i.e., of type $\geq$ or $\leq$), the words "positive" and "negative" may be replaced by "non-negative" and "non-positive" respectively.

We shall, in the future, deal mainly with loose inequalities of the form $a \leq b$ or $a \geq b$. The reason for this is that physical, practical constraints are generally of this form; the amount of goods produced by a factory can be positive or zero, but not negative, which leads to a constraint of the form $x \geq 0$, or the total number of hours worked by an employee cannot (in certain cases) be more than 8 hours, which leads to a constraint of the form $t \leq 8$. In both cases we see that the loose inequalities appear naturally.

We are now in a position to study the graph of a linear inequality. We can do this best by considering an example first. Let us take the inequality

$$3x + 2y \leq 6$$

which differs from a linear equation only in that the inequality symbol appears rather than the equals sign. Following the rules (1.4.8) to (1.4.12), we can solve for y in terms of x: first, we add $-3x$ to both sides:

$$2y \leq 6 - 3x$$

and then we multiply both sides by 1/2. Since 1/2 is positive, this preserves the inequality:

$$y \leq 3 - \frac{3}{2}x$$

If we consider, instead of the inequality $3x + 2y \leq 6$, the corresponding linear equation

$$3x + 2y = 6$$

(whose graph is shown in Figure I.4.1), and solve for y in terms of x, we obtain

$$y = 3 - \frac{3}{2}x$$

Suppose, now, that the point P, with coordinates (x_1, y_1) lies on this line, i.e., we have

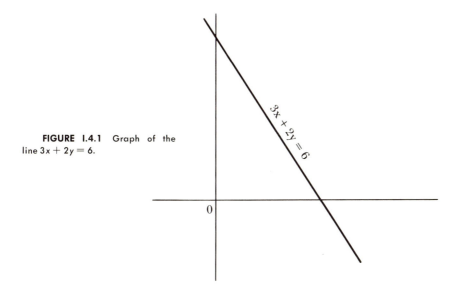

FIGURE I.4.1 Graph of the line $3x + 2y = 6$.

$$y_1 = 3 - \frac{3}{2}x_1$$

Let Q be another point with coordinates (x_1, y_2), such that $y_2 < y_1$. This can best be expressed by saying that Q lies *below* P (since it has the same abscissa but a smaller ordinate). We see then, that

$$y_2 < 3 - \frac{3}{2}x_1$$

so that the coordinates of Q satisfy the linear inequality. But P was an arbitrary point on the line, and Q was any point below P. We find, thus, that the linear inequality $3x + 2y \leq 6$ is satisfied by all the points *on* or *below* the line $3x + 2y = 6$. The converse is also true: if a point satisfies the given linear inequality, it must lie on or below the line (Figure I.4.2). The points below the line satisfy the *strict* inequality $3x + 2y < 6$, while those on the line, of course, satisfy the equation.

In a similar manner, we can consider the reverse inequality

$$3x + 2y \geq 6$$

If we do so, we find that this inequality is satisfied by all the points on or *above* the line $3x + 2y = 6$; the *strict* inequality $3x + 2y > 6$ is satisfied by the points above the line only.

In general, linear inequalities behave more or less in this manner. If we consider the general linear equation (1.3.5)

$$ax + by = c$$

FINITE MATHEMATICS

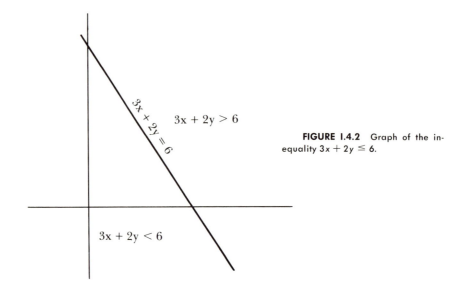

$3x + 2y > 6$

$3x + 2y = 6$

$3x + 2y < 6$

FIGURE 1.4.2 Graph of the inequality $3x + 2y \leq 6$.

we know that its graph is a straight line. This line divides the remainder of the Cartesian plane into two parts, or half-planes. In one of these half-planes, the inequality

1.4.13 $ax + by > c$

holds; in the other half-plane, it is the opposite inequality

1.4.14 $ax + by < c$

that holds. If we consider the *closed* half-planes, i.e., the half-planes along with the line that is their common boundary, we see that one closed half-plane is the graph of

1.4.15 $ax + by \geq c$

while the other closed half-plane is the graph of

1.4.16 $ax + by \leq c$

It is, naturally, important to determine which half-plane corresponds to each of the two inequalities (1.4.15) and (1.4.16). Generally speaking, this can best be done by solving for y. Given the inequality (1.4.15), we reduce it to the form

$$by \geq c - ax$$

Now, if b is positive, we can divide b (or equivalently, multiply by its reciprocal $1/b$) and obtain

$$y \geq \frac{c}{b} - \frac{a}{b} x$$

while, if b is negative, we obtain

$$y \leq \frac{c}{b} - \frac{a}{b} x$$

since in this case multiplication by $1/b$ reverses the inequality. For positive b, the inequality (1.4.15) corresponds to the half-plane *above* the line (1.3.5), while (1.4.16) corresponds to the half-plane below the line. For negative b, however, we find that (1.4.15) corresponds to the half-plane *below* the line, and (1.4.16) to that above the line.

In the event that $b = 0$, of course, this analysis falls through. In fact, (1.3.5) reduces now to the form

$$ax = c$$

and this line is vertical (parallel to the y-axis) so that two half-planes cannot be classified as being above or below the line. Here we can only talk about the half-plane to the right or left of the line. It is not difficult to see that, if a is positive, then $ax \leq c$ corresponds to the left half-plane, while $ax \geq c$ corresponds to the right half-plane. If a is negative, the roles are reversed, and this time $ax \geq c$ corresponds to the left half-plane.

I.4.1 *Example.* Give the graph of the inequality

$$2x - 3y \geq 12$$

In this case, solution for y gives us

$$y \leq -4 + \frac{2}{3} x$$

which means that we want the half-plane below the line. The line itself is most easily plotted if we take its intercepts, which are $x_0 = 6$ and $y_0 = -4$ (Figure I.4.3). The graph of the inequality is, then, the area on and below this line.

I.4.2 *Example.* Find the graph of the inequality

$$-3x \leq 15$$

This inequality reduces to

$$x \geq -5$$

so we want the half-plane to the right of the vertical line $x = -5$.

FINITE MATHEMATICS

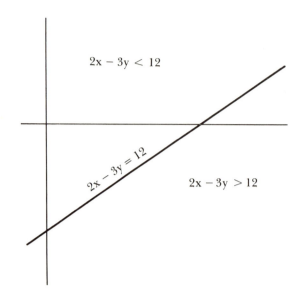

$2x - 3y < 12$

$2x - 3y = 12$

$2x - 3y > 12$

FIGURE I.4.3 Example I.4.1.

PROBLEMS ON INEQUALITIES

1. On graph paper, show the points which satisfy the following inequalities:

(a) $3x + 4y \leq 12$

(b) $4x - 2y \geq 10$

(c) $x + 2y \leq 14$

(d) $-x + 3y \geq 11$

(e) $x + 4y < 10$

(f) $x - 2y > 8$

(g) $x \geq 6$

(h) $3x - y \geq 3$

(i) $x - 3y < 12$

(j) $4x + 2y > 8$

5. SYSTEMS OF EQUATIONS AND INEQUALITIES

We have, so far, studied linear equations and linear inequalities separately. Quite often, however, we are given a *system* of two or more such equations and inequalities. The idea, then, is to look for numbers that satisfy all these equations or inequalities simultaneously. Equivalently, we look for points that lie on the graphs of all these equations (or inequalities) simultaneously.

Let us consider the simplest such system, one consisting of two equations:

1.5.1 $$a_1 x + b_1 y = c_1$$

1.5.2 $$a_2 x + b_2 y = c_2$$

Each of these equations has, as its graph, a straight line. We know that two lines can be parallel or intersecting. If parallel they have no points in common; if they intersect they have exactly one point in common. It is also possible that the two equations represent the same straight line (we say that the two lines are *coincident*) and, in this case, all the points on the line lie on both graphs.

We see, therefore, that there are three possibilities for the system of linear equations (1.5.1) to (1.5.2), corresponding to the three possibilities for the pair of lines. If the lines are parallel, the system does not have a solution: there is no pair (x,y) that satisfies both these equations, and we say that the system is *inconsistent*. If the lines intersect, there is a unique pair (x,y) that satisfies both equations; this pair is called, quite naturally, the *solution* of the system. If the lines are coincident, the system will have an infinity of solutions.

Let us consider now the actual method of solution of systems of equations. Suppose we are given two equations such as (1.5.1) and (1.5.2). Let us assume that these two lines have a point (x_0, y_0) in common (it may, of course, happen that the lines have no point in common; in such a case, we shall see that this assumption leads to a contradiction). This means that the two equations, (1.5.1) and (1.5.2), will both hold if we substitute the values (x_0, y_0). Now, we can multiply the first equation by any number r, and the second equation by any number, s, obtaining

$$ra_1 x + rb_1 y = rc_1$$
$$sa_2 x + sb_2 y = sc_2$$

and we can add these two equations, obtaining a third equation

1.5.3 $$(ra_1 + sa_2)x + (rb_1 + sb_2)y = rc_1 + sc_2$$

which will hold whenever (1.5.1) and (1.5.2) both hold. That is, it will hold if we substitute the values (x_0, y_0) into the equation. Thus (1.5.3), if the left side does not vanish, is the equation of a line passing through the point of intersection (x_0, y_0) of the lines with equations (1.5.1) and (1.5.2).

The equation (1.5.3), which is obtained by multiplying (1.5.1) and (1.5.2) each by a constant and adding them is said to be a *linear combination* of (1.5.1) and (1.5.2).

We now see that a linear combination of (1.5.1) and (1.5.2) is the equation of another line that will pass through the point of intersection of these two lines. If, moreover, (1.5.1) and (1.5.2) have different slopes, so that they have a unique point in common, it is not too difficult to see that the expression

1.5.4
$$m = -\frac{ra_1 + sa_2}{rb_1 + sb_2}$$

which is the slope of (1.5.3), can be given any value desired by a judicious choice of the two constants, r and s. Thus any one of the lines passing through the point of intersection of (1.5.1) and (1.5.2) may be obtained as a linear combination of their equations, i.e., in the form (1.5.3). In particular, the two lines passing through this point, parallel to the x-axis and y-axis respectively, may be obtained in this manner. The importance of this, of course, is that, in the equation of a line parallel to the x-axis, the variable x does not appear; in the equation of a line parallel to the y-axis, the variable y does not appear. Such an equation can then be solved for the single variable that appears in it; this is the process of *elimination*, which the reader doubtless encountered in high school algebra. Once the value of one of the unknowns is found, this value is substituted in one of the original equations; the other variable is then obtained directly.

1.5.1 Example. Find the solution of the system

$$3x + 2y = 12$$
$$2x - y = 1$$

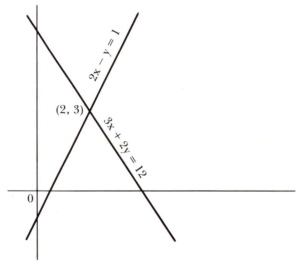

FIGURE I.5.1 Example I.5.1.

We look for a linear combination of these equations that represents a horizontal line (i.e., parallel to the x-axis). This has slope 0, and so is obtained by choosing r and s so that the numerator in the right side of (1.5.4) is equal to 0. This is most easily done by setting $r = a_2 = 2$ and $s = -a_1 = -3$; thus we have the two equations

$$6x + 4y = 24$$
$$-6x + 3y = -3$$

Adding these two equations together we obtain

$$7y = 21$$

an equation from which x has been eliminated. This of course gives us $y = 3$; substitution of this value into the first of the original equations gives us

$$3x + 2 \cdot 3 = 12$$

or

$$3x = 6$$

which has the solution $x = 2$. Thus $(2,3)$ is the solution of the system.

1.5.2 *Example.* Find the solution of the system

$$x + 2y = 4$$
$$2x - y = 3$$

Again, we eliminate x by setting $r = 2$, $s = -1$. This gives us

$$2x + 4y = 8$$
$$-2x + y = -3$$

which upon addition reduces to

$$5y = 5$$

or $y = 1$. Substituting in the first equation, we have

$$x + 2 \cdot 1 = 4$$

or $x = 2$. Thus $(2,1)$ is the solution of the system.

This method of solving systems of equations will of course break down if the two equations do not have a unique solution. In fact, the two lines (1.5.1) and (1.5.2) will fail to have a unique point in common

ANALYTIC GEOMETRY

FINITE MATHEMATICS

if they have the same slope (in which case they are either parallel or coincident). Suppose, then, that these two lines have the common slope k; this means $a_1 = -kb_1$ and $a_2 = -kb_2$. But then, from (1.5.4)

$$m = +\frac{rkb_1 + skb_2}{rb_1 + sb_2} = k$$

and so any linear combination must also have slope k. But the method of elimination consists precisely in finding a linear combination with slope 0 (or with infinite slope). It follows that the method of elimination cannot work for such systems; if we let the numerator in (1.5.4) vanish, we find that the denominator vanishes simultaneously, and (1.5.3) reduces either to an absurdity of the form $0 = 1$ (when the lines are parallel) or to an identity $0 = 0$ (when the lines are coincident).

1.5.3 Example. Find the solution of the system

$$3x + 2y = 5$$
$$9x + 6y = 10$$

In this case we can eliminate x by letting $r = 3$, $s = -1$. We obtain

$$9x + 6y = 15$$
$$-9x - 6y = -10$$

Addition of these two equations, however, shows us that in eliminating x we have also eliminated y, giving rise to the absurdity $0 = 5$. We can only conclude that the system is infeasible: the two lines are parallel.

1.5.4 Example. Find the solution of the system

$$6x - 4y = 8$$
$$-9x + 6y = -12$$

In this case, we eliminate x by setting $r = 3$ and $s = 2$, obtaining

$$18x - 12y = 24$$
$$-18x + 12y = -24$$

Addition of these two equations leads to the identity $0 = 0$. This means that the two equations are equivalent. The two lines are coincident and there is no *unique* solution; there is, rather, an infinity (all the points on a line). We may, in such cases, leave the system in the form of *one* of the original equations; we may also solve for one of the variables in terms of the other. Thus, we could solve for y:

$$y = \frac{3}{2}x - 2$$

and state that this is the general solution in the sense that, for any (arbitrary) value of x, the value of y given here will satisfy the system. Similarly, we could solve for x:

$$x = \frac{4}{3} + \frac{2}{3}y$$

and treat this also as the general solution.

For systems of more than two equations, the procedure is similar, though with certain differences. If we consider the geometry of the situation, we see that, in general, three lines in the plane form a triangle (Figure I.5.2); it is only in special cases that they have a point

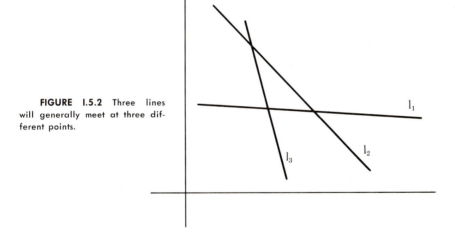

FIGURE I.5.2 Three lines will generally meet at three different points.

in common (Figure I.5.3). Sometimes it may even happen that all three lines are coincident, so that the system may have an infinity of solutions. In general, however, a system of three or more equations in the two unknowns, x and y, will have no solution.

The method used for solving such systems is, generally, to find the point of intersection of two of the lines. Once this point has been found, it is simply a matter of checking whether the remaining lines of the system all pass through this point. If they do, well and good; if not, the system has no solution. A systematic procedure would run as follows:

Solve the first two equations in the system as has just been explained. There are three possibilities: the two equations may have a unique solution, or no solution at all, or an infinity of solutions.

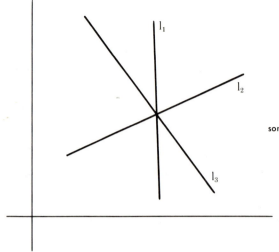

FIGURE I.5.3 Three lines will sometimes meet at a single point.

If the two equations have a unique solution, substitute this solution in the remaining equations of the system to check whether it satisfies them all. If it does, it is the unique solution of the system. If not, the system has no solution.

If the two equations have no common solution, the system has no solution.

If the two equations have an infinity of solutions, they represent coincident lines and are equivalent. This means one of the two equations may be discarded, giving rise to a smaller system. We then repeat the procedure with the smaller system.

I.5.5 Example. Solve the system

$$3x + y = 5$$
$$x - 2y = -3$$
$$2x + 2y = 6$$
$$x + 2y = 5$$

In this system, we take the first two equations:

$$3x + y = 5$$
$$x - 2y = -3$$

and solve by the elimination procedure: choosing $r = 1$, $s = -3$ to give

$$3x + y = 5$$
$$-3x + 6y = 9$$

and adding these,

$$7y = 14$$

gives us $y = 2$; substitution in the first equation gives

$$3x + 2 = 5$$

or $x = 1$. We then substitute $(1,2)$, the unique solution of the first two equations, in the remaining equations:

$$2 \cdot 1 + 2 \cdot 2 = 6$$
$$1 + 2 \cdot 2 = 5$$

Since both of these are correct, it follows that $(1,2)$ is the solution of the system.

1.5.6 Example. Solve the system

$$3x + y = 5$$
$$x - 2y = -3$$
$$2x + y = 6$$

In this case, we once again consider the first two equations, which have the unique solution $(1,2)$. We substitute this into the third equation, obtaining

$$2 \cdot 1 + 2 = 6$$

This is false, and it follows that the system has no solution.

1.5.7 Example. Solve the system

$$x - 2y = 3$$
$$-2x + 4y = -6$$
$$x + y = 6$$

In this case, the first two equations

$$x - 2y = 3$$
$$-2x + 4y = -6$$

are seen to be equivalent. Thus, we may discard one of them, obtaining the reduced system

$$x - 2y = 3$$
$$x + y = 6$$

which can be seen (subtracting one equation from the other) to have

FINITE MATHEMATICS

34

the solution $x = 5$, $y = 1$. Thus, $(5,1)$ is the unique solution to the system.

1.5.8 Example. Solve the system

$$x + 2y = 6$$
$$2x + 4y = 10$$
$$x - 3y = 7$$
$$3x + y = 5$$

In this case, the first two equations

$$x + 2y = 6$$
$$2x + 4y = 10$$

are seen to represent parallel lines. Since they have no common solution, it follows that the system has no solutions at all.

PROBLEMS ON SYSTEMS OF EQUATIONS

1. Solve the following systems of simultaneous equations. Give all solutions:

(a) $x + 2y = 5$
     $2x + y = 4$

(b) $x - 3y = 7$
     $x + y = 11$

(c) $2x + 3y = 7$
     $x - 2y = -1$

(d) $3x - 2y = 5$
     $x + y = 5$

(e) $2x + 4y = 8$
     $x + 2y = 4$

(f) $x + 3y = 6$
     $-x + 2y = -1$

(g) $x - 4y = 3$
     $2x + y = -3$

(h) $4x + 6y = 8$
     $6x + 9y = 10$

(i) $-2x + 3y = 5$
     $x - 2y = -4$

(j) $3x - 4y = 6$
     $x + y = 9$

(k) $3x + 2y = 5$
     $6x + 4y = 10$

(l) $2x + 4y = 5$
     $3x + 6y = 7$

(m) $x + 2y = 5$
     $3x + y = 5$
     $-x + 4y = 7$
     $x - 2y = -3$

(n) $-2x + 4y = 6$
     $-3x + 6y = 9$
     $x + 2y = 1$
     $-x + 3y = 4$

(o) $x - y = 3$
     $2x + y = 6$
     $-x - y = 2$
     $x + 2y = 3$

(p) $x + 3y = 4$
$3x + y = 2$
$x - 4y = 3$
$x + 3y = 2$

(q) $3x + 4y = 10$
$-x + 2y = 0$
$2x + y = 5$
$x - 3y = -1$

(r) $-x + 2y = 6$
$x + 3y = 9$
$2x + y = 3$
$x - 2y = -6$

2. Show that (a) the medians and (b) the altitudes of the triangle with vertices $(5,1)$, $(2,4)$, and $(1,0)$ intersect at one point.

The procedure with systems of linear inequalities is generally different. In fact, each linear inequality determines, not a line, but the entire half-space on one side of a line. Thus the solution to a system of linear inequalities will generally be, not a unique point, but rather, a figure bounded by straight lines, i.e., a polygon or some similar figure.

In the case of two inequalities, each is determined by a line. Generally, they are two intersecting lines; these divide the plane into four regions, labeled A, B, C, and D in Figure I.5.4. The general solution consists of all the points in one of these regions. For instance, the system of inequalities

$$x - y \geq 0$$
$$x - 2y \leq 0$$

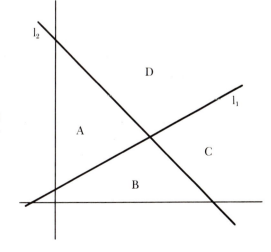

FIGURE I.5.4 The lines l_1 and l_2 divide the plane into four regions.

FINITE MATHEMATICS

may be seen to be equivalent to the relation (1.2.2). Its solution is the shaded area in Figure I.2.2, which we repeat as Figure I.5.5.

Special cases arise, of course. If the lines are parallel, the solution may be the entire, infinite strip between two lines (Figure I.5.6).

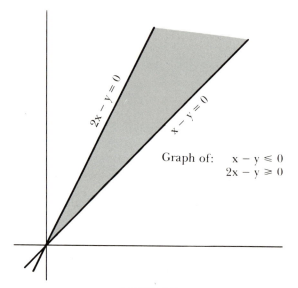

Graph of: $\quad x - y \leq 0$
$\quad\quad\quad\quad 2x - y \geq 0$

FIGURE I.5.5

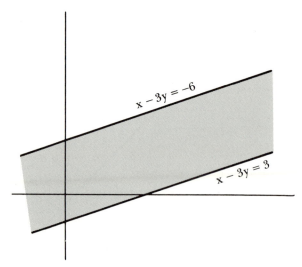

FIGURE I.5.6 The shaded area is the solution of the system

$$x - 3y \leq 3$$
$$x - 3y \geq -6$$

It may also happen that the two inequalities are contradictory as in the system

$$x + 2y \geq 5$$
$$x + 2y \leq 3$$

In this case, we want the points above the first line and below the second line. But the first line is parallel to and above the second, so there can be no solution.

It may even happen that the two inequalities give a single line as solution, thus:

$$x + 2y \geq 5$$
$$x + 2y \leq 5$$

These two lines are coincident. We want those points that are on or above the line, and also those on or below the line. But in this case all the points must lie on the line $x + 2y = 5$.

For larger systems of inequalities, we proceed similarly. Consider the system

$$x + y \leq 4$$
$$x + 2y \leq 5$$
$$x \geq 0$$
$$y \geq 0$$

Let us take the first of these inequalities,

$$x + y \leq 4$$

and multiply the fourth by -1, which gives us

$$-y \leq 0$$

These two inequalities, being of the same type, can be added to give

$$x \leq 4$$

Since we already have $x \geq 0$, we know that x is free to vary in the interval from 0 to 4, i.e.,

$$0 \leq x \leq 4$$

For any value x in this range, y can vary within a certain range that depends on x: we must have, from the inequalities, $y \geq 0$, and, moreover,

$$y \leq 4 - x$$

$$y \leq \frac{5}{2} - \frac{1}{2}x$$

These are two inequalities that must be satisfied by y. Depending on the value of x, one of them will be stronger than the other: thus, if $x = 1$, the inequalities are

$$y \leq 3$$
$$y \leq 2$$

and it is clear that the second is stronger than the first. On the other hand, for $x = 4$, the inequalities become

$$y \leq 0$$

$$y \leq \frac{1}{2}$$

and in this case it is clear that the first is stronger than the second. The transition between the two cases occurs when the two inequalities are the same, i.e., when

$$4 - x = \frac{5}{2} - \frac{1}{2}x$$

which is solved to give us $x = 3$. For x greater than 3, it is the first inequality that dominates, while for x smaller than 3, the second inequality dominates. Thus the general solution to the system of inequalities might be:

$$0 \leq x \leq 4$$

$$\text{If } 0 \leq x \leq 3 \text{ then } 0 \leq y \leq \frac{5}{2} - \frac{1}{2}x$$

$$\text{If } 3 \leq x \leq 4 \text{ then } 0 \leq y \leq 4 - x$$

This is a general solution of the system; it remains to be asked, however, whether we have gained much by expressing it in this form. There is the advantage of being able to obtain a point in the set directly: we can, say, choose $x = 2$ (this satisfies the first line of the solution); according to the second line, we must then have

$$0 \leq y \leq \frac{3}{2}$$

so that (2,1) is a solution—as may be checked directly.

Against this advantage, we have lost much of the conciseness of the original system of inequalities, and this after much work. Thus we shall not, in general, attempt to give the solutions to systems of inequalities in such form.

It is much more interesting, from a practical point of view, to know what the constraint set "looks like." We can say that it is a quadrilateral bounded by the four lines that determine the inequalities; to avoid ambiguities, we may give the four vertices or *extreme points* of the quadrilateral.

Each of the four vertices of the quadrilateral is at the intersection of two of the bounding lines of the quadrilateral: in other words, each of the vertices can be found by solving a system of two equations in the unknowns x and y: the equations are simply the inequalities that determine the constraint set, with the inequality sign replaced by an equals sign. In this example the four vertices will be the solutions of the systems

$$x + y = 4$$
$$x + 2y = 5$$

which gives us the point (3,1);

$$x + y = 4$$
$$y = 0$$

which gives (4,0);

$$x + 2y = 5$$
$$x = 0$$

which gives (0,5/2); and

$$x = 0$$
$$y = 0$$

which, of course, gives (0,0). The graph of the system of four inequalities is best characterized in this form: it is the quadrilateral with the vertices at (0,0), (0,5/2), (4,0), and (3,1) as shown in Figure I.5.7.

It may be noted that we did not consider all the points of intersection of pairs of the lines bounding the quadrilateral. We could, for

FINITE MATHEMATICS

40

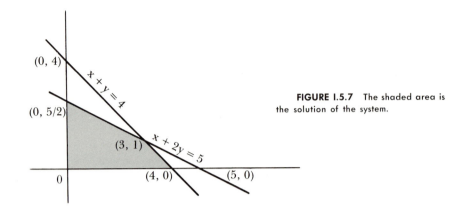

FIGURE I.5.7 The shaded area is the solution of the system.

example, have taken the two lines

$$x + 2y = 5$$
$$y = 0$$

and solved to obtain (5,0). This is not, however, a vertex of the graph, for the simple reason that it does not satisfy the inequality

$$x + y \leq 4$$

which must be satisfied by all points of the figure (including, of course, the vertices). Similarly, the system

$$x + y = 4$$
$$x = 0$$

gives the point (0,4), which does not satisfy the inequality

$$x + 2y \leq 5$$

and is not, therefore, in the figure.

Points obtained by treating two of the inequalities as equations and solving the resulting system need not be vertices of the graph, as we have no guarantee that they will satisfy the other inequalities. If they do satisfy them, however, they will be vertices. The rule is: in a system of loose linear inequalities (in two variables), the vertices of the graph can be found by treating two of the inequalities as equations and solving. If this system of two equations has a unique solution, this is a vertex of the graph if, and only if, it satisfies the remaining inequalities.

I.5.9 Example. Give the graph of the system

$$x + y \leq 5$$
$$x \geq 0$$
$$x \leq 3$$
$$y \geq 0.$$

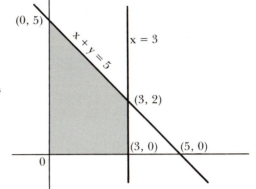

FIGURE I.5.8 The shaded area is the solution of Example I.5.9.

From Figure I.5.8 we see that the set is a quadrilateral. To find its vertices, we take the inequalities, two at a time, and treat them as equations. The first and second inequalities give us

$$x + y = 5$$
$$x = 0$$

which has solution (0,5). This satisfies the other two inequalities. The first and third inequalities give

$$x + y = 5$$
$$x = 3$$

with solution (3,2). Again, this satisfies the remaining inequalities. The first and fourth inequalities give

$$x + y = 5$$
$$y = 0$$

with solution (5,0). This does not, however, satisfy the inequality

$$x \leq 3$$

and is therefore not a vertex of the graph.

The second and third inequalities give

$$x = 0$$
$$x = 3$$

which is clearly not feasible.

The second and fourth give the point (0,0). This satisfies the other inequalities.

Finally, the third and fourth inequalities give the point (3,0). It may be checked that it satisfies the other inequalities.

Of the six pairs of inequalities, then, we see that four give the points (0,5), (3,2), (0,0), and (3,0), which satisfy the remaining inequalities and are therefore vertices of the graph. Of the two other pairs, one gives a point that fails to satisfy the inequalities, while the other pair fails to have a solution.

This, then, characterizes the graph: it is, simply, a quadrilateral with vertices at (0,5), (3,2), (0,0), and (3,0).

1.5.10 Example. We give an example now to show how practical problems give rise to systems of linear inequalities (and, possibly, equations).

A mixture must be made with foods A and B. Each unit of food A weighs 5 gm. and contains 1 gm. of protein. Each unit of food B weighs 3 gm. and contains 0.5 gm. protein. The mixture must weigh 60 gm. or less and contain at least 8 gm. protein.

We can summarize the data of the problem, if we wish, in the form of a table (Table I.5.1).

TABLE I.5.1

Food	Units	Unit Weight	Protein/Unit
A	x	5	1
B	y	3	0.5
Total		≤ 60	≥ 8

Letting x and y be the number of units of food A and B to be put into the mixture, we find that the weight of food A is $5x$ grams, while the weight of food B is $3y$ grams. Thus we obtain the relation

$$5x + 3y \leq 60$$

In a similar way, the need for at least 8 grams of protein can be restated as

$$x + 0.5y \geq 8$$

Two additional inequalities arise due to the fact that the amounts used must be positive or zero. These are the so-called *non-negativity constraints.* The problem can be stated in the form

$$5x + 3y \leq 60$$
$$x + 0.5y \geq 8$$
$$x \geq 0$$
$$y \geq 0$$

We proceed then to solve the system as we did with Example I.5.9. The first two inequalities give the system

$$5x + 3y = 60$$
$$x + 0.5y = 8$$

with solution $(-12,40)$, which does not satisfy the constraint $x \geq 0$. The first and third inequalities give

$$5x + 3y = 60$$
$$x = 0$$

with solution $(0,20)$, which does satisfy the remaining inequalities. The first and fourth inequalities give the point $(12,0)$; the second and third inequalities, $(0,16)$; and the second and fourth, $(8,0)$. All these points satisfy the inequalities and are therefore vertices of the set. On the other hand, the third and fourth inequalities give $(0,0)$, which does not satisfy the inequality $x + 0.5y \geq 8$ and thus is not a vertex. This set can, therefore, be characterized as the quadrilateral with vertices $(0,20)$, $(12,0)$, $(0,16)$, and $(8,0)$.

PROBLEMS ON SYSTEMS OF LINEAR INEQUALITIES

1. On graph paper, show the sets of points corresponding to the following systems of linear inequalities. Find the extreme points of these sets.

(a)
$$x + 2y \leq 12$$
$$3x + y \leq 9$$
$$x \geq 0$$
$$y \geq 0$$

(b)
$$2x + y \leq 20$$
$$2x + 3y \leq 24$$
$$x - y \geq 5$$
$$x \geq 0$$
$$y \geq 0$$

(c)
$$3x + y \geq 16$$
$$x + 2y \leq 20$$
$$x \geq 0$$
$$y \geq 0$$

(d)
$$2x + y \geq 20$$
$$2x + 3y \geq 24$$
$$x - y \geq 5$$
$$x \geq 0$$
$$y \geq 0$$

(e)
$$3x + 4y \leq 16$$
$$2x + y \leq 8$$
$$x - y \leq 2$$
$$x \geq 0$$
$$y \geq 0$$

2. A nut company has 6000 lb. of mixed nuts and 4000 lb. of peanuts. The company sells three products: an expensive mixture containing 5 lb. of mixed nuts for each pound of peanuts, a cheaper mixture containing equal parts of mixed nuts and peanuts, and peanuts without any other nuts. Give a system of linear inequalities expressing the amounts which the company can make of each of these products.

6. HIGHER DIMENSIONS

We have seen that the set of real numbers can be put into a natural correspondence with the set of points on a line; the set of pairs of real numbers with the set of points in a plane. We related one-dimensional geometry with the analysis of one variable, two-dimensional geometry with the analysis of two variables. Similarly, we can relate solid (three-dimensional) geometry with the analysis of three real variables. It is then only a short step to talking about space of four, five, six, or any number of dimensions, even though there is no physical interpretation of this idea. The justification is, simply, that it allows us to speak of the relations among four or more variables in the highly descriptive language of geometry, though it be only by generalization of the concepts that we have studied in solid geometry. Let us look, first, at the three-dimensional case. Once again a point, O, is chosen; through this point (called, as usual, the origin) three mutually perpendicular lines are then taken and called respectively, the x-axis, the y-axis, and the z-axis (collectively, the coordinate axes).

Three points, I, I', and I'', are then taken, equidistant from O, on the three axes. These serve to give us, first, the unit of distance (the common length of the three segments $\overline{OI}$, $\overline{OI'}$, and $\overline{OI''}$) and the "positive" direction on each of the three axes.

In Figure I.6.1, the x-axis and the y-axis are assumed to lie in the plane of the page. The z-axis, however, is supposed to come *out* of the page (toward the reader). Such a coordinate system is known as a *right-handed system*. This nomenclature is derived from the fact that a right-handed screw, rotated from the direction of positive x toward the direction of positive y (counterclockwise in this case), will move in the direction of positive z. We could just as easily have chosen the point I'' in such a way that the positive z-direction was *into* the book (away from the reader), a *left-handed system*. The principal reason for choosing a right-handed system is convenience: most authors assume right-handed systems, and a left-handed system, while perfectly consistent, clashes with established use. What this means is immediately apparent to anyone who has tried to screw a left-handed bolt into a right-handed nut.

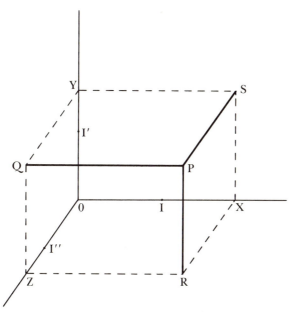

FIGURE I.6.1   Location of a point in three-dimensional space.

The three coordinate axes, taken two at a time, determine three planes called the coordinate planes. The x- and y-axes determine a plane called the xy-plane, the x- and z-axes determine the xz-plane, and the y- and z-axes, the yz-plane.

To each point in space, we can assign three numbers, the directed (perpendicular) distances to that point from the three coordinate planes. The distance to the yz-plane is called the x-coordinate, that to the xz-plane the y-coordinate, and that to the xy-plane the z-coordinate. Thus, in Figure I.6.1, to the point P are assigned, as coordinates, the distances $\overline{QP}$, $\overline{RP}$, and $\overline{SP}$, respectively. Here Q is the point of intersection of a line through P parallel to the x-axis, with the yz-plane. The points R and S are obtained analogously.

Conversely, any triple of numbers (x,y,z) corresponds to a point in space. The point (a,b,c) can be obtained by moving a units in the x-direction, then b units in the y-direction, and finally, c units in the z-direction. (The order in which these three motions are carried out is unimportant. We could just as well move, first, b units in the y-direction, followed by a units in the x-direction, and c units in the z-direction. The same point would be obtained, thanks to the geometric theorem which states that opposite sides of a rectangle are equal.)

The general first-degree equation

1.6.1 $$ax + by + cz = d$$

represents, not a line, but a plane in space. (This can best be seen if we consider that the equation $x = 0$ represents the yz-plane.) Nevertheless, it is still called a linear equation.

The general loose linear inequality

1.6.2
$$ax + by + cz \geq d$$

will represent one of the two *half-spaces* bounded by the plane of equation (1.6.1). That is, it will represent all the space that lies on one side of the plane (1.6.1), plus the plane itself. The corresponding strict inequality

1.6.3
$$ax + by + cz > d$$

will represent the half-space, minus the bounding plane.

The solution of systems of linear equations can best be understood if we look at certain geometric considerations. Normally, two planes intersect along a line, while three planes intersect at a point. Four planes normally form a tetrahedron and therefore have no points in common. Analytically, this means that two equations (in three variables) generally have an infinity of solutions, while three equations have a unique solution, and four or more equations none at all. This is, of course, only the general case; it may happen that a system of only two equations has no solutions (parallel planes), just as it may happen that a larger system has a solution (several planes passing through a common point) or even an infinity of solutions (if the planes are coaxal, i.e., have a line in common). It may happen, for that matter, that all the planes are coincident.

Such systems are solved in essentially the same way as systems of equations in two unknowns, i.e., by elimination and substitution. Naturally, it is a considerably longer process, since now there are more unknowns to eliminate and substitute.

1.6.1 Example. Solve the system

$$x + 2y + 3z = 9$$
$$2x + y + 2z = 7$$
$$x + y + 3z = 8$$

We start by eliminating x. This can be done by multiplying the first equation by 2, the second by -1, and adding:

$$2x + 4y + 6z = 18$$
$$-2x - y - 2z = -7$$

to obtain

$$3y + 4z = 11$$

Similarly, we can multiply the second equation by -1, the third equation by 2, and add

$$\begin{aligned} -2x - y - 2z &= -7 \\ 2x + 2y + 6z &= 16 \end{aligned}$$

obtaining

$$y + 4z = 9$$

We have now a system of two equations

$$\begin{aligned} 3y + 4z &= 11 \\ y + 4z &= 9 \end{aligned}$$

in the two unknowns, y and z. We then proceed to solve this system by subtracting the second equation from the first, to obtain

$$2y = 2$$

or $y = 1$. This value is then substituted to give us

$$3 + 4z = 11$$

which reduces to $z = 2$. Finally, the values $y = 1$, $z = 2$ are substituted in one of the original equations, giving,

$$x + 2 + 6 = 9$$

or, equivalently, $x = 1$. Thus (1,1,2) is the (unique) solution to the system.

1.6.2 Example. Find the solution of the system

$$\begin{aligned} 3x + 4y + z &= 11 \\ -x - 2y + 2z &= 3 \\ -x + 2y + z &= 8 \end{aligned}$$

In this case, we can eliminate x from the first and second equations if we multiply the second by 3 and add:

$$\begin{aligned} 3x + 4y + z &= 11 \\ -3x - 6y + 6z &= 9 \end{aligned}$$

to obtain

$$-2y + 7z = 20$$

Similarly, adding the first equation to three times the third

$$3x + 4y + z = 11$$
$$-3x + 6y + 3z = 24$$

we obtain

$$10y + 4z = 35$$

We now have a system

$$-2y + 7z = 20$$
$$10y + 4z = 35$$

from which we can eliminate y by multiplying the first equation by 5:

$$-10y + 35z = 100$$
$$10y + 4z = 35$$

to get

$$39z = 135$$

or $z = 45/13$. This is then substituted in the previous equations, obtaining $y = 55/26$ and $x = -10/13$. Thus $(-10/13, 55/26, 45/13)$ is the solution of the system.

1.6.3 Example. Find the solution of the system

$$x + 2y - z = 6$$
$$2x - y + 2z = 4$$
$$3x - 4y + 5z = 3$$

In this case, we can eliminate x by subtracting the second equation from two times the first, obtaining

$$5y - 4z = 8$$

and subtracting the third equation from three times the first, which gives us

$$10y - 8z = 15$$

Because the resulting system of two equations is, however, infeasible, the original system is also infeasible.

I.6.4 *Example.* Solve the system

$$
\begin{aligned}
x + 2y - z &= 6 &\quad (e_1)\\
2x - y + 2z &= 4 &\quad (e_2)\\
3x - 4y + 5z &= 2 &\quad (e_3)
\end{aligned}
$$

This system is the same as that of Example I.6.3 except for the right-hand side of the third equation. Proceeding as before, we obtain the two equations

$$
\begin{aligned}
5y - 4z &= 8 &\quad (e_4)\\
10y - 8z &= 16 &\quad (e_5)
\end{aligned}
$$

which are equivalent, i.e., represent the same plane in space. We must see, now, exactly what this means. The equation (e_4) was obtained as a linear combination of (e_1) and (e_2). This means (as in the case of systems of equations of two unknowns), that every solution of (e_1) and (e_2) must also solve (e_4). Thus the plane of (e_4) must contain the line of intersection of the planes (e_1) and (e_2). Similarly, the plane (e_5) will contain the line of intersection of (e_1) and (e_3). But because (e_4) and (e_5) are the same plane, the two lines at which (e_1) intersects (e_2) and (e_3) will both lie in the plane (e_4).

Although two distinct lines can lie *at most* in one common plane, these two lines are seen to lie in both (e_1) and (e_4). The implication is that either the two lines are the same, or the two planes (e_1) and (e_4) are the same. But, since it is easy to see that (e_1) and (e_4) are different planes, e.g., $(2,2,0)$ satisfies (e_1) but not (e_4), it follows that the two lines are the same.

Thus (e_1) intersects (e_2) and (e_3) along the same line. This line is common to all three planes, and therefore every point on the line satisfies all three equations; there is an infinity of solutions. We obtain these by letting one of the variables, say, z, be arbitrary, and then solving (e_4) for y (in terms of z):

$$
y = \frac{8}{5} + \frac{4}{5}z \qquad (e_6)
$$

This value is then substituted in (e_1) to give us

$$
x = \frac{14}{5} - \frac{3}{5}z \qquad (e_7)
$$

These two expressions, (e_6) and (e_7), together with the statement that z is arbitrary, form the general solution of the system. Of course, this is not the only form in which the solution can be given; we might just as easily have made y arbitrary and obtained

$$
z = -2 + \frac{5}{4}y
$$

FINITE MATHEMATICS

$$x = 4 - \frac{3}{4}y$$

as the solution. We could also solve for y and z in terms of an arbitrary x. In any of these three cases, the resulting expressions would be called the general solution of the original system of equations.

I.6.5 Example. Solve the system

$$\begin{aligned} x + 2y + 3z &= 9 \\ 2x + y + 2z &= 7 \\ x + y + 3z &= 8 \\ x - y + 2z &= 4 \end{aligned}$$

In a system such as this, we solve the sub-system consisting of the first three equations, and, if this system has a unique solution, we check to see whether this satisfies the remaining equations. (The procedure is much the same as when handling systems of three or more equations in two unknowns.)

From Example I.6.1, the first three equations have solution $(1,1,2)$. Since this satisfies the fourth equation as well, viz., $1 - 1 + 2 \cdot 2 = 4$, we conclude that $(1,1,2)$ is the solution of this system.

I.6.6 Example. Solve the system

$$\begin{aligned} x + 2y - z &= 6 \\ 2x - y + 2z &= 4 \\ 3x - 4y + 5z &= 2 \\ x + y - z &= 2 \end{aligned}$$

Once again, we solve the first three equations; as in Example I.6.4, we find that they do not have a unique solution, but rather, a general solution consisting of an entire line. We may actually do two things in this case. One possibility is to discard one of the first three equations, reducing to a system of three equations that we solve in the usual manner. Another possibility is to take the general solution of Example I.6.4 and introduce it into the fourth equation here; we do this since it turns out to be considerably shorter (after all, most of the work has already been done). We obtain the single equation

$$\left(4 - \frac{3}{4}y\right) + y - \left(-2 + \frac{5}{4}y\right) = 2$$

which reduces to

$$6 - y = 2$$

or $y = 4$. Introducing this into the solution obtained for Example

I.6.4 (with y arbitrary) we obtain $x = 1$, $z = 3$. Thus $(1,4,3)$ is the solution of this system.

I.6.7 Example. Solve the system

$$x + 2y + 3z = 9$$
$$2x + y + 2z = 7$$
$$x + y + 3z = 8$$
$$x - y + 2z = 6$$

In this case, the first three equations give the solution $(1,1,2)$. Introducing these values into the fourth equation, we obtain,

$$1 - 1 + 2 \cdot 2 = 6$$

which is patently false. We conclude that this system has no solution.

For systems of linear inequalities, the procedure could be more or less as in the two-variable case. Again, we could try to give the solution set explicitly. It may be more interesting, however, to look for the extreme points (vertices) of the set, which the inequalities determine. This is considerably complicated by the fact that we cannot draw figures to help us determine the way in which the planes intersect.

I.6.8 Example. A factory produces widgets, gadgets, and flibbers, using two machines called a bender and a twister. Each widget requires two hours in the bender and one hour in the twister. A gadget requires three hours in the bender and two hours in the twister, while a flibber requires one hour in the bender and two in the twister. The bender can be used at most 50 hours per week, while the twister can be used at most 40 hours per week. What combinations of the products can be made?

If we denote the number of widgets, gadgets, and flibbers made in a week by x, y, and z, respectively, the data of the problem can be written in the form of a table (Table I.6.1).

TABLE I.6.1

Product	Number	Hr. in Bender	Twister
Widgets	x	2	1
Gadgets	y	3	2
Flibbers	z	1	2
Total		≤ 50	≤ 40

This table can be used to give the constraints of the problem: there are two constraints corresponding to the availability of the machines, and three non-negativity constraints due to the fact that negative amounts of a product are not possible. Thus we obtain the system

$$
\begin{aligned}
2x + 3y + z &\le 50 && (i_1) \\
x + 2y + 2z &\le 40 && (i_2) \\
x &\ge 0 && (i_3) \\
y &\ge 0 && (i_4) \\
z &\ge 0 && (i_5)
\end{aligned}
$$

The set of possible combinations forms a polyhedron in space; it will be adequately described if we give its extreme points (vertices). In a manner analogous to that used in the two-variable case, we obtain the extreme points by taking three of the inequalities, solving them as equations, and then checking that the point obtained satisfies the remaining inequalities. The inequalities (i_1), (i_2), and (i_3) give us the system of equations

$$
\begin{aligned}
2x + 3y + z &= 50 \\
x + 2y + 2z &= 40 \\
x &= 0
\end{aligned}
$$

The x is immediately eliminated, the resulting system of two equations has solution $y = 15$, $z = 5$, and these three planes intersect at $(0,15,5)$. It is easily checked that this point satisfies the other inequalities.

There are 10 combinations of three inequalities from among these five. Taking one combination at a time, we put our results in table form (Table I.6.2).

TABLE I.6.2

Inequalities	Point	Check
1,2,3	(0,15,5)	Yes
1,2,4	(20,0,10)	Yes
1,2,5	(−20,30,0)	No
1,3,4	(0,0,50)	No
1,3,5	(0,50/3,0)	Yes
1,4,5	(25,0,0)	Yes
2,3,4	(0,0,20)	Yes
2,3,5	(0,20,0)	No
2,4,5	(40,0,0)	No
3,4,5	(0,0,0)	Yes

Here we find that the set of possible combination is a five-faced figure having the six points (0,15,5), (20,0,10), (0,50/3,0), (25,0,0), (0,0,20), and (0,0,0) as its vertices.

Unfortunately, not every such problem will have a solution as easily characterized as this. The trouble is that there is no guarantee that the desired set will be bounded. If not bounded, it is not a polyhedron and cannot be characterized in terms of its vertices. An example of this danger follows.

1.6.9 *Example.* Find the solution of the system

$$
\begin{aligned}
2x + 3y + \ z &\geq 50 \\
x + 2y + 2z &\geq 40 \\
x\ \ \ \ \ \ \ \ \ \ &\geq 0 \\
y\ \ \ \ \ \ &\geq 0 \\
z &\geq 0
\end{aligned}
$$

This system is the same as that in Example I.6.8, except that the first two inequalities have been reversed. If we proceed as before, although we will obtain the same 10 points of intersection for the bounding planes, the process of checking whether they satisfy the remaining inequalities will give different results. Indeed, we will find that the vertices in this case are the points (0,15,5), (20,0,10), (0,0,50), (0,20,0), and (40,0,0).

We are tempted to describe the solution set of this problem as the polyhedron with these vertices. This is not, however, the case. We have here a five-faced figure with only five vertices. Such a polyhedron must be a pyramid with a quadrilateral base, but in such a pyramid, four of the faces pass through the same point. This is clearly not true here, since each combination of three faces gives a different point. It follows that this figure cannot be a polyhedron. It is, in fact, unbounded and, hence, not characterized by its vertices. Our method fails us.

For the case of four or more unknowns, we will briefly describe the behavior of such systems here. A method of solution for these is given in Chapters II and III.

In general, when dealing with many unknowns, we label them $x_1, x_2, \ldots, x_n$. The general first-degree equation in n unknowns has the form

1.6.4 $$a_1x_1 + a_2x_2 + \ldots + a_nx_n = b$$

and represents a hyperplane (i.e., an $(n-1)$-dimensional subset of n-dimensional space).

A system of m such equations, known as an m *by* n *system*, has the general form

1.6.5

$$a_{11}x_1 + a_{12}x_2 + \ldots + a_{1n}x_n = b_1$$
$$a_{21}x_1 + a_{22}x_2 + \ldots + a_{2n}x_n = b_2$$
$$\cdots\cdots\cdots\cdots\cdots\cdots\cdots\cdots$$
$$a_{m1}x_1 + a_{m2}x_2 + \ldots + a_{mn}x_n = b_m$$

in which the coefficients are now given double subscripts to denote the equation and variable, respectively.

That is, a_{ij} is the coefficient of the variable x_j in the ith equation, while b_i is the constant term in the ith equation. The 2×3 system

$$3x_1 + 2x_2 - x_3 = 5$$
$$4x_1 - 5x_2 + 2x_3 = 6$$

is an example of the more general system

$$a_{11}x_1 + a_{12}x_2 + a_{13}x_3 = b_1$$
$$a_{21}x_1 + a_{22}x_2 + a_{23}x_3 = b_2$$

in which

$$a_{11} = 3 \qquad a_{12} = 2 \qquad a_{13} = -1 \qquad b_1 = 5$$
$$a_{21} = 4 \qquad a_{22} = -5 \qquad a_{23} = 2 \qquad b_2 = 6$$

A system such as (1.6.5) usually has solutions for $m \leq n$, but not for $m > n$. More exactly, the usual case is that a system has a solution set of dimension $n - m$, whenever $m < n$. If $m = n$, the system normally has a unique solution, whereas for $m > n$, there usually is no solution.

This is, however, only the usual case. It may be that a system of two equations in many unknowns has no solutions (parallel hyperplanes), just as it may happen that a system with more equations than unknowns has one or more solutions. We have seen all this in dealing with two- and three-variable systems.

The system of linear inequalities

1.6.6

$$a_{11}x_1 + a_{12}x_2 + \ldots + a_{1n}x_n \leq b_1$$
$$a_{21}x_1 + a_{22}x_2 + \ldots + a_{2n}x_n \leq b_2$$
$$\cdots\cdots\cdots\cdots\cdots\cdots\cdots\cdots$$
$$a_{m1}x_1 + a_{m2}x_2 + \ldots + a_{mn}x_n \leq b_n$$

generally represents a solid figure in n-dimensional space, bounded only by hyperplanes. Often it will be a hyperpolyhedron, i.e., the generalization of a polyhedron to n-dimensional space. Sometimes, however, this figure is not bounded. The considerations are quite similar to those in the three-variable case.

The general method of solution of a system such as (1.6.5) is much the same as in the two- or three-variable case, by elimination and substitution. Errors are, however, likely to occur (i.e., a particular equation might be disregarded) unless a very systematic procedure is used. Such a procedure is studied in Chapter II.

If we attempt to solve a system such as (1.6.6) as we did in the three-variable case, the same difficulties will arise, complicated by the fact that if m and n are large the number of combinations of n inequalities from the full set of m can be enormous. Generally, we shall concern ourselves with finding one particular vertex (extreme point) of the solution set—usually, the one that maximizes or minimizes some objective such as profits or costs. A procedure for this is discussed fully in Chapter III.

PROBLEMS ON HIGHER DIMENSIONS

Solve the following systems of simultaneous equations.

(a) $\begin{aligned} 3x + 4y - 2z &= 9 \\ x - 2y + z &= -2 \\ 2x + 3y + z &= 4 \end{aligned}$

(b) $\begin{aligned} 5x + 2y - 3z &= -3 \\ x - 3y - z &= 2 \\ 3x + y + z &= 4 \end{aligned}$

(c) $\begin{aligned} x - 2y + 2z &= 5 \\ 2x + y - 2z &= 6 \\ -x - 3y + 4z &= 2 \end{aligned}$

(d) $\begin{aligned} x - 2y + 2z &= 5 \\ 2x + y - 2z &= 6 \\ -x - 3y + 4z &= -1 \end{aligned}$

(e) $\begin{aligned} -2x + 4y + z &= -1 \\ x - y + z &= 3 \\ 2y + z &= 3 \end{aligned}$

(f) $\begin{aligned} 3x + y - 2z &= -3 \\ x + 2y + 3z &= -4 \\ 2x + y + 3z &= 1 \\ x - y - 2z &= 3 \end{aligned}$

CHAPTER II

VECTORS AND MATRICES

1. THE IDEA OF ABSTRACTION

Up to now, most of the mathematics that we have studied has dealt with the system of numbers — the natural numbers, the integers, the fractions, and finally, the real numbers — and its structure — its behavior with respect to the elementary operations of arithmetic, its order properties, and such. This is undoubtedly the natural process, for our mathematics is an outgrowth of the old sciences of arithmetic and geometry, most of whose concern was, precisely, with counting and measuring ideas that can only be explained in terms of the number system. Even today, most of the concrete applications of mathematics are, therefore, in terms of numbers.

It is the mathematician's desire, however, to look for ever more general mathematical structures, seeking always to understand the logical basis of mathematics — to uncover, as it were, the skeleton of the mathematical systems he knows — so as to adapt it to other related systems. This is the process of mathematical abstraction; in this manner the various branches of modern pure mathematics have been born.

Because mathematics evolved from arithmetic, the mathematical system that has been studied in greatest detail is that of the real numbers. It follows that most of the new branches of mathematics have been created by abstracting certain of the properties of the real number system and studying them in a sort of vacuum. Thus, the idea of "closeness" between numbers (or between points in Euclidean space, related to the numbers through the medium of analytic geometry) has been generalized — abstracted — to develop the science of *topology*. Alternatively, the ordering of the real numbers can be abstracted and generalized to give the complicated science of *cardinal* and *ordinal* numbers and *order-types*. It is also possible —

and this is our interest now—to abstract the idea of a *binary operation* (i.e., a rule for composing a pair of elements into a third element of the system), which appears amid the real numbers in the form of addition and the other fundamental arithmetic operations. This abstraction gives rise to the science of *modern algebra.*

We are interested in a branch of modern algebra, called *linear algebra,* that deals with certain generalizations of numbers, called *vectors* and *matrices.* The fact that they are generalizations means that some, but not all, of the properties of the number system will hold. We start by defining these concepts.

II.1.1   Definition.   A *matrix* is a rectangular array of numbers, arranged in m rows and n columns.

II.1.2   *Example.*   The following are all examples of matrices:

(a) $\begin{pmatrix} 5 & 1 & 2 \\ 2 & 3 & 7 \end{pmatrix}$ (b) $\begin{pmatrix} 1 \\ 5 \\ -1 \end{pmatrix}$ (c) $\begin{pmatrix} 1 & 3 & 8 & 4 \\ 2 & 6 & -5 & 1 \\ -7 & 1 & 3 & 4 \end{pmatrix}$

The numbers, m and n, of rows and columns in a matrix, are called the *dimensions* of the matrix. If a matrix has m rows and n columns it is called an $m \times n$ *matrix.* For instance, in Example II.1.2, (a) is a 2×3 matrix, (b) is 3×1, and (c) is 3×4.

The usual procedure, in dealing with matrices, is to represent all the numbers in the matrix, called *entries* of the matrix, by the same letter with different subscripts. Thus we have, as the "general" 2×3 matrix:

$$\begin{pmatrix} a_{11} & a_{12} & a_{13} \\ a_{21} & a_{22} & a_{23} \end{pmatrix}$$

or, more generally still, for the general $m \times n$ matrix:

2.1.1

$$A = \begin{pmatrix} a_{11} & a_{12} & \ldots & a_{1n} \\ a_{21} & a_{22} & \ldots & a_{2n} \\ \ldots & \ldots & \ldots & \ldots \\ a_{m1} & a_{m2} & \ldots & a_{mn} \end{pmatrix}$$

Note that the first subscript refers to the row, while the second subscript refers to the column. Thus a_{ij} is the entry in the ith row and jth column of the matrix A. The shorter, symbolic notation (a_{ij}) is also in common use for the matrix A given by (2.1.1).

II.1.3   Definition.   Two matrices, $A = (a_{ij})$ and $B = (b_{ij})$, are equal if they have the same dimensions and, for each i and j, $a_{ij} = b_{ij}$.

Equality of matrices is defined by saying that two matrices are equal only if they are identical in the sense of having the same shape and having corresponding entries equal. This means, among other things, that the single matrix equation $A = B$ is equivalent to the mn simultaneous equations $a_{ij} = b_{ij}$.

Of the possible dimensions for matrices, two cases of special interest arise.

II.1.4 Definition.   An $m \times n$ matrix is said to be *square* if $m = n$. An $n \times n$ matrix is also called an *nth-order* matrix.

II.1.5 Definition.   A $1 \times n$ matrix is called a *row vector*, or *row n-vector*. An $m \times 1$ matrix is called a *column vector*, or *column m-vector*.

A vector, as used here, is simply a special type of matrix. It may also be defined as an *n*-tuple of numbers, arranged either in a row, as, say

$$(5, 1, 4, 6, 2)$$

or in a column, as

$$\begin{pmatrix} 3 \\ 1 \\ 4 \\ 1 \end{pmatrix}$$

The entries in a vector are also called *components*. We shall usually denote a vector by a boldface letter, say $\mathbf{b} = (b_1, b_2, \ldots, b_m)$.

2. ADDITION AND SCALAR MULTIPLICATION

As already mentioned, matrices are an algebraic generalization of the concept of a number. This means that some operations can be performed on matrices. We introduce here two such operations: the first is that of addition.

II.2.1 Definition.   Let $A = (a_{ij})$ and $B = (b_{ij})$ be $m \times n$ matrices. Then by their sum, $A + B$, we mean the $m \times n$ matrix, $C = (c_{ij})$, where

2.2.1 $$c_{ij} = a_{ij} + b_{ij}$$

For example, we have the following additions:

$$\begin{pmatrix} 3 & 5 & 1 \\ 2 & 6 & 8 \end{pmatrix} + \begin{pmatrix} 1 & 4 & 6 \\ 2 & -2 & 3 \end{pmatrix} = \begin{pmatrix} 4 & 9 & 7 \\ 4 & 4 & 11 \end{pmatrix}$$

$$\begin{pmatrix} 1 & 3 \\ 1 & -4 \\ 2 & 5 \end{pmatrix} + \begin{pmatrix} -2 & 1 \\ 6 & 2 \\ 4 & -1 \end{pmatrix} = \begin{pmatrix} -1 & 4 \\ 7 & -2 \\ 6 & 4 \end{pmatrix}$$

$$(1, 5, 7, 2) + (2, 4, 6, 8) = (3, 9, 13, 10)$$

Addition is carried out in this way for matrices of equal dimensions by the simple expedient of adding corresponding entries in the two matrices. If two matrices have different dimensions, their sum is not defined, i.e., it is not possible to add them.
Thus, the sums

$$\begin{pmatrix} 1 & 3 & 5 \\ 2 & 6 & 4 \end{pmatrix} + \begin{pmatrix} -1 & 2 \\ 3 & 14 \\ 5 & -6 \end{pmatrix}$$

$$\begin{pmatrix} 1 & 3 \\ 5 & 8 \end{pmatrix} + \begin{pmatrix} 1 & -2 & 4 \\ 6 & 1 & -2 \end{pmatrix}$$

$$(1, 3, 1) + (2, 5)$$

are not defined; they *do not exist.*

The second operation to be considered is that of multiplication by a number. In the context of vectors and matrices, numbers are called *scalars,* so that this operation is called *scalar multiplication.*

II.2.2 Definition. Let s be a scalar, and let $A = (a_{ij})$ be an $m \times n$ matrix. Then by their product sA we shall mean the $m \times n$ matrix $B = (b_{ij})$ given by

2.2.2
$$b_{ij} = sa_{ij}$$

Scalar multiplication is accomplished by multiplying each entry in the matrix by the scalar.
For example,

$$3 \begin{pmatrix} 2 & 1 & 4 \\ 6 & -5 & 1 \end{pmatrix} = \begin{pmatrix} 6 & 3 & 12 \\ 18 & -15 & 3 \end{pmatrix}$$

$$-4 \begin{pmatrix} 2 & 1 & -2 & 3 \\ -1 & 4 & -1 & 0 \end{pmatrix} = \begin{pmatrix} -8 & -4 & 8 & -12 \\ 4 & -16 & 4 & 0 \end{pmatrix}$$

$$2 \begin{pmatrix} 1 \\ 4 \\ 5 \end{pmatrix} = \begin{pmatrix} 2 \\ 8 \\ 10 \end{pmatrix}$$

We shall use the symbol θ to represent a matrix all of whose entries are equal to zero. There are, of course, many such matrices — one

FINITE MATHEMATICS

for each pair of dimensions. This ambiguity is not very important, inasmuch as we will generally be able to tell, from the context, in which of all these matrices θ is needed. For instance, if we see the expression $A + \theta$, we shall know that the θ here must have the same dimensions as A, since matrices cannot be added unless they have the same dimensions.

It is not too difficult to see that addition and scalar multiplication satisfy the following laws:

2.2.3 $$A + B = B + A$$

2.2.4 $$A + (B + C) = (A + B) + C$$

2.2.5 $$A + \theta = A$$

2.2.6 $$A + (-1)A = \theta$$

2.2.7 $$r(sA) = (rs)A$$

2.2.8 $$r(A + B) = rA + rB$$

2.2.9 $$(r + s)A = rA + sA$$

2.2.10 $$1A = A$$

The law (2.2.3) is called the *commutative law* for addition. Laws (2.2.4) and (2.2.7) are the *associative laws* for addition and scalar multiplication, respectively. Law (2.2.5) states that θ is an *additive identity*, while (2.2.6) says that $(-1)A$, which is more briefly written $-A$, is an *additive inverse* for A. The two laws (2.2.8) and (2.2.9) are the two *distributive laws*; finally, (2.2.10) states that the scalar 1 is an identity for scalar multiplication.

It is a well-known fact, of course, that the real numbers satisfy all these laws; it is, indeed, due to this fact that the matrices themselves satisfy them. Let us consider (2.2.4). This law states that, given three matrices A, B, and C (of the same dimensions, of course), we will obtain the same result if we first add A and B together and then add their sum to C as if we first add B and C together and then add A to *their* sum.

We prove this in the case of three 2×2 matrices, $A = (a_{ij})$, $B = (b_{ij})$, and $C = (c_{ij})$, as follows:

$$A + B = \begin{pmatrix} a_{11} & a_{12} \\ a_{21} & a_{22} \end{pmatrix} + \begin{pmatrix} b_{11} & b_{12} \\ b_{21} & b_{22} \end{pmatrix} = \begin{pmatrix} a_{11} + b_{11} & a_{12} + b_{12} \\ a_{21} + b_{21} & a_{22} + b_{22} \end{pmatrix}$$

so

$$(A + B) + C = \begin{pmatrix} a_{11} + b_{11} & a_{12} + b_{12} \\ a_{21} + b_{21} & a_{22} + b_{22} \end{pmatrix} + \begin{pmatrix} c_{11} & c_{12} \\ c_{21} & c_{22} \end{pmatrix}$$

$$= \begin{pmatrix} a_{11} + b_{11} + c_{11} & a_{12} + b_{12} + c_{12} \\ a_{21} + b_{21} + c_{21} & a_{22} + b_{22} + c_{22} \end{pmatrix}$$

On the other hand,

$$B + C = \begin{pmatrix} b_{11} & b_{12} \\ b_{21} & b_{22} \end{pmatrix} + \begin{pmatrix} c_{11} & c_{12} \\ c_{21} & c_{22} \end{pmatrix} = \begin{pmatrix} b_{11} + c_{11} & b_{12} + c_{12} \\ b_{21} + c_{21} & b_{22} + c_{22} \end{pmatrix}$$

and so

$$A + (B + C) = \begin{pmatrix} a_{11} & a_{12} \\ a_{21} & a_{22} \end{pmatrix} + \begin{pmatrix} b_{11} + c_{11} & b_{12} + c_{12} \\ b_{21} + c_{21} & b_{22} + c_{22} \end{pmatrix}$$

$$= \begin{pmatrix} a_{11} + b_{11} + c_{11} & a_{12} + b_{12} + c_{12} \\ a_{21} + b_{21} + c_{21} & a_{22} + b_{22} + c_{22} \end{pmatrix}$$

We see that

$$(A + B) + C = A + (B + C)$$

For matrices of other dimensions, a similar proof using similar reasoning may be given. An alternative method, which the reader is invited to try, proves that the "typical" element (i.e., the element in the ith row and jth column) in each of the two matrices $(A + B) + C$ and $A + (B + C)$ is the same.

The other laws may all be proved in much the same way by using the fact that they hold for real numbers.

We may also, if we wish, define the operation of subtraction for matrices by analogy with the operation of subtraction for real numbers. In fact, if a and b are numbers, we know that $b - a$ is the number x, which satisfies the equation

$$x + a = b$$

In a similar way, for matrices A and B, the matrix $B - A$ would be an X such that

$$X + A = B$$

We notice first of all that this is possible only if A and B have the same dimensions. If this is the case, it is easy to see that $X = B + (-1)A$ satisfies the equation. In fact, the typical entry in $B + (-1)A$ is $b_{ij} - a_{ij}$; this means that the typical entry in $(B + (-1)A) + A$ is $(b_{ij} - a_{ij}) + a_{ij}$, or b_{ij}.

We now have

$$B + (-1)A + A = B$$

Thus, we define subtraction by the natural expression

2.2.11 $$B - A = B + (-1)A$$

II.2.3 Examples. Some examples of matrix addition and scalar multiplication are:

62

$$\begin{pmatrix} 5 & 1 \\ 2 & 4 \end{pmatrix} - 2\begin{pmatrix} 1 & -1 \\ 2 & 6 \end{pmatrix} + 3\begin{pmatrix} 1 & 2 \\ -5 & 4 \end{pmatrix}$$

$$= \begin{pmatrix} 5-2(1)+3(1) & 1-2(-1)+3(2) \\ 2-2(2)+3(-5) & 4-2(6) & +3(4) \end{pmatrix} = \begin{pmatrix} 6 & 9 \\ -17 & 4 \end{pmatrix}$$

$$\begin{pmatrix} 1 & 3 & 5 \\ 2 & 4 & 7 \end{pmatrix} - \begin{pmatrix} 6 & 1 & -8 \\ 2 & 6 & 2 \end{pmatrix} = \begin{pmatrix} -5 & 2 & 13 \\ 0 & -2 & 5 \end{pmatrix}$$

$$\begin{pmatrix} 1 \\ 2 \\ -1 \end{pmatrix} + 5\begin{pmatrix} 6 \\ -2 \\ 3 \end{pmatrix} = \begin{pmatrix} 31 \\ -8 \\ 14 \end{pmatrix}$$

PROBLEMS ON ADDITION AND SCALAR MULTIPLICATION

1. Perform the indicated vector operations.

(a) $(3,5,1) + (2,8,-4)$

(b) $\begin{pmatrix} 5 \\ 1 \end{pmatrix} + \begin{pmatrix} 6 \\ -3 \end{pmatrix} - \begin{pmatrix} -2 \\ 4 \end{pmatrix}$

(c) $2(1,3,6,7) - 4(2,5,-2,2)$

(d) $\begin{pmatrix} 3 \\ 6 \\ 1 \end{pmatrix} + 4\begin{pmatrix} 1 \\ -2 \\ 0 \end{pmatrix} - 3\begin{pmatrix} 2 \\ -5 \\ -3 \end{pmatrix}$

(e) $2(1,5) + 4(-2,1) - 3(1,-4) + (0,2)$

(f) $5(1,3,-5) + 3(-2,6,2) - 4(1,1,-1)$

(g) $(1,2,4) + 3(1,6,2) - 2(-2,4,1)$

(h) $\begin{pmatrix} 3 \\ 5 \end{pmatrix} + \begin{pmatrix} 2 \\ 8 \end{pmatrix} - 3\begin{pmatrix} 1 \\ -3 \end{pmatrix} + 4\begin{pmatrix} 1 \\ -2 \end{pmatrix}$

(i) $2\begin{pmatrix} 1 \\ 2 \\ 4 \\ 1 \end{pmatrix} - 3\begin{pmatrix} -2 \\ 1 \\ -5 \\ 6 \end{pmatrix} + 4\begin{pmatrix} 1 \\ 3 \\ -7 \\ 2 \end{pmatrix}$

(j) $(1,6,3) + 2(-2,4,2) - 3(1,5,-8)$

2. Solve the following vector equations.

(a) $2\begin{pmatrix} x \\ y \end{pmatrix} + 4\begin{pmatrix} 1 \\ -2 \end{pmatrix} = \begin{pmatrix} 10 \\ 2 \end{pmatrix}$

(b) $5\begin{pmatrix} 2x \\ 3y \end{pmatrix} + 3\begin{pmatrix} 4 \\ -2 \end{pmatrix} = \begin{pmatrix} -8 \\ 9 \end{pmatrix}$

(c) $\begin{pmatrix} x \\ 2y \end{pmatrix} + \begin{pmatrix} -y \\ 3x \end{pmatrix} = \begin{pmatrix} 1 \\ 8 \end{pmatrix}$

(d) $\begin{pmatrix} x \\ 3y \\ 2x \end{pmatrix} + \begin{pmatrix} 2y \\ -4x \\ y \end{pmatrix} = \begin{pmatrix} 4 \\ 17 \\ -1 \end{pmatrix}$

(e) $x(1,5) + y(-2,1) = (-5,-3)$

(f) $x\begin{pmatrix} 3 \\ -2 \end{pmatrix} - y\begin{pmatrix} 2 \\ 4 \end{pmatrix} = \begin{pmatrix} -7 \\ -22 \end{pmatrix}$

(g) $x\begin{pmatrix} 2 \\ 1 \\ 4 \end{pmatrix} + y\begin{pmatrix} 1 \\ -2 \\ 2 \end{pmatrix} = \begin{pmatrix} 7 \\ 1 \\ 14 \end{pmatrix}$

(h) $x\begin{pmatrix} 2 \\ 1 \\ 5 \end{pmatrix} + y\begin{pmatrix} 4 \\ 2 \\ -3 \end{pmatrix} = \begin{pmatrix} 12 \\ 7 \\ 4 \end{pmatrix}$

(i) $x(3,2,4) + y(6,4,1) = (9,6,-2)$

(j) $x\begin{pmatrix} 2 \\ 1 \\ 4 \end{pmatrix} + y\begin{pmatrix} 1 \\ -3 \\ -2 \end{pmatrix} + z\begin{pmatrix} 1 \\ 5 \\ 1 \end{pmatrix} = \begin{pmatrix} 8 \\ 14 \\ 11 \end{pmatrix}$

(k) $x\begin{pmatrix} 1 \\ -1 \\ 2 \end{pmatrix} + y\begin{pmatrix} 2 \\ 1 \\ 4 \end{pmatrix} + z\begin{pmatrix} 0 \\ -1 \\ 1 \end{pmatrix} = \begin{pmatrix} 3 \\ 3 \\ 3 \end{pmatrix}$

3. In each of the following problems, two vectors **a** and **b** are given. In which cases is it possible to express every vector **c** of the same dimensions in the form $x\mathbf{a} + y\mathbf{b}$? (In other words, in which cases will the vector equation $x\mathbf{a} + y\mathbf{b} = \mathbf{c}$ have solutions for all values of **c**?)

(a) $\mathbf{a} = \begin{pmatrix} 1 \\ 0 \end{pmatrix}$ $\mathbf{b} = \begin{pmatrix} 0 \\ 1 \end{pmatrix}$

(b) $\mathbf{a} = \begin{pmatrix} 2 \\ 1 \end{pmatrix}$ $\mathbf{b} = \begin{pmatrix} 3 \\ 2 \end{pmatrix}$

(c) $\mathbf{a} - (1,4)$ $\mathbf{b} = (2,8)$

(d) $\mathbf{a} = (6,-4)$ $\mathbf{b} = (-9,6)$

(e) $\mathbf{a} = (3,2)$ $\mathbf{b} = (3,-2)$

4. Perform the following matrix operations.

(a) $\begin{pmatrix} 2 & 1 \\ 6 & -3 \end{pmatrix} + \begin{pmatrix} 1 & 4 \\ 1 & 8 \end{pmatrix} - \begin{pmatrix} 2 & 2 \\ -4 & -1 \end{pmatrix}$

(b) $5\begin{pmatrix} 1 & 8 \\ 6 & -3 \\ 2 & -4 \end{pmatrix} - 3\begin{pmatrix} 1 & 3 \\ 2 & -2 \\ -5 & 3 \end{pmatrix} + 4\begin{pmatrix} 1 & 3 \\ 2 & 0 \\ 1 & 0 \end{pmatrix}$

(c) $5\begin{pmatrix} 3 & 4 & 1 & 2 \\ 6 & 5 & -4 & 0 \\ -5 & 3 & 0 & 1 \end{pmatrix} + 2\begin{pmatrix} 1 & -3 & -5 & 7 \\ 2 & -4 & 6 & -2 \\ -1 & 4 & 0 & 1 \end{pmatrix}$

(d) $2\begin{pmatrix} 1 & 5 \\ 2 & -2 \\ -3 & 6 \\ 1 & 0 \end{pmatrix} + 3\begin{pmatrix} 2 & -3 \\ 0 & 4 \\ -1 & 2 \\ 5 & 1 \end{pmatrix} - 4\begin{pmatrix} 1 & 5 \\ -2 & 4 \\ -6 & 7 \\ 1 & -3 \end{pmatrix}$

(e) $-2\begin{pmatrix} 1 & 5 & 8 \\ 3 & 4 & 2 \\ 2 & -1 & 6 \end{pmatrix} + 3\begin{pmatrix} 1 & 2 & 5 \\ 6 & -1 & 7 \\ -4 & 2 & 8 \end{pmatrix} + \begin{pmatrix} 1 & 3 & 0 \\ 2 & -4 & 2 \\ 1 & 3 & 1 \end{pmatrix}$

3. SCALAR PRODUCTS OF VECTORS

Before considering any further binary operations on matrices, we introduce a *unitary* operation, a rule that assigns to each matrix another matrix. This is the operation of *transposition*.

II.3.1 Definition. Let $A = (a_{ij})$ be an $m \times n$ matrix. Then the *transpose* of A is the $n \times m$ matrix $B = (b_{ij})$ defined by $b_{ij} = a_{ji}$.

We shall use the notation A^t for the transpose of A. Heuristically, the two matrices A and A^t are related in that each row of A is transformed into a column of A^t, and the converse. Examples of this follow.

II.3.2 Example.

$$\begin{pmatrix} 2 & 5 \\ 1 & -1 \\ 3 & 2 \end{pmatrix}^t = \begin{pmatrix} 2 & 1 & 3 \\ 5 & -1 & 2 \end{pmatrix}$$

$$\begin{pmatrix} 1 & 3 & 1 \\ 2 & 6 & 7 \\ 4 & 1 & 5 \end{pmatrix}^t = \begin{pmatrix} 1 & 2 & 4 \\ 3 & 6 & 1 \\ 1 & 7 & 5 \end{pmatrix}$$

$$(1,5,6,2)^t = \begin{pmatrix} 1 \\ 5 \\ 6 \\ 2 \end{pmatrix}$$

It is easy to see that

2.3.1 $$(A + B)^t = A^t + B^t$$

2.3.2 $$(rA)^t = rA^t$$

2.3.3 $$A^{tt} = A$$

The two laws (2.3.1) and (2.3.2) state that the transposition preserves the operations of addition and scalar multiplication. Law (2.3.3) states that the transpose of the transpose is the original matrix.

The theory of matrices and their algebra was first developed in detail by two Englishmen, Arthur Cayley (1821-1895) and James Sylvester (1814-1897). It was their observation that certain physical processes could be described by matrices. Quite often it was desirable to consider the *resultant* of two such processes (i.e., the result of carrying these out one after the other). In obtaining the resultant, the following definition was found of considerable value.

II.3.3 Definition. Let $x = (x_1, \ldots, x_n)$ be a row vector, and $y = (y_1, \ldots, y_n)^t$ be a column vector. Then the product xy is the scalar defined by

2.3.4 $$xy = x_1 y_1 + x_2 y_2 + \ldots + x_n y_n$$

or, using the summation symbol (see Appendix),

2.3.5 $$xy = \sum_{i=1}^{n} x_i y_i$$

Thus, the vectors x and y are multiplied by multiplying their corresponding components and adding the products obtained. We note that this is done only if the first vector is a row vector, the second is a column vector, and both have the same number of components. We add, further, that the product we have obtained here is xy; it is *not* yx. The product yx exists, but we will not define it as yet.

A possible heuristic interpretation would be as follows: let x be a "supply" vector, that is, let x_i represent the amount available of a certain commodity C_i. Let y be a "price" vector, that is, let y_i be the price of a unit amount of the commodity C_i. Then the total value of the supply $x = (x_i, \ldots, x_n)$ is precisely the product xy.

II.3.4 Example. We give, now, some examples of vector multiplication.

(a) $$(3, 5, 1) \begin{pmatrix} 2 \\ 4 \\ -2 \end{pmatrix} = 3(2) + 5(4) + 1(-2) = 24$$

(b) $$(6,1)\ (2,3)^t = 6(2) + 1(3) = 15$$

(c) $$(1,-2) \begin{pmatrix} 2 \\ 1 \end{pmatrix} = 1(2) - 2(1) = 0$$

(d) $(0,0,0) \begin{pmatrix} 1 \\ 4 \\ 8 \end{pmatrix} = 0(1) + 0(4) + 0(8) = 0$

It may be seen that the following laws hold for vector multiplication. Here, x and x' are row vectors, y and y' are column vectors, and s is a scalar:

2.3.6 $(sx)y = x(sy) = s(xy)$

2.3.7 $(x + x')y = xy + x'y$

2.3.8 $x(y + y') = xy + xy'$

2.3.9 $xy = y^t x^t$

The law (2.3.6) is called the homogeneity law; (2.3.7) and (2.3.8) are distributive laws, similar to (2.2.8) and (2.2.9). The law (2.3.9) is a modification of the commutative law $ab = ba$ which, we know, holds for multiplication of real numbers. Note that, while we have not, as yet, defined the product yx, the product $y^t x^t$ is well defined, since y^t is a row vector and x^t is a column vector.

Certain observations may be made concerning laws (2.3.6) to (2.3.9). We may, in (2.3.6), let $s = 0$. In this case, we find that the product of any vector with the vector 0 (that is, the vector of which all the components are equal to zero) will, if defined, be the scalar 0. This corresponds to our usual idea that the product of any number by 0 is always equal to 0. This property of multiplication of real numbers carries over to the multiplication of vectors. On the other hand, the converse property of real numbers, that if the product ab of two numbers is equal to zero, then at least one of the two factors, a and b, must be equal to zero, does not hold for vectors. We can see this in Example II.3.4(c), in which the product of (1,–2) and (2,1)^t is 0. Vectors such as these are called *orthogonal* vectors.

We note, finally, that the product (2.3.4) is quite generally called the *scalar product* or *inner product* of the two vectors, x and y.

4.   MATRIX MULTIPLICATION

We are now in a position to define multiplication of matrices. We shall do this in terms of the scalar products of vectors. In effect, to form the product AB of two matrices A and B, we would like to form the scalar products of all the rows of A with all the columns of B. These products (if they exist) can then be put into a matrix array in a very natural manner.

This, then, is the way in which we define the product of the two matrices, A and B. Each entry in the product matrix AB should be the scalar product of one of the rows of A with one of the columns of B. Now this is possible only if the rows of A and the columns of B have the same number of components. But the number of components in each row of A is equal to the number of columns in A, while the number of components in a column of B is equal to the number of rows in B. Thus, we find that *the product AB can be defined only if A has as many columns as B has rows.*

II.4.1   Definition.   Let $A = (a_{ij})$ be an $m \times n$ matrix, and let $B = (b_{jk})$ be an $n \times p$ matrix. The product AB of these two matrices is the $m \times p$ matrix $C = (c_{ik})$, in which

2.4.1
$$c_{ik} = \sum_{j=1}^{n} a_{ij} b_{jk}$$

In other words, the entry in the ith row and kth column of C is the scalar product of the ith row of A and kth column of B.

II.4.2   Examples.   Some examples of matrix multiplication are:

(a) $\begin{pmatrix} 6 & 1 \\ 2 & 4 \\ -1 & 3 \end{pmatrix} \begin{pmatrix} 1 & 5 \\ 4 & 8 \end{pmatrix} = \begin{pmatrix} 6 \cdot 1 + 1 \cdot 4 & 6 \cdot 5 + 1 \cdot 8 \\ 2 \cdot 1 + 4 \cdot 4 & 2 \cdot 5 + 4 \cdot 8 \\ -1 \cdot 1 + 3 \cdot 4 & -1 \cdot 5 + 3 \cdot 8 \end{pmatrix} = \begin{pmatrix} 10 & 38 \\ 18 & 42 \\ 11 & 19 \end{pmatrix}$

(b) $\begin{pmatrix} 3 & 5 \\ 1 & 4 \end{pmatrix} \begin{pmatrix} 2 & -3 \\ 1 & 8 \end{pmatrix} = \begin{pmatrix} 3 \cdot 2 + 5 \cdot 1 & 3(-3) + 5 \cdot 8 \\ 1 \cdot 2 + 4 \cdot 1 & 1(-3) + 4 \cdot 8 \end{pmatrix} = \begin{pmatrix} 11 & 31 \\ 6 & 29 \end{pmatrix}$

(c) $\begin{pmatrix} 2 & -3 \\ 1 & 8 \end{pmatrix} \begin{pmatrix} 3 & 5 \\ 1 & 4 \end{pmatrix} = \begin{pmatrix} 2 \cdot 3 - 3 \cdot 1 & 2 \cdot 5 - 3 \cdot 4 \\ 1 \cdot 3 + 8 \cdot 1 & 1 \cdot 5 + 8 \cdot 4 \end{pmatrix} = \begin{pmatrix} 3 & -2 \\ 11 & 37 \end{pmatrix}$

(d) $\begin{pmatrix} 1 & 5 & -1 \\ 2 & -4 & 3 \end{pmatrix} \begin{pmatrix} 6 & 2 \\ 1 & 4 \\ 1 & 8 \end{pmatrix} = \begin{pmatrix} 1 \cdot 6 + 5 \cdot 1 - 1 \cdot 1 & 1 \cdot 2 + 5 \cdot 4 - 1 \cdot 8 \\ 2 \cdot 6 - 4 \cdot 1 + 3 \cdot 1 & 2 \cdot 2 - 4 \cdot 4 + 3 \cdot 8 \end{pmatrix}$

$$= \begin{pmatrix} 10 & 14 \\ 11 & 12 \end{pmatrix}$$

(e) $\begin{pmatrix} 2 & -1 & 8 \\ 6 & 2 & -1 \end{pmatrix} \begin{pmatrix} 1 \\ 5 \\ 2 \end{pmatrix} = \begin{pmatrix} 2 \cdot 1 - 1 \cdot 5 + 8 \cdot 2 \\ 6 \cdot 1 + 2 \cdot 5 - 1 \cdot 2 \end{pmatrix} = \begin{pmatrix} 13 \\ 14 \end{pmatrix}$

The following rules may be seen to hold for matrix multiplication:

2.4.2
$$A(B + C) = AB + AC$$

2.4.3
$$(A + B)C = AC + BC$$

2.4.4
$$(sA)B = A(sB) = s(AB)$$

FINITE MATHEMATICS

2.4.5 $$A(BC) = (AB)C$$

2.4.6 $$(AB)^t = B^t A^t$$

The laws (2.4.2) and (2.4.3) are distributive laws; (2.4.4) and (2.4.5) are associative laws for matrix multiplication. Law (2.4.6) is, like (2.3.9), a modification of the commutative law for multiplication. The commutative law itself, that is, $AB = BA$, does not, in general, hold for matrix multiplication (though in certain cases it may be true). A counterexample is provided by Example II.4.2 (b) and (c).

Of the laws (2.4.2) to (2.4.6), the distributive laws are easily proved; they depend directly on the distributive laws for real numbers. The typical entry in the matrix AB is given by the expression

$$\sum_{j=1}^{n} a_{ij} b_{jk}$$

while that in AC is given by

$$\sum_{j=1}^{n} a_{ij} c_{jk}$$

Now, the typical entry in $A(B + C)$ is

$$\sum_{j=1}^{n} a_{ij}(b_{jk} + c_{jk})$$

and, by the distributive law for real numbers, it follows that the typical entry in $A(B + C)$ is the sum of the corresponding entries in AB and AC. Law (2.4.3) is proved similarly.

The law (2.4.6) is also easy to prove; it depends on the fact that the operation of transposition transforms each row into a column, and the converse. Thus the entry in the ith row and kth column of $(AB)^t$ is the same as the entry in the kth row and ith column of AB, and is therefore the scalar product of the kth row of A and ith column of B. But these are, respectively, the kth column of A^t and ith row of B^t. Application of (2.3.9) will then yield (2.4.6).

The associative law, (2.4.5), is somewhat less obvious. We shall prove this law in the case of 2×2 matrices $A = (a_{ij})$, $B = (b_{ij})$, and $C = (c_{ij})$. We have

$$AB = \begin{pmatrix} a_{11} & a_{12} \\ a_{21} & a_{22} \end{pmatrix} \begin{pmatrix} b_{11} & b_{12} \\ b_{21} & b_{22} \end{pmatrix} = \begin{pmatrix} a_{11}b_{11} + a_{12}b_{21} & a_{11}b_{12} + a_{12}b_{22} \\ a_{21}b_{11} + a_{22}b_{21} & a_{21}b_{12} + a_{22}b_{22} \end{pmatrix}$$

and so

$$(AB)C = \begin{pmatrix} a_{11}b_{11} + a_{12}b_{21} & a_{11}b_{12} + a_{12}b_{22} \\ a_{21}b_{11} + a_{22}b_{21} & a_{21}b_{12} + a_{22}b_{22} \end{pmatrix} \begin{pmatrix} c_{11} & c_{12} \\ c_{21} & c_{22} \end{pmatrix}$$

$$= \begin{pmatrix} (a_{11}b_{11} + a_{12}b_{21})c_{11} + (a_{11}b_{12} + a_{12}b_{22})c_{21} \\ (a_{21}b_{11} + a_{22}b_{21})c_{11} + (a_{21}b_{12} + a_{22}b_{22})c_{21} \end{pmatrix}$$

$$\begin{pmatrix} (a_{11}b_{11} + a_{12}b_{21})c_{12} + (a_{11}b_{12} + a_{12}b_{22})c_{22} \\ (a_{21}b_{11} + a_{22}b_{21})c_{12} + (a_{21}b_{12} + a_{22}b_{22})c_{22} \end{pmatrix}$$

On the other hand,

$$BC = \begin{pmatrix} b_{11} & b_{12} \\ b_{21} & b_{22} \end{pmatrix} \begin{pmatrix} c_{11} & c_{12} \\ c_{21} & c_{22} \end{pmatrix} = \begin{pmatrix} b_{11}c_{11} + b_{12}c_{21} & b_{11}c_{12} + b_{12}c_{22} \\ b_{21}c_{11} + b_{22}c_{21} & b_{21}c_{12} + b_{22}c_{22} \end{pmatrix}$$

and so

$$A(BC) = \begin{pmatrix} a_{11} & a_{12} \\ a_{21} & a_{22} \end{pmatrix} \begin{pmatrix} b_{11}c_{11} + b_{12}c_{21} & b_{11}c_{12} + b_{12}c_{22} \\ b_{21}c_{11} + b_{22}c_{21} & b_{21}c_{12} + b_{22}c_{22} \end{pmatrix}$$

$$= \begin{pmatrix} a_{11}(b_{11}c_{11} + b_{12}c_{21}) + a_{12}(b_{21}c_{11} + b_{22}c_{21}) \\ a_{21}(b_{11}c_{11} + b_{12}c_{21}) + a_{22}(b_{21}c_{11} + b_{22}c_{21}) \end{pmatrix}$$

$$\begin{pmatrix} a_{11}(b_{11}c_{12} + b_{12}c_{22}) + a_{12}(b_{21}c_{12} + b_{22}c_{22}) \\ a_{21}(b_{11}c_{12} + b_{12}c_{22}) + a_{22}(b_{21}c_{12} + b_{22}c_{22}) \end{pmatrix}$$

It may now be seen directly (by expanding the products) that the two matrices $(AB)C$ and $A(BC)$ are equal. This being so, we will frequently dispense with the parentheses and write ABC instead of $(AB)C$ or $A(BC)$.

The proof for matrices of other dimensions is entirely similar (though of course somewhat longer), and the reader is invited to try it.

II.4.3 Examples. The following examples illustrate these laws. Let

$$A = \begin{pmatrix} 5 & 1 \\ 3 & 2 \end{pmatrix} \quad B = \begin{pmatrix} -1 & 2 \\ 6 & 4 \end{pmatrix} \quad C = \begin{pmatrix} 2 & 3 \\ 1 & 0 \end{pmatrix}$$

Then

$$AB = \begin{pmatrix} 1 & 14 \\ 9 & 14 \end{pmatrix} \quad AC = \begin{pmatrix} 11 & 15 \\ 8 & 9 \end{pmatrix}$$

so that

$$AB + AC = \begin{pmatrix} 12 & 29 \\ 17 & 23 \end{pmatrix}$$

On the other hand,

$$B + C = \begin{pmatrix} 1 & 5 \\ 7 & 4 \end{pmatrix}$$

and so

$$A(B + C) = \begin{pmatrix} 5 & 1 \\ 3 & 2 \end{pmatrix} \begin{pmatrix} 1 & 5 \\ 7 & 4 \end{pmatrix} = \begin{pmatrix} 12 & 29 \\ 17 & 23 \end{pmatrix}$$

so that $A(B + C) = AB + AC$.

Similarly, we have

$$BC = \begin{pmatrix} 0 & -3 \\ 16 & 18 \end{pmatrix}$$

so that

$$A(BC) = \begin{pmatrix} 5 & 1 \\ 3 & 2 \end{pmatrix} \begin{pmatrix} 0 & -3 \\ 16 & 18 \end{pmatrix} = \begin{pmatrix} 16 & 3 \\ 32 & 27 \end{pmatrix}$$

while

$$(AB)C = \begin{pmatrix} 1 & 14 \\ 9 & 14 \end{pmatrix} \begin{pmatrix} 2 & 3 \\ 1 & 0 \end{pmatrix} = \begin{pmatrix} 16 & 3 \\ 32 & 27 \end{pmatrix}$$

and so $A(BC) = (AB)C$.

Finally,

$$B^t A^t = \begin{pmatrix} -1 & 6 \\ 2 & 4 \end{pmatrix} \begin{pmatrix} 5 & 3 \\ 1 & 2 \end{pmatrix} = \begin{pmatrix} 1 & 9 \\ 14 & 14 \end{pmatrix}$$

and so $B^t A^t = (AB)^t$.

5. INVERSE MATRICES

We have defined (for certain cases) the operations of addition, subtraction, and multiplication of matrices. The question naturally arises whether some type of matrix division can be defined. In fact it cannot, but we shall see that the treatment of this question leads to very important results.

As usual, we proceed by analogy with the real number system. Where real numbers are concerned, the quotient a/b is defined as the number x, which is the solution of the equation

2.5.1 $$bx = a$$

or

2.5.2 $$xb = a$$

assuming that this solution is unique. We know that multiplication of real numbers satisfies the commutative law, so that (2.5.1) and (2.5.2) are equivalent equations; they will have a unique solution whenever $b \neq 0$. Thus, the quotient a/b is well defined except when the denominator b is equal to zero; i.e., division, except by zero, is a well-defined operation.

By analogy with this notion, we might try to define the quotient A/B as the solution of the equations

2.5.3 $$BX = A$$

or

2.5.4
$$XB = A$$

Two difficulties arise. One is that multiplication of matrices is not commutative, and so the two equations (2.5.4) and (2.5.3) will not, usually, be equivalent. The other difficulty is that, even if we consider only one of these two equations, there is no guarantee of a solution, much less of a unique solution. A deeper study of this question is necessary.

As we know, division by the real number b is equivalent to multiplication by its reciprocal, $1/b$. The reciprocal $1/b$ is in turn defined as the solution of the equation

$$bx = 1$$

We see that for real numbers, the number 1 plays an important role in the theory of multiplication and division. We look, therefore, for a matrix that will play a similar role. We find, in fact, that there are several such matrices: a unit matrix of order n for each value of n.

II.5.1 Definition. The *unit matrix* of order n is the $n \times n$ matrix $I_n = (\delta_{ij})$, where

2.5.5
$$\delta_{ij} = \begin{cases} 0 & \text{if } i \neq j \\ 1 & \text{if } i = j \end{cases}$$

In other words, I_n is a square matrix that has 1's on the main diagonal, and zeros elsewhere:

$$I_4 = \begin{pmatrix} 1 & 0 & 0 & 0 \\ 0 & 1 & 0 & 0 \\ 0 & 0 & 1 & 0 \\ 0 & 0 & 0 & 1 \end{pmatrix}$$

and

$$I_2 = \begin{pmatrix} 1 & 0 \\ 0 & 1 \end{pmatrix}$$

There are, as we have said, an infinity of unit matrices, one for each value of n. When there is any ambiguity, a subscript is used to show the order of the matrix; in other cases, the order is clear from the context, and we omit the subscript.

As just mentioned, the unit matrices play a role analogous to that of the number 1 in ordinary multiplication. What this role is can best be seen from the following theorem.

II.5.2 Theorem. Let A be an $m \times n$ matrix. Then

$$I_m A = A I_n = A$$

In other words, multiplication by a unit matrix, if it can be carried out at all, leaves any matrix unchanged.

Proof. We shall prove Theorem II.5.2 for the case of 2×2 matrices. Consider $A = (a_{ij})$ and I_2. We have

$$AI_2 = \begin{pmatrix} a_{11} & a_{12} \\ a_{21} & a_{22} \end{pmatrix} \begin{pmatrix} 1 & 0 \\ 0 & 1 \end{pmatrix} = \begin{pmatrix} a_{11} \cdot 1 + a_{12} \cdot 0 & a_{11} \cdot 0 + a_{12} \cdot 1 \\ a_{21} \cdot 1 + a_{22} \cdot 0 & a_{21} \cdot 0 + a_{22} \cdot 1 \end{pmatrix}$$

$$= \begin{pmatrix} a_{11} & a_{12} \\ a_{21} & a_{21} \end{pmatrix} = A$$

Hence $AI = A$. The proof that $IA = A$ is quite similar.

The fundamental property of the unit matrices is that they serve as multiplicative identities, much as the matrices θ, considered earlier, serve as additive identities.

If we multiply the identity matrix I by a scalar s, we obtain a matrix, sI, all of whose entries are zero, except for the entries on the main diagonal, which are all equal to s. Such a matrix is called a *scalar matrix*.

If we apply (2.4.4) to a scalar matrix sI, we find that

$$(sI)A = s(IA) = sA$$

and

$$A(sI) = s(AI) = sA$$

so that multiplication, on either side, by the scalar matrix sI is equivalent to multiplication by the scalar s. This is, of course, the reason for the name.

We proceed now, to define the *inverse* of a matrix. By analogy with real numbers, we have

II.5.3 Definition. By the *inverse* of a matrix A, we shall mean a matrix A^{-1}, satisfying the equations

$$AA^{-1} = A^{-1}A = I$$

If a matrix, A, has an inverse, we say that it is *invertible* or *non-singular*. A matrix that does not have an inverse is said to be *singular*.

It may be seen that the matrix A can have an inverse only if it is a square matrix; the matrix AA^{-1} will have as many rows as A, while $A^{-1}A$ will have as many columns as A. Hence A must have the same dimensions as the unit matrix I, which is, by definition, a square. It follows, then, that A^{-1} is also a square matrix of the same order as A. Thus, all non-singular matrices are square. The converse, however, is not true: a square matrix might be singular. The matrix θ, whatever its dimensions, will always be singular, since we know that $\theta X = \theta$ for any matrix X.

While we have no guarantee that a matrix will have an inverse, we see that, if the matrix does have an inverse, it will be unique. Suppose, for instance, that A had two inverses, B and C. Consider, then, the triple product $C(AB)$ or $(CA)B$. We have, by the associative law,

$$C(AB) = (CA)B$$

But, since $AB = I$,

$$C(AB) = CI = C$$

while, since $CA = I$,

$$(CA)B = IB = B$$

Therefore, $C = B$, and the inverse is unique. We are, hence, justified in talking about *the* inverse of A, and using the symbol A^{-1} to denote this inverse.

The following laws may be seen to hold for matrix inverses (we assume A,B are invertible).

2.5.4 $$(A^{-1})^{-1} = A$$

2.5.5 $$(sA)^{-1} = \frac{1}{s} A^{-1} \quad \text{if } s \neq 0$$

2.5.6 $$(A^t)^{-1} = (A^{-1})^t$$

2.5.7 $$(AB)^{-1} = B^{-1}A^{-1}$$

All these laws are quite easily proved; for instance, (2.5.4), which states that the inverse of the inverse is the original matrix, follows directly from the definition. Law (2.5.7) is proved by observing that

$$(AB)(B^{-1}A^{-1}) = A(BB^{-1})A^{-1} = AIA^{-1} = AA^{-1} = I$$

and

$$(B^{-1}A^{-1})(AB) = B^{-1}(A^{-1}A)B = B^{-1}IB = B^{-1}B = I$$

so that $B^{-1}A^{-1}$ is indeed the inverse of AB. Law (2.5.6) is proved by applying (2.4.6) and using the fact that $I^t = I$.

The actual problem of finding the inverse of a square matrix is quite complicated, and we shall defer it until a subsequent section of this chapter. We will, however, give some examples of matrix inverses; it is not difficult to check that a pair of matrices are inverses of each other, since all we have to do is perform a pair of matrix multiplications and see that the product is indeed an identity matrix.

II.5.4 Example. Show that the two matrices

$$A = \begin{pmatrix} 5 & 1 \\ 4 & 1 \end{pmatrix} \quad B = \begin{pmatrix} 1 & -1 \\ -4 & 5 \end{pmatrix}$$

are inverses of each other.

This is done by constructing the two products, AB and BA. We have

$$AB = \begin{pmatrix} 5 \cdot 1 + 1(-4) & 5(-1) + 1 \cdot 5 \\ 4 \cdot 1 + 1(-4) & 4(-1) + 1 \cdot 5 \end{pmatrix} = \begin{pmatrix} 1 & 0 \\ 0 & 1 \end{pmatrix}$$

and

$$BA = \begin{pmatrix} 1 \cdot 5 - 1 \cdot 4 & 1 \cdot 1 - 1 \cdot 1 \\ -4 \cdot 5 + 5 \cdot 4 & -4 \cdot 1 + 5 \cdot 1 \end{pmatrix} = \begin{pmatrix} 1 & 0 \\ 0 & 1 \end{pmatrix}$$

so that $AB = BA = I$, and we find that A and B are indeed inverses.

II.5.5 Example. Show that the two matrices

$$A = \begin{pmatrix} 1 & 1 & -3 \\ 2 & 5 & 1 \\ 1 & 3 & 2 \end{pmatrix} \quad B = \begin{pmatrix} 7 & -11 & 16 \\ -3 & 5 & -7 \\ 1 & -2 & 3 \end{pmatrix}$$

are inverses of each other.

Once again, we compute AB and BA:

$$AB = \begin{pmatrix} 1 \cdot 7 + 1(-3) - 3 \cdot 1 & 1(-11) + 1 \cdot 5 - 3(-2) & 1 \cdot 16 + 1(-7) + 2 \cdot 3 \\ 2 \cdot 7 + 5(-3) + 1 \cdot 1 & 2(-11) + 5 \cdot 5 + 1(-2) & 2 \cdot 16 + 5(-7) + 1 \cdot 3 \\ 1 \cdot 7 + 3(-3) + 2 \cdot 1 & 1(-11) + 3 \cdot 5 + 2(-2) & 1 \cdot 16 + 3(-7) + 2 \cdot 3 \end{pmatrix}$$

$$= \begin{pmatrix} 1 & 0 & 0 \\ 0 & 1 & 0 \\ 0 & 0 & 1 \end{pmatrix}$$

and

$$BA = \begin{pmatrix} 7 \cdot 1 - 11 \cdot 2 + 16 \cdot 1 & 7 \cdot 1 - 11 \cdot 5 + 16 \cdot 3 & 7(-3) - 11 \cdot 1 + 16 \cdot 2 \\ -3 \cdot 1 + 5 \cdot 2 - 7 \cdot 1 & -3 \cdot 1 + 5 \cdot 5 - 7 \cdot 3 & -3(-3) + 5 \cdot 1 - 7 \cdot 2 \\ 1 \cdot 1 - 2 \cdot 2 + 3 \cdot 1 & 1 \cdot 1 - 2 \cdot 5 + 3 \cdot 3 & 1(-3) - 2 \cdot 1 + 3 \cdot 2 \end{pmatrix}$$

$$= \begin{pmatrix} 1 & 0 & 0 \\ 0 & 1 & 0 \\ 0 & 0 & 1 \end{pmatrix}$$

and A and B are inverses.

II.5.6 Example. Find the inverse of the matrix

$$A = \begin{pmatrix} a & b \\ c & d \end{pmatrix}$$

if it has one.

We are looking, here, for a matrix

$$X = \begin{pmatrix} x & y \\ z & w \end{pmatrix}$$

that will satisfy $AX = XA = I$. Let us consider the equation $AX = I$. We have

$$AX = \begin{pmatrix} ax + bz & ay + bw \\ cx + dz & cy + dw \end{pmatrix} = \begin{pmatrix} 1 & 0 \\ 0 & 1 \end{pmatrix}$$

The two matrices will be equal if, and only if, corresponding entries are equal. We therefore obtain a system of four equations in the unknowns x, y, z, w:

$$ax + bz = 1$$
$$cx + dz = 0$$
$$ay + bw = 0$$
$$cy + dw = 1$$

This system can be split into two systems of two equations each: the first two equations, with the variables x and z only,

$$ax + bz = 1$$
$$cx + dz = 0$$

can be solved by the methods discussed in Chapter I, giving

$$x = \frac{d}{ad - bc}, \quad z = \frac{-c}{ad - bc}$$

as solution, assuming, of course, that $ad - bc \neq 0$. The last two equations, in the variables y and w, will have a solution

$$y = \frac{-b}{ad - bc}, \quad w = \frac{a}{ad - bc}$$

Let us write $\Delta = ad - bc$. This quantity is called the *determinant* of the matrix A, and is often denoted by the symbol $|A|$, or det A. We find, thus, that the matrix equation $AX = I$ has a solution if $\Delta \neq 0$; this solution is

$$X = \begin{pmatrix} \dfrac{d}{\Delta} & \dfrac{-b}{\Delta} \\ \dfrac{-c}{\Delta} & \dfrac{a}{\Delta} \end{pmatrix}$$

Before stating that X is the inverse of A, we must still check that $XA = I$. We have

$$XA = \begin{pmatrix} \dfrac{ad - bc}{\Delta} & \dfrac{bd - bd}{\Delta} \\ \dfrac{-ac + ac}{\Delta} & \dfrac{-bc + ad}{\Delta} \end{pmatrix} = \begin{pmatrix} 1 & 0 \\ 0 & 1 \end{pmatrix}$$

and so we can write $X = A^{-1}$.

II.5.7 *Example.*   Find the inverses of the matrices

$$A = \begin{pmatrix} 5 & 2 \\ 3 & 1 \end{pmatrix} \quad B = \begin{pmatrix} 1 & 4 \\ 2 & 8 \end{pmatrix} \quad C = \begin{pmatrix} 2 & 3 \\ 5 & 1 \end{pmatrix}$$

We can apply Example II.5.6. For matrix A, we find

$$\Delta = 5 \cdot 1 - 3 \cdot 2 = -1$$

and so we will have

$$A^{-1} = \begin{pmatrix} \dfrac{1}{-1} & \dfrac{-2}{-1} \\ \dfrac{-3}{-1} & \dfrac{5}{-1} \end{pmatrix} = \begin{pmatrix} -1 & 2 \\ 3 & -5 \end{pmatrix}$$

For B, we find that $\Delta = 1 \cdot 8 - 2 \cdot 4 = 0$; this means that B is singular and does not have an inverse.

For C, we have

$$\Delta = 2 \cdot 1 - 5 \cdot 3 = -13$$

and so

$$C^{-1} = \begin{pmatrix} \dfrac{-1}{13} & \dfrac{3}{13} \\ \dfrac{5}{13} & -\dfrac{2}{13} \end{pmatrix}$$

It was mentioned that not all square matrices are invertible, i.e., some, such as the matrix θ, are singular. The problem of deciding which are invertible is solved in the next two sections of this chapter; here we give the following negative criterion:

II.5.8 **Theorem.**   If $AB = \theta$, but $B \neq \theta$, then A is singular.

Proof.   Suppose A is invertible. Then

$$A^{-1}(AB) = A^{-1}\theta = \theta$$

But

$$(A^{-1}A)B = IB = B$$

By the associative law, $A^{-1}(AB) = (A^{-1}A)B$, meaning that $B = \theta$. We had assumed, however, that $B \neq \theta$. This contradiction proves that A^{-1} cannot exist.

II.5.9 Example. Find the inverses of

$$A = \begin{pmatrix} 1 & 3 & 1 \\ 2 & 6 & 2 \\ 5 & 1 & 4 \end{pmatrix} \quad B = \begin{pmatrix} 5 & 1 \\ 10 & 2 \end{pmatrix}$$

if they exist.

We can apply Theorem II.5.8 here. In fact, we can see that

$$\begin{pmatrix} 1 & 3 & 1 \\ 2 & 6 & 2 \\ 5 & 1 & 4 \end{pmatrix} \begin{pmatrix} 11 \\ 1 \\ -14 \end{pmatrix} = \begin{pmatrix} 0 \\ 0 \\ 0 \end{pmatrix}$$

so that we have $AS = \theta$ with $S \neq \theta$. Similarly,

$$\begin{pmatrix} 5 & 1 \\ 10 & 2 \end{pmatrix} \begin{pmatrix} 1 \\ -5 \end{pmatrix} = \begin{pmatrix} 0 \\ 0 \end{pmatrix}$$

and we have $BT = \theta$ with $T \neq \theta$. It follows that both A and B are singular matrices: their inverses do not exist.

PROBLEMS ON INVERSE MATRICES

1. Perform the following operations.

(a) $(3, 1, 5)\,(2,4,6)^t$

(b) $(2,4) \begin{pmatrix} 1 \\ -2 \end{pmatrix}$

(c) $(1,5,3,6) \begin{pmatrix} 2 \\ 6 \\ 1 \\ -8 \end{pmatrix}$

(d) $\begin{pmatrix} 5 \\ -1 \\ 3 \end{pmatrix}^t \begin{pmatrix} 2 \\ 6 \\ 5 \end{pmatrix}$

(e) $\begin{pmatrix} 2 & 6 & 3 \\ 3 & -1 & 8 \end{pmatrix} \begin{pmatrix} 1 & 5 & 4 \\ -3 & 1 & 2 \\ 0 & -6 & 3 \end{pmatrix}$

FINITE MATHEMATICS

78

(f) $\begin{pmatrix} 1 & 0 & 3 \\ 0 & 1 & 0 \\ 0 & 0 & 1 \end{pmatrix} \begin{pmatrix} 1 & 5 \\ 2 & 4 \\ 6 & -3 \end{pmatrix}$

(g) $\begin{pmatrix} 2 & 1 & 3 \\ 4 & 6 & -2 \\ 5 & 3 & 3 \end{pmatrix} \begin{pmatrix} 1 & 4 & -2 \\ 6 & 0 & 1 \\ 2 & -5 & 0 \end{pmatrix}$

(h) $\begin{pmatrix} 1 & 4 & -2 \\ 6 & 0 & 1 \\ 2 & -5 & 0 \end{pmatrix} \begin{pmatrix} 2 & 1 & 3 \\ 4 & 6 & -2 \\ 5 & 3 & 3 \end{pmatrix}$

(i) $\begin{pmatrix} 1 & 6 & 2 \\ 4 & 0 & -5 \\ -2 & 1 & 0 \end{pmatrix} \begin{pmatrix} 2 & 4 & 5 \\ 1 & 6 & 3 \\ 3 & -2 & 3 \end{pmatrix}$

(j) $\begin{pmatrix} -1 & 3 & 5 \\ 2 & -1 & -2 \\ 0 & 5 & 8 \end{pmatrix} \begin{pmatrix} 1 & 3 & -1 \\ -8 & -24 & 8 \\ 5 & 15 & -5 \end{pmatrix}$

(k) $\begin{pmatrix} 1 & 3 & 6 & 4 \\ 2 & -4 & 1 & 3 \\ 5 & 1 & 8 & 4 \end{pmatrix} \begin{pmatrix} 1 \\ 2 \\ -4 \\ 6 \end{pmatrix}$

(l) $\begin{pmatrix} 2 & 5 & 8 \\ 6 & 1 & 2 \end{pmatrix} \begin{pmatrix} 1 & 3 & 1 \\ -2 & 4 & 2 \end{pmatrix}^t$

(m) $\begin{pmatrix} 2 & 5 & 8 \\ 6 & 1 & 2 \end{pmatrix}^t \begin{pmatrix} 1 & 3 & 1 \\ -2 & 4 & 2 \end{pmatrix}$

(n) $\begin{pmatrix} 5 & 1 & 6 \\ 3 & -5 & 1 \end{pmatrix} \begin{pmatrix} 1 & 4 & 3 & 1 \\ 2 & -4 & 3 & 2 \\ 0 & 5 & 6 & 8 \end{pmatrix}$

(o) $\begin{pmatrix} 1 & 3 \\ 2 & 4 \end{pmatrix} \begin{pmatrix} 1 & 5 \\ 6 & -1 \end{pmatrix}$

(p) $\begin{pmatrix} 2 & 6 \\ 3 & 1 \end{pmatrix} \begin{pmatrix} 1 & 3 \\ 4 & 8 \end{pmatrix}$

(q) $\begin{pmatrix} -1 & 5 \\ 2 & 6 \end{pmatrix} \begin{pmatrix} 1 & 4 \\ 6 & 2 \end{pmatrix}$

(r) $\begin{pmatrix} 1 & 5 \\ -2 & 4 \\ 1 & -6 \end{pmatrix} \begin{pmatrix} 1 & 4 \\ 1 & 6 \end{pmatrix}$

(s) $(1,5,3)\ (1,6,-8)^t$

(t) $\begin{pmatrix} 1 \\ 4 \\ 2 \end{pmatrix} (1,-3,4)$

2. Verify the associative law, $A(BC) = (AB)C$, by computing the following products in two different ways.

(a) $\begin{pmatrix} 1 & 5 \\ 2 & -4 \end{pmatrix} \begin{pmatrix} 3 \\ -2 \end{pmatrix} (1,3,1)$

(b) $\begin{pmatrix} 1 & -2 \\ 6 & 4 \end{pmatrix} \begin{pmatrix} 0 & 5 \\ 1 & -8 \end{pmatrix} \begin{pmatrix} -3 & 1 \\ 2 & 4 \end{pmatrix}$

(c) $\begin{pmatrix} 1 & 3 \\ 2 & 4 \\ 1 & -6 \end{pmatrix} \begin{pmatrix} 1 & 5 & 1 & 8 \\ 2 & -6 & -1 & 4 \end{pmatrix} \begin{pmatrix} 1 & -3 & 2 \\ 6 & 1 & -5 \\ 2 & 1 & 2 \\ 1 & 0 & 1 \end{pmatrix}$

(d) $\begin{pmatrix} 2 & 5 \\ 1 & 4 \end{pmatrix} \begin{pmatrix} 2 & 6 & 3 \\ 1 & -7 & 4 \end{pmatrix} \begin{pmatrix} 3 & 1 \\ 2 & -4 \\ 1 & 8 \end{pmatrix}$

(e) $\begin{pmatrix} 2 & 4 \\ 1 & 6 \end{pmatrix} \begin{pmatrix} 2 & 1 \\ 1 & -8 \end{pmatrix} \begin{pmatrix} 3 & 2 & 5 \\ 1 & 6 & 8 \end{pmatrix}$

(f) $(1,3,4) \begin{pmatrix} 1 & 6 & -2 \\ 2 & 4 & -3 \\ -1 & 0 & 2 \end{pmatrix} \begin{pmatrix} 1 \\ 5 \\ -2 \end{pmatrix}$

3. Solve the following equations.

(a) $\begin{pmatrix} 3 & 1 \\ 1 & 5 \end{pmatrix} \begin{pmatrix} x \\ y \end{pmatrix} = \begin{pmatrix} 7 \\ 7 \end{pmatrix}$

(b) $\begin{pmatrix} 4 & 2 \\ 6 & 0 \end{pmatrix} \begin{pmatrix} x \\ y \end{pmatrix} = \begin{pmatrix} 2 \\ 3 \end{pmatrix}$

(c) $\begin{pmatrix} 1 & 5 \\ 2 & 4 \\ 4 & 2 \end{pmatrix} \begin{pmatrix} x \\ y \end{pmatrix} = \begin{pmatrix} 3 \\ 0 \\ -6 \end{pmatrix}$

(d) $\begin{pmatrix} 2 & 1 \\ 3 & 5 \end{pmatrix} \begin{pmatrix} x & y \\ z & w \end{pmatrix} = \begin{pmatrix} 7 & 0 \\ 0 & 7 \end{pmatrix}$

(e) $\begin{pmatrix} 1 & 2 \\ 2 & 3 \end{pmatrix} \begin{pmatrix} x & y \\ z & w \end{pmatrix} = \begin{pmatrix} 3 & 5 \\ 1 & 4 \end{pmatrix}$

(f) $\begin{pmatrix} 1 & 3 \\ 2 & 5 \end{pmatrix} \begin{pmatrix} x & y \\ z & w \end{pmatrix} = \begin{pmatrix} 2 & 4 \\ 3 & 1 \end{pmatrix}$

6. ROW OPERATIONS AND THE SOLUTION OF SYSTEMS OF LINEAR EQUATIONS

Let us consider a system of two linear equations in three unknowns:

$$2x + 5y - 4z = 6$$
$$x - 2y + 2z = 3$$

It may be seen that this can be rewritten in matrix notation as

$$\begin{pmatrix} 2 & 5 & -4 \\ 1 & -2 & 2 \end{pmatrix} \begin{pmatrix} x \\ y \\ z \end{pmatrix} = \begin{pmatrix} 6 \\ 3 \end{pmatrix}$$

More generally, the system of m linear equations in n unknowns,

2.6.1
$$\begin{aligned} a_{11}x_1 + a_{12}x_2 + \ldots + a_{1n}x_n &= c_1 \\ a_{21}x_1 + a_{22}x_2 + \ldots + a_{2n}x_n &= c_n \\ \cdots\cdots\cdots\cdots\cdots\cdots\cdots\cdots\cdots\cdots \\ a_{m1}x_1 + a_{m2}x_2 + \ldots + a_{mn}x_n &= c_m \end{aligned}$$

can be rewritten in the form

2.6.2
$$\begin{pmatrix} a_{11} & a_{12} & \ldots & a_{1n} \\ a_{21} & a_{22} & \ldots & a_{2n} \\ & \cdots\cdots\cdots & \\ a_{m1} & a_{m2} & \ldots & a_{mn} \end{pmatrix} \begin{pmatrix} x_1 \\ x_2 \\ \cdot \\ \cdot \\ \cdot \\ x_n \end{pmatrix} = \begin{pmatrix} c_1 \\ c_2 \\ \cdot \\ \cdot \\ \cdot \\ c_m \end{pmatrix}$$

or, more simply, in the form

2.6.3
$$A \mathbf{x} = \mathbf{c}$$

where A is the matrix of coefficients, $\mathbf{x}$ is the column vector of variables, and $\mathbf{c}$ is the vector of constant terms. Let K be any $m \times m$ matrix. If we multiply both sides of (2.6.3) by K, we obtain, by the associative law,

2.6.4
$$(KA)\, \mathbf{x} = K\mathbf{c}$$

and it is clear that (2.6.3) implies (2.6.4), i.e., any vector $\mathbf{x}$ that satisfies (2.6.3) will also satisfy (2.6.4). The converse is not, generally, true: the solutions of (2.6.4) need not be solutions to (2.6.3). If, however, K is invertible, we may multiply both sides of (2.6.4) by K^{-1}:

$$K^{-1}(KA)\, \mathbf{x} = K^{-1}K\mathbf{c}$$

which reduces to (2.6.3). Thus, in this case, (2.6.4) implies (2.6.3), and so the two matrix equations, (2.6.3), and (2.6.4), are equivalent: any vector that solves the one will also solve the other. Now, it must be admitted that (2.6.4) has approximately the same form as (2.6.3): for an arbitrary, non-singular matrix K, we have no reason to believe that solution of the matrix equation (2.6.4), will be any easier than that of (2.6.3). We shall see, however, that the matrix K can be chosen in such a way that the equation (2.6.4) is either trivially easy to solve or else shows clearly that no solution is possible. More exactly, we shall show the existence of sequence, $K_1, K_2, \ldots, K_p$, of non-singular matrices such that the equation

$$K_p \ldots K_2 K_1 A x = K_p \ldots K_2 K_1 c$$

can be easily solved if it has any solution at all.

Let us consider the following operations that may be performed on any matrix:

2.6.5 Multiply every entry in the kth row by the non-zero scalar, r; i.e., replace a_{kj} by ra_{kj}.

2.6.6 Multiply each entry in the lth row by r and add to the corresponding entry in the kth row; i.e., replace a_{kj} by $a_{kj} + ra_{lj}$ (where $k \neq l$).

These two operations are called *fundamental row* operations; examples of them are:

II.6.1 Examples

(a) $\begin{pmatrix} 1 & 5 & 2 \\ -1 & 4 & 6 \\ 2 & 1 & 8 \end{pmatrix} \rightarrow \begin{pmatrix} 1 & 5 & 2 \\ 3(-1) & 3 \cdot 4 & 3 \cdot 6 \\ 2 & 1 & 8 \end{pmatrix} = \begin{pmatrix} 1 & 5 & 2 \\ -3 & 12 & 18 \\ 2 & 1 & 8 \end{pmatrix}$

In this example, the second row was multiplied by the scalar 3.

(b) $\begin{pmatrix} 1 & -1 & 6 \\ 2 & 4 & 5 \\ 3 & 1 & 8 \end{pmatrix} \rightarrow \begin{pmatrix} 1+2\cdot2 & -1+2\cdot4 & 6+2\cdot5 \\ 2 & 4 & 5 \\ 3 & 1 & 8 \end{pmatrix} = \begin{pmatrix} 3 & 7 & 16 \\ 2 & 4 & 5 \\ 3 & 1 & 8 \end{pmatrix}$

Here, the second row was multiplied by 2, and added to the first row.

(c) $\begin{pmatrix} 1 & 4 & 7 \\ 1 & 4 & 6 \\ 2 & 1 & 8 \end{pmatrix} \rightarrow \begin{pmatrix} 1 & 4 & 7 \\ 1-1 & 4-4 & 6-7 \\ 2 & 1 & 8 \end{pmatrix} = \begin{pmatrix} 1 & 4 & 7 \\ 0 & 0 & -1 \\ 2 & 1 & 8 \end{pmatrix}$

In this example, the first row was multiplied by -1, and added to the second row, i.e., the first row was subtracted from the second.

It is our desire to use these row operations to solve systems of linear equations. To do this, we shall show, first, that these operations are always equivalent to multiplication on the left by a non-singular matrix K; and second, that the matrix K depends only on the form of the operation, and not on the particular matrix A upon which we operate. This second property means that multiplication by K will perform the same operation on the vector c as on the matrix A.

Consider, then, operation (2.6.5). It may be seen that the corresponding matrix is $T = (t_{ij})$, defined by

2.6.7
$$t_{ij} = \begin{cases} r & \text{if } i=j=k \\ 1 & \text{if } i=j \neq k \\ 0 & \text{if } i \neq j \end{cases}$$

(T is almost an identity matrix, the only difference being that its (k,k)th entry is r instead of 1.) The matrix T is non-singular; its inverse is $T^{-1} = (t'_{ij})$, where

2.6.8
$$t'_{ij} = \begin{cases} 1/r & \text{if } i=j=k \\ 1 & \text{if } i=j \neq k \\ 0 & \text{if } i \neq j \end{cases}$$

That T will, in fact, cause the operation (2.6.5) can be proved in a manner similar to the proof, given earlier, that the matrix I is a multiplicative identity. That (2.6.8) does, in fact, define the inverse matrix is most easily seen when we consider that it induces the inverse operation, i.e., it causes the kth row to be *divided* by the scalar r.

Next let us consider operation (2.6.6). It may be seen that this is induced by the matrix $I + S$, in which I is the identity matrix, and $S = (s_{ij})$ is defined by

2.6.9
$$s_{ij} = \begin{cases} r & \text{if } i=k, j=l \\ 0 & \text{otherwise} \end{cases}$$

(S has all entries equal to zero except for the (k,l)th entry, which is equal to r.) It is not difficult to see, since $k \neq l$, that $SS = \theta$, and so

$$\begin{aligned}(I+S)(I-S) &= II + SI - IS + SS \\ &= I + S - S + \theta \\ &= I\end{aligned}$$

Similarly, we can see that $(I-S)(I+S) = I$, so that $I - S = (I+S)^{-1}$, and $I + S$ is non-singular.

We are now in a position to apply these elementary row operations to the solution of systems of linear equations. The general idea is to perform the operations on the matrix of coefficients until a very simple type of matrix is obtained; we try to change A, if possible, into an identity matrix. If A is singular, this will not be feasible, so we have to content ourselves with something that looks as much like an identity matrix as possible.

We have seen that each of these row operations is equivalent to multiplication by a non-singular matrix K. The vector of constant terms, $\mathbf{c}$, must simultaneously be multiplied by the same K. But K induces the same row operation on all matrices (so long as they have the proper number of rows so that multiplication is possible). Hence we need only perform the same row operations on the vector c as we do on the matrix A.

Since the same row operations are to be performed on both A and c, it is generally more convenient to deal with the *augmented* matrix

$$2.6.10 \quad (A \mid c) = \begin{pmatrix} a_{11}\, a_{12} \cdots\cdots\cdots\cdots a_{1n} & c_1 \\ a_{21}\, a_{22} \cdots\cdots\cdots\cdots a_{2n} & c_2 \\ \cdots\cdots\cdots\cdots\cdots\cdots\cdots & \cdots \\ a_{m1} a_{m2} \cdots\cdots\cdots\cdots a_{mn} & c_m \end{pmatrix}$$

It is then simply a question of performing the row operations on $(A \mid c)$; this guarantees that they be performed simultaneously on A and c.

II.6.2 Example. Solve the system of equations

$$\begin{aligned} x + 2y - z + w &= 5 \\ 2x - y + w &= 6 \\ x - y + z + 3w &= 9 \\ 4y - z - 2w &= 1 \end{aligned}$$

This system may be written, using matrix notation, in the form

$$\begin{pmatrix} 1 & 2 & -1 & 1 \\ 2 & -1 & 0 & 1 \\ 1 & -1 & 1 & 3 \\ 0 & 4 & -1 & -2 \end{pmatrix} \begin{pmatrix} x \\ y \\ z \\ w \end{pmatrix} = \begin{pmatrix} 5 \\ 6 \\ 9 \\ 1 \end{pmatrix}$$

We obtain, thus, the augmented matrix

$$\begin{pmatrix} 1 & 2 & -1 & 1 & 5 \\ 2 & -1 & 0 & 1 & 6 \\ 1 & -1 & 1 & 3 & 9 \\ 0 & 4 & -1 & -2 & 1 \end{pmatrix}$$

We would like, if possible, to change this matrix, by means of row operations, until the first four columns form an identity matrix. The best method is, generally, to take the columns one at a time, changing the "main diagonal" entry to a 1 and getting rid of any other non-zero entries in the column. In the first column, here, we find that the first entry is, as desired, a 1. The entries in the second and third rows are not zero, so we must eliminate them. To eliminate the 2 in the second row, we can multiply the first row by -2 and add to the second row; we also subtract the first row from the third, obtaining

$$\begin{pmatrix} 1 & 2 & -1 & 1 & 5 \\ 2-2 & -1-4 & 0+2 & 1-2 & 6-10 \\ 1-1 & -1-2 & 1+1 & 3-1 & 9-5 \\ 0 & 4 & -1 & -2 & 1 \end{pmatrix} = \begin{pmatrix} 1 & 2 & -1 & 1 & 5 \\ 0 & -5 & 2 & -1 & -4 \\ 0 & -3 & 2 & 2 & 4 \\ 0 & 4 & -1 & -2 & 1 \end{pmatrix}$$

We have, thus, reduced the first column to the required form. We now attack the second column: we must obtain here a 1 in the second row and zeros elsewhere, *being careful meanwhile that the first column remains as it is.* (This means that we cannot, say, multiply the first row by 3 and add to the second row; while this would accomplish the feat of putting a 1 in the second row, second column, it would also put a 3 in the second row, first column.)

There are many ways of putting a 1 in the second row, second column. One possibility is to multiply the third row by -2 and add to the second row. Another possibility is to multiply the second row by $-1/5$. The former procedure has the advantage of avoiding fractions, at least for the moment. The latter has the advantage of being somewhat more systematic. Let us, then, carry out this latter procedure; we obtain the matrix

$$\left(\begin{array}{cccc|c} 1 & 2 & -1 & 1 & 5 \\ 0 & 1 & -2/5 & 1/5 & 4/5 \\ 0 & -3 & 2 & 2 & 4 \\ 0 & 4 & -1 & -2 & 1 \end{array}\right)$$

We now use the second row to remove all the other non-zero entries in this column. We can multiply the second row by -2, and add to the first row; by 3, and add to the third row; by -4, and add to the fourth row. Carrying out all these operations, we will have

$$\left(\begin{array}{cccc|c} 1 & 2-2 & -1+4/5 & 1-2/5 & 5-8/5 \\ 0 & 1 & -2/5 & 1/5 & 4/5 \\ 0 & -3+3 & 2-6/5 & 2+3/5 & 4+12/5 \\ 0 & 4-4 & -1+8/5 & -2-4/5 & 1-16/5 \end{array}\right) = \left(\begin{array}{cccc|c} 1 & 0 & -1/5 & 3/5 & 17/5 \\ 0 & 1 & -2/5 & 1/5 & 4/5 \\ 0 & 0 & 4/5 & 13/5 & 32/5 \\ 0 & 0 & 3/5 & -14/5 & -11/5 \end{array}\right)$$

We proceed to the third column. We can multiply the third row by 5/4:

$$\left(\begin{array}{cccc|c} 1 & 0 & -1/5 & 3/5 & 17/5 \\ 0 & 1 & -2/5 & 1/5 & 4/5 \\ 0 & 0 & 1 & 13/4 & 8 \\ 0 & 0 & 3/5 & -14/5 & -11/5 \end{array}\right)$$

and now we use the third row to remove the other non-zero entries in the third column, obtaining

$$\left(\begin{array}{cccc|c} 1 & 0 & -1/5+1/5 & 3/5+13/20 & 17/5+8/5 \\ 0 & 1 & -2/5+2/5 & 1/5+3/10 & 4/5+16/5 \\ 0 & 0 & 1 & 13/4 & 8 \\ 0 & 0 & 3/5-3/5 & -14/5-39/20 & -11/5-24/5 \end{array}\right) = \left(\begin{array}{cccc|c} 1 & 0 & 0 & 5/4 & 5 \\ 0 & 1 & 0 & 3/2 & 4 \\ 0 & 0 & 1 & 13/4 & 8 \\ 0 & 0 & 0 & -19/4 & -7 \end{array}\right)$$

We next multiply the bottom row by $-4/19$:

$$\begin{pmatrix} 1 & 0 & 0 & 5/4 & 5 \\ 0 & 1 & 0 & 3/2 & 4 \\ 0 & 0 & 1 & 13/4 & 8 \\ 0 & 0 & 0 & 1 & 28/19 \end{pmatrix}$$

and use this row to eliminate the other fourth-column entries:

$$\begin{pmatrix} 1 & 0 & 0 & 0 & 60/19 \\ 0 & 1 & 0 & 0 & 34/19 \\ 0 & 0 & 1 & 0 & 61/19 \\ 0 & 0 & 0 & 1 & 28/19 \end{pmatrix}$$

We have now accomplished what we set out to do: the matrix of coefficients has become an identity matrix, and the original matrix equation, $A\mathbf{x} = \mathbf{c}$, has been reduced to the form

$$I\begin{pmatrix} x \\ y \\ z \\ w \end{pmatrix} = \begin{pmatrix} 60/19 \\ 34/19 \\ 61/19 \\ 28/19 \end{pmatrix}$$

But the matrix I will not change anything. This last equation gives us, directly, the solution of the original system. It is, precisely, $(x, y, z, w) = (60/19, 34/19, 61/19, 28/19)$. That this satisfies the original system may be checked directly.

II.6.3 Example. Solve the system

$$\begin{aligned} x - 2y + z + 2w &= 18 \\ 2x + y + 4z - w &= 16 \\ x + y - z + w &= 4 \\ x + 3y - 2z + w &= 1 \end{aligned}$$

As before, we form the augmented matrix

$$\begin{pmatrix} 1 & -2 & 1 & 2 & 18 \\ 2 & 1 & 4 & -1 & 16 \\ 1 & 1 & -1 & 1 & 4 \\ 1 & 3 & -2 & 1 & 1 \end{pmatrix}$$

We use the first row to remove the other non-zero entries in the first column:

$$\begin{pmatrix} 1 & -2 & 1 & 2 & 18 \\ 0 & 5 & 2 & -5 & -20 \\ 0 & 3 & -2 & -1 & -14 \\ 0 & 5 & -3 & -1 & -17 \end{pmatrix}$$

FINITE MATHEMATICS

Next, we divide the second row by 5 and use this row to remove the non-zero entries in the second column:

$$\begin{pmatrix} 1 & 0 & 9/5 & 0 & | & 10 \\ 0 & 1 & 2/5 & -1 & | & -4 \\ 0 & 0 & -16/5 & 2 & | & -2 \\ 0 & 0 & -5 & 4 & | & 3 \end{pmatrix}$$

We multiply the third row by $-5/16$ and use it to clear the third column:

$$\begin{pmatrix} 1 & 0 & 0 & 9/8 & | & 71/8 \\ 0 & 1 & 0 & -3/4 & | & -17/4 \\ 0 & 0 & 1 & -5/8 & | & 5/8 \\ 0 & 0 & 0 & 7/8 & | & 49/8 \end{pmatrix}$$

Finally, we multiply the last row by $8/7$ and clear the fourth column,

$$\begin{pmatrix} 1 & 0 & 0 & 0 & | & 1 \\ 0 & 1 & 0 & 0 & | & 1 \\ 0 & 0 & 1 & 0 & | & 5 \\ 0 & 0 & 0 & 1 & | & 7 \end{pmatrix}$$

This gives us the solution; it is $(x, y, z, w) = (1, 1, 5, 7)$. Again, we can check directly by substituting these values in the original equations.

7. SOLUTION OF GENERAL $m \times n$ SYSTEMS OF EQUATIONS

Naturally, it is not always possible to solve systems of equations as we did in Examples II.6.2 and II.6.3. Even if the matrix of coefficients is square, we have no guarantee that it will be invertible. We can see what happens in such cases from the following examples.

II.7.1 Example. Solve the system

$$\begin{aligned} x + 2y - z &= 6 \\ 3x - y + z &= 8 \\ 5x + 3y - z &= 20 \end{aligned}$$

In this case, we form the augmented matrix

$$\begin{pmatrix} 1 & 2 & -1 & | & 6 \\ 3 & -1 & 1 & | & 8 \\ 5 & 3 & -1 & | & 20 \end{pmatrix}$$

and use the first row to clear the first column:

$$\begin{pmatrix} 1 & 2 & -1 & | & 6 \\ 0 & -7 & 4 & | & -10 \\ 0 & -7 & 4 & | & -10 \end{pmatrix}$$

It may be seen that the second and third rows have become identical; this means that one of the equations is redundant, so that, with only two independent equations, we cannot hope to have a unique solution. If we continue with our procedure, multiplying the second row by $-1/7$ and clearing the second column, we obtain

$$\begin{pmatrix} 1 & 0 & 1/7 & | & 22/7 \\ 0 & 1 & -4/7 & | & 10/7 \\ 0 & 0 & 0 & | & 0 \end{pmatrix}$$

Consider this new matrix: we would like to get a 1 in the third row, third column. This cannot be done by multiplying the third row, since we now have a zero in that position. We could do this by using one of the *other* rows, say, by multiplying the first row by 7 and adding to the third row. If we do this, however, we obtain a 7 in the third row, first column, thus undoing some of our previous work. We find, in fact, that we can go no further in simplifying the coefficient matrix. Our final result is the system

$$x + 1/7z = 22/7$$
$$y - 4/7z = 10/7$$

The third equation, of course, reduces to the identity $0 = 0$, and can be dispensed with. We can, then, solve for x and y in terms of z:

$$x = 22/7 - 1/7z$$
$$y = 10/7 + 4/7z$$

The variable z is arbitrary. This is the general solution, or, more exactly, one form of the general solution. We could, of course, solve for any two of the variables in terms of the third. Let us suppose that we want to solve for y and z in terms of x. We can do this by obtaining a second-order identity matrix in the columns corresponding to y and z, rather than in those corresponding to x and y. Starting from the last of the foregoing matrices, we can multiply the first row by 7:

$$\begin{pmatrix} 7 & 0 & 1 & | & 22 \\ 0 & 1 & -4/7 & | & 10/7 \\ 0 & 0 & 0 & | & 0 \end{pmatrix}$$

We then clear the third column by means of the first row:

$$\begin{pmatrix} 7 & 0 & 1 & | & 22 \\ 4 & 1 & 0 & | & 14 \\ 0 & 0 & 0 & | & 0 \end{pmatrix}$$

We could, of course, have discarded the third row a long time ago, reducing to the 2×4 matrix

$$\begin{pmatrix} 7 & 0 & 1 & | & 22 \\ 4 & 1 & 0 & | & 14 \end{pmatrix}$$

Now, the two columns corresponding to y and z do not form an identity matrix, but this is not too important. In fact, these two columns form what is called a *permutation* matrix, a matrix that can be transformed into an identity matrix by simply interchanging some of the rows. As such, the foregoing matrix is sufficient for our purposes. Indeed, the equations we now have are

$$\begin{aligned} 7x \quad\;\; + z &= 22 \\ 4x + y \quad\;\; &= 14 \end{aligned}$$

or, solving for y and z in terms of x,

$$\begin{aligned} y &= 14 - 4x \\ z &= 22 - 7x \end{aligned}$$

which, for arbitrary x, is another form of the general solution to the system.

II.7.2 Example. Solve the system

$$\begin{aligned} x + 2y - z &= 6 \\ 3x - y + z &= 8 \\ 5x + 3y - z &= 22 \end{aligned}$$

Again we form the augmented matrix

$$\begin{pmatrix} 1 & 2 & -1 & | & 6 \\ 3 & -1 & 1 & | & 8 \\ 5 & 3 & -1 & | & 22 \end{pmatrix}$$

and use the first row to clear the first column:

$$\begin{pmatrix} 1 & 2 & -1 & | & 6 \\ 0 & -7 & 4 & | & -10 \\ 0 & -7 & 4 & | & -8 \end{pmatrix}$$

We then divide the second row by -7, and clear the second column, obtaining

$$\begin{pmatrix} 1 & 0 & 1/7 & | & 22/7 \\ 0 & 1 & -4/7 & | & 10/7 \\ 0 & 0 & 0 & | & 2 \end{pmatrix}$$

Consider, now, the third row: it represents the contradiction $0 = 2$. This means that the system has no solution. In general, a system will be infeasible (i.e., have no solution) if a row is obtained in which all the entries except the last are zero. On the other hand, as we saw from Example II.7.1, a row that has all its entries (including the last one) equal to zero may be discarded and does not affect the system.

For systems that have fewer equations than variables, the procedure will generally be as in Example II.7.1: we try to obtain an identity matrix (or a permutation matrix) in the columns corresponding to some of the variables; this solves the system for the corresponding variables in terms of the others.

II.7.3 Example. Solve the system

$$2x + y - 4z = 10$$
$$x - y + 2z = 6$$

We can, here, attempt to solve the system for any two of the variables in terms of the third. (In this case, our attempts will be successful, but in some cases they need not be.) Let us, then, solve for x and z in terms of y. We will work on the matrix in such a way that the columns corresponding to x and z form an identity (or permutation) matrix. The augmented matrix is

$$\begin{pmatrix} 2 & 1 & -4 & | & 10 \\ 1 & -1 & 2 & | & 6 \end{pmatrix}$$

We can divide the first row by 2 and clear the first column:

$$\begin{pmatrix} 1 & 1/2 & -2 & | & 5 \\ 0 & -3/2 & 4 & | & 1 \end{pmatrix}$$

Next we divide the second row by 4 and clear the third column to obtain

$$\begin{pmatrix} 1 & -1/4 & 0 & | & 11/2 \\ 0 & -3/8 & 1 & | & 1/4 \end{pmatrix}$$

so that the desired solution is

$$x = 11/2 + 1/4y$$
$$z = 1/4 + 3/8y$$

Let us suppose, now that we want to solve for y and z in terms of x. In this last matrix, we multiply the first row by -4:

$$\begin{pmatrix} -4 & 1 & 0 & | & -22 \\ 0 & -3/8 & 1 & | & 1/4 \end{pmatrix}$$

and use the first row to clear the second column:

$$\begin{pmatrix} -4 & 1 & 0 & | & -22 \\ -3/2 & 0 & 1 & | & -8 \end{pmatrix}$$

thus obtaining the solution

$$y = -22 + 4x$$
$$z = -8 + 3/2x$$

II.7.4 Example. Solve the system

$$2x + y + 4z = 10$$
$$x - y + 2z = 6$$

Suppose we want to solve this system for x and z in terms of y. We take the augmented matrix

$$\begin{pmatrix} 2 & 1 & 4 & | & 10 \\ 1 & -1 & 2 & | & 6 \end{pmatrix}$$

and, as before, divide the first row by 2, and clear the first column:

$$\begin{pmatrix} 1 & 1/2 & -2 & | & 5 \\ 0 & -3/2 & 0 & | & 1 \end{pmatrix}$$

We would like to clear the third column now, but it has become impossible. In fact, we cannot solve for x and z in terms of y. The reason for this is best seen if we note that the second row represents the equation

$$-\frac{3}{2}y = 1$$

which tells us that $y = -2/3$. In effect, we have found that y is not arbitrary; if we were to give x and z in terms of y, we would get only one solution. But the problem has an infinity of solutions.

While it is not possible, here, to solve for x and z in terms of y, we may, if we wish, solve for x and y in terms of z. In the last matrix given, we may multiply the second row by $-2/3$, and clear the second column:

$$\begin{pmatrix} 1 & 0 & -2 & | & 16/3 \\ 0 & 1 & 0 & | & -2/3 \end{pmatrix}$$

which gives us

$$x = \frac{16}{3} + 2z$$

$$y = \frac{-2}{3}$$

Note that x is given in terms of z; y, however, has only one value.

In case we have more equations than variables, the usual procedure is to try to obtain an identity matrix in the first n rows and n columns; below this, we try to get nothing but zeros. When this has been done, we might find that the last $m - n$ rows contain zeros only; if so, we have obtained the solution. On the other hand, it may be that one of the last $m - n$ rows has a non-zero entry in the right-hand column, but all zeros otherwise. In this case, the problem has no solution, as this row represents the contradiction $0 = 1$.

II.7.5 *Example*. Solve the system

$$\begin{aligned}
x + y + z &= 9 \\
2x - y + 2z &= 15 \\
x - y + 2z &= 12 \\
x + 2y - z &= 0
\end{aligned}$$

We take the augmented matrix

$$\left(\begin{array}{ccc|c}
1 & 1 & 1 & 9 \\
2 & -1 & 2 & 15 \\
1 & -1 & 2 & 12 \\
1 & -2 & -1 & 0
\end{array}\right)$$

and use the first row to clear the first column:

$$\left(\begin{array}{ccc|c}
1 & 1 & 1 & 9 \\
0 & -3 & 0 & -3 \\
0 & -2 & 1 & 3 \\
0 & 1 & -2 & -9
\end{array}\right)$$

We divide the second row by -3, and use it to clear the second column:

$$\left(\begin{array}{ccc|c}
1 & 0 & 1 & 8 \\
0 & 1 & 0 & 1 \\
0 & 0 & 1 & 5 \\
0 & 0 & -2 & -10
\end{array}\right)$$

Finally, we can use the third row to clear the second column, obtaining the matrix

$$\left(\begin{array}{ccc|c}
1 & 0 & 0 & 3 \\
0 & 1 & 0 & 1 \\
0 & 0 & 1 & 5 \\
0 & 0 & 0 & 0
\end{array}\right)$$

which gives us the solution

$$x = 3$$
$$y = 1$$
$$z = 5$$

Note that the last row consists only of zeros: it may be disregarded.

II.7.6   Example.   Solve the system

$$
\begin{aligned}
x + y - z &= 1 \\
2x + y + z &= 9 \\
x - 2y + z &= 1 \\
x + 2y - 2z &= 0 \\
x - y + z &= 2
\end{aligned}
$$

Let us once again take the augmented matrix

$$
\left(
\begin{array}{rrr|r}
1 & 1 & -1 & 1 \\
2 & 1 & 1 & 9 \\
1 & -2 & 1 & 1 \\
1 & 2 & -2 & 0 \\
1 & -1 & 1 & 2
\end{array}
\right)
$$

Once again, we use the first row to clear the first column:

$$
\left(
\begin{array}{rrr|r}
1 & 1 & -1 & 1 \\
0 & -1 & 3 & 7 \\
0 & -3 & 2 & 0 \\
0 & 1 & -1 & -1 \\
0 & -2 & 2 & 1
\end{array}
\right)
$$

We multiply the second row by -1, and clear the second column:

$$
\left(
\begin{array}{rrr|r}
1 & 0 & 2 & 8 \\
0 & 1 & -3 & -7 \\
0 & 0 & -7 & -21 \\
0 & 0 & 2 & 6 \\
0 & 0 & -4 & -13
\end{array}
\right)
$$

Finally, we divide the third row by -7, and clear the third column:

$$
\left(
\begin{array}{rrr|r}
1 & 0 & 0 & 2 \\
0 & 1 & 0 & 2 \\
0 & 0 & 1 & 3 \\
0 & 0 & 0 & 0 \\
0 & 0 & 0 & 1
\end{array}
\right)
$$

Consider this matrix: the first three rows seem to give us the solution $(x, y, z) = (2, 2, 3)$. The last row, however, represents the contradiction $0 = 1$. We conclude that the system is infeasible, since the unique solution to the first three equations fails to satisfy the fifth equation.

PROBLEMS ON SOLUTION OF GENERAL $m \times n$ SYSTEMS OF EQUATIONS

1. Solve the following systems of simultaneous equations.

(a) $3x + 5y = 6$
$x - 2y = 2$

(b) $3x + 5y = 4$
$x + 2y = 7$

(c) $5x + 2y + 4z = 7$
$3x - 2y + z = 6$
$x + y + z = 4$

(d) $2x - 6y + z = -11$
$x + 2y - z = 2$
$-x + y + 2z = 12$

(e) $x - 2y + 3z - w = 12$
$2x + y + z - 2w = -6$
$-x + y + 2z + w = 2$
$-y + z + w = 1$

(f) $3x + y + z = 2$
$x - 2y + 2z = 5$
$2x - y - z = -7$
$5x + 3y + z = 2$

(g) $x + 2y + 4z + w = 6$
$3x + 2y - 3z - w = 5$
$-x + 6y + 2z + 3w = 8$

(h) $2x + 3y + z - 2w = 0$
$x - 2y + 2z - w = 9$
$3x - y + 3z + w = 6$
$-x + y - z + 3w = 2$

(i) $3x + 2y + 6z + w = 12$
$x - 2y + 3z + 2w = 4$
$-2x + 4y - 2z - w = -1$
$x + y - 2z + 3w = 3$

(j) $2x + 4y + 2z - w = 18$
$x + 3y - 2z + w = 13$
$-x + 2y + 2w = 28$
$3x + y + 4z = 1$

(k) $x + 2y + 6z - w = 5$
$3x + y + 4z - 2w = 7$
$x - y + 7z + w = 9$
$-x + 2y + 3z - 2w = 5$

(l) $x - 3y + 5z - w = -3$
$2x + y + 4z + w = 11$
$-x + y + z - 3w = -7$
$3x + 6y + 2z - 4w = 5$
$2x + 4y + z - 7w = -10$
$x + y + z - w = 1$

(m) $x - 2y + 3z + w = 6$
$2x + y + 6z - 2w = 1$
$5x + 2y - 3z - 2w = -4$
$4x + y - 2z - w = -1$
$6x + 2y + z + w = 10$

(n) $2x + 3y + z - 5w = 8$
$-x + 3y + 7z + 5w = 30$
$x - 2y + 4z - 6w = 7$
$-2x + y + z + 8w = 11$

(o) $3x + y + 2z - w = 6$
$-x - y + 2z - 2w = 5$
$x + 4y + 6z - 3w = 8$

8. COMPUTATION OF THE INVERSE MATRIX

In Section 4 of this chapter, we saw that we can, in effect, solve a matrix equation $A\mathbf{x} = \mathbf{c}$ by using the inverse matrix; the solution is $\mathbf{x} = A^{-1}\mathbf{c}$. On the other hand, it may be noticed that, while all the row operations gave us this solution, the matrix A^{-1} was never actually computed. It is true, of course, that this was not necessary in the context of such problems. We may, however, conceive of a situation in which computation of A^{-1} may actually help us to avoid a great amount of work.

Let us suppose, for instance, that we are given several systems of equations:

2.8.1 $A\mathbf{x} = \mathbf{c}$

2.8.2 $A\mathbf{x} = \mathbf{c}'$

2.8.3 $A\mathbf{x} = \mathbf{c}''$
$\cdots\cdots\cdots$

2.8.4 $A\mathbf{x} = \mathbf{c}^{(p)}$

in which, in each case, the matrix of coefficient A, remains the same, but the vector of constant terms varies. This may happen, in a physical case, where the matrix A represents the physical characteristics of a system, while the vectors $\mathbf{c}, \mathbf{c}', \mathbf{c}'', \ldots, \mathbf{c}^{(p)}$ represent the several outputs that one may wish to obtain from the system. The variables $\mathbf{x}$ would then be the inputs necessary to obtain the desired outputs.

Now, we may solve each of the systems (2.8.1), (2.8.2), (2.8.3), ..., (2.8.4) by the method of Section II.6, if we wish. On the other hand, if there are many of them, this will mean carrying out the same set of row operations several times over. It becomes much more practical, in this case, to compute the matrix A^{-1}; it is then quite simple to generate the solutions $A^{-1}\mathbf{c}, A^{-1}\mathbf{c}', \ldots, A^{-1}\mathbf{c}^{(p)}$ of the several systems. This procedure also has the advantage of allowing us to see how the solution vector, $\mathbf{x}$, depends on the vector of constant terms, $\mathbf{c}$.

We wish, then, a procedure for inverting matrices. Our method will once again employ the fundamental row operations (2.6.5) and (2.6.6). As we have seen, a non-singular matrix A may, by means of these operations, be transformed into the identity matrix I. But each row operation corresponds to multiplication by a matrix K. What these operations do, in effect, is to provide us with a sequence of matrices, $K_1, K_2, \ldots, K_q$, such that

2.8.5 $$K_q \ldots K_2 K_1 A = I$$

But this means that

2.8.6 $$K_q \ldots K_2 K_1 = A^{-1}$$

Now, precisely because I is the multiplicative identity, we have

2.8.7 $$A^{-1} = K_q \ldots K_2 K_1 I$$

This means that the inverse matrix, A^{-1}, may be obtained by subjecting I to those row operations that changed A into I.

II.8.1 Example. Invert the matrix

$$A = \begin{pmatrix} 3 & 1 \\ 5 & 2 \end{pmatrix}$$

We could, of course, invert the matrix directly by using the results of Section II.5. Let us, however, invert it by the method of row operations. We form the double matrix

$$(A \mid I) = \begin{pmatrix} 3 & 1 & 1 & 0 \\ 5 & 2 & 0 & 1 \end{pmatrix}$$

In this matrix, we divide the first row by 3:

$$\begin{pmatrix} 1 & 1/3 & \bigm| & 1/3 & 0 \\ 5 & 2 & \bigm| & 0 & 1 \end{pmatrix}$$

and use this row to clear the first column:

$$\begin{pmatrix} 1 & 1/3 & \bigm| & 1/3 & 0 \\ 0 & 1/3 & \bigm| & -5/3 & 1 \end{pmatrix}$$

Next we multiply the second row by 3, and use it to clear the second column:

$$\begin{pmatrix} 1 & 0 & \bigm| & 2 & -1 \\ 0 & 1 & \bigm| & -5 & 3 \end{pmatrix}$$

We have now, by means of row operations, reduced the double matrix, $(A \mid I)$ to the form $(I \mid B)$. But this B is precisely the desired inverse matrix. In other words,

$$A^{-1} = \begin{pmatrix} 2 & -1 \\ -5 & 3 \end{pmatrix}$$

which, of course, coincides with the results of Section II.5.

II.8.2 *Example.* Invert the matrix

$$A = \begin{pmatrix} 1 & -2 & 1 \\ 2 & 1 & 4 \\ 1 & 1 & -1 \end{pmatrix}$$

Once again, we form the double matrix

$$\begin{pmatrix} 1 & -2 & 1 & \bigm| & 1 & 0 & 0 \\ 2 & 1 & 4 & \bigm| & 0 & 1 & 0 \\ 1 & 1 & -1 & \bigm| & 0 & 0 & 1 \end{pmatrix}$$

We use the first row to clear the first column:

$$\begin{pmatrix} 1 & -2 & 1 & \bigm| & 1 & 0 & 0 \\ 0 & 5 & 2 & \bigm| & -2 & 1 & 0 \\ 0 & 3 & -2 & \bigm| & -1 & 0 & 1 \end{pmatrix}$$

We then divide the second row by 5, and clear the second column:

$$\begin{pmatrix} 1 & 0 & 9/5 & \bigm| & 1/5 & 2/5 & 0 \\ 0 & 1 & 2/5 & \bigm| & -2/5 & 1/5 & 0 \\ 0 & 0 & -16/5 & \bigm| & 1/5 & -3/5 & 1 \end{pmatrix}$$

Finally, we multiply the third row by $-5/16$, and use this row to clear the third column:

$$\begin{pmatrix} 1 & 0 & 0 & | & 5/16 & 1/16 & 9/16 \\ 0 & 1 & 0 & | & -3/8 & 1/8 & 1/8 \\ 0 & 0 & 1 & | & -1/16 & 3/16 & -5/16 \end{pmatrix}$$

This gives us, then, the inverse matrix; it is the right half of this double matrix, or, more simply,

$$A^{-1} = \frac{1}{16}\begin{pmatrix} 5 & 1 & 9 \\ -6 & 2 & 2 \\ -1 & 3 & -5 \end{pmatrix}$$

It may be checked directly that this is, indeed, the inverse matrix.

II.8.3 Example. Find the solution of the system

$$\begin{aligned} x + 3y - z + 4w &= a \\ -x + 2y + z - w &= b \\ 2x - y + w &= c \\ x + 2z + 3w &= d \end{aligned}$$

in which a, b, c, d are *parameters*, i.e., constants that may be changed at the option of the person handling the system.

Since a solution is wanted in terms of general values for the parameters, we invert the matrix of coefficients. Once again, we take the double matrix

$$\begin{pmatrix} 1 & 3 & -1 & 4 & | & 1 & 0 & 0 & 0 \\ -1 & 2 & 1 & -1 & | & 0 & 1 & 0 & 0 \\ 2 & -1 & 0 & 1 & | & 0 & 0 & 1 & 0 \\ 1 & 0 & 2 & 3 & | & 0 & 0 & 0 & 1 \end{pmatrix}$$

and we use the first row to clear the first column. The intermediate steps follow without further comment:

$$\begin{pmatrix} 1 & 3 & -1 & 4 & | & 1 & 0 & 0 & 0 \\ 0 & 5 & 0 & 3 & | & 1 & 1 & 0 & 0 \\ 0 & -7 & 2 & -7 & | & -2 & 0 & 1 & 0 \\ 0 & -3 & 3 & -1 & | & -1 & 0 & 0 & 1 \end{pmatrix}$$

$$\begin{pmatrix} 1 & 0 & -1 & 11/5 & | & 2/5 & -3/5 & 0 & 0 \\ 0 & 1 & 0 & 3/5 & | & 1/5 & 1/5 & 0 & 0 \\ 0 & 0 & 2 & -15/4 & | & -3/5 & 7/5 & 1 & 0 \\ 0 & 0 & 3 & 4/5 & | & -2/5 & 3/5 & 0 & 1 \end{pmatrix}$$

$$\begin{pmatrix} 1 & 0 & 0 & 4/5 & | & 1/10 & 1/10 & 1/2 & 0 \\ 0 & 1 & 0 & 3/5 & | & 1/5 & 1/5 & 0 & 0 \\ 0 & 0 & 1 & -7/5 & | & -3/10 & 7/10 & 1/2 & 0 \\ 0 & 0 & 0 & 5 & | & 1/2 & -3/2 & -3/2 & 1 \end{pmatrix}$$

$$\begin{pmatrix} 1 & 0 & 0 & 0 & | & 1/50 & 17/50 & 37/50 & -4/25 \\ 0 & 1 & 0 & 0 & | & 7/50 & 19/50 & 9/50 & -3/25 \\ 0 & 0 & 1 & 0 & | & -4/25 & 7/25 & 2/25 & 7/25 \\ 0 & 0 & 0 & 1 & | & 1/10 & -3/10 & -3/10 & 1/5 \end{pmatrix}$$

The right side of this last matrix gives us the inverse matrix; the solution to the problem is then simply given by

$$\begin{pmatrix} x \\ y \\ z \\ w \end{pmatrix} = \frac{1}{50} \begin{pmatrix} 1 & 17 & 37 & -8 \\ 7 & 19 & 9 & -6 \\ -8 & 14 & 2 & 14 \\ 5 & -15 & -15 & 10 \end{pmatrix} \begin{pmatrix} a \\ b \\ c \\ d \end{pmatrix}$$

or, equivalently,

$$\begin{aligned} x &= 1/50a + 17/50b + 37/50c - 4/25d \\ y &= 7/50a + 19/50b + 9/50c - 3/25d \\ z &= -4/25a + 7/25b + 1/25c + 7/25d \\ w &= 1/10a - 3/10b - 3/10c + 1/5d \end{aligned}$$

PROBLEMS ON COMPUTATION OF INVERSE MATRICES

1. Find the inverses of the following matrices.

(a) $\begin{pmatrix} 2 & 5 \\ 6 & 1 \end{pmatrix}$

(b) $\begin{pmatrix} 3 & 4 \\ 7 & 8 \end{pmatrix}$

(c) $\begin{pmatrix} 1 & -3 \\ 2 & 5 \end{pmatrix}$

(d) $\begin{pmatrix} 1 & 6 \\ 1 & 5 \end{pmatrix}$

(e) $\begin{pmatrix} -1 & 3 \\ -2 & 4 \end{pmatrix}$

(f) $\begin{pmatrix} 3 & 1 & -5 \\ 2 & 1 & 7 \\ 8 & 3 & -2 \end{pmatrix}$

(g) $\begin{pmatrix} 3 & 6 & 1 \\ 2 & 4 & 8 \\ 5 & 1 & 9 \end{pmatrix}$

(h) $\begin{pmatrix} 6 & 4 & 2 \\ 3 & 1 & 5 \\ 3 & 3 & -3 \end{pmatrix}$

(j) $\begin{pmatrix} 5 & 1 & 4 & 2 \\ 3 & 2 & 7 & 1 \\ 2 & -1 & -3 & 4 \\ 1 & -4 & -13 & 2 \end{pmatrix}$

(i) $\begin{pmatrix} 2 & 4 & 7 & 1 \\ 6 & 3 & -2 & 4 \\ -5 & 4 & 8 & 1 \\ 2 & 6 & 3 & -7 \end{pmatrix}$

2. Consider an economy in which n basic commodities are produced. Each commodity is produced through a process that uses (as input) the other commodities in the economy. More precisely, the ith process $(i=1,\ldots,n)$, when operating at unit intensity, uses as input a_{ij} units of the jth commodity $(j=1,\ldots,n)$ and produces as output 1 unit of the ith commodity.

(a) Show that, if the ith process operates at intensity x_i, then the total production of the economy is given by the vector

$$\mathbf{x} = (x_1, x_2, \ldots, x_n)$$

and the total consumption, for production purposes, is given by the vector $\mathbf{x}A$, where

$$A = \begin{pmatrix} a_{11} & a_{12} & \cdots & a_{1n} \\ a_{21} & a_{22} & \cdots & a_{2n} \\ \cdots\cdots\cdots\cdots \\ \cdots\cdots\cdots\cdots \\ a_{n1} & a_{n2} & \cdots & a_{nn} \end{pmatrix}$$

164090

is the *technological matrix* of the economy. Show that the net production of the economy is given by the vector

$$\mathbf{y} = \mathbf{x}(I - A)$$

(b) Under the natural assumption that the sum of the entries in any row of A be smaller than 1, the matrix $I - A$ will be invertible. Thus, if the demand $\mathbf{y}$ is given, the production intensity vector $\mathbf{x}$ must be chosen to satisfy

$$\mathbf{x}(I - A) = \mathbf{y}$$

100

What is x, then, in terms of y?

(c) Consider an economy with two basic products: transportation and fuel. The fuel must be taken from its natural deposit (mine, or oil well) to a processing plant, whereas the transportation needs fuel to run. Assume that each ton of fuel requires 100 T.-mi. of transportation, while each ton-mile of transportation consumes 2 lb. of fuel. Choosing suitable units, construct a technological matrix for this economy. What intensity vector should be used, if there is an outside demand for 1000 T. of fuel and 25,000 T.-mi. of transportation?

3. In the economy of the previous question, let p_i be the price of 1 unit of the ith commodity, and q_i be the total value added in producing 1 unit of that commodity.

(a) Show that the two vectors

$$\mathbf{p} = \begin{pmatrix} p_1 \\ p_2 \\ \vdots \\ p_n \end{pmatrix} \quad \mathbf{q} = \begin{pmatrix} q_1 \\ q_2 \\ \vdots \\ q_n \end{pmatrix}$$

are related by the equation

$$(I - A)\mathbf{p} = \mathbf{q}$$

(b) In the foregoing example, what should the price of fuel and the freight rates be, given that the value added by the fuel industry is 5¢ per pound, while the value added by transportation is 10¢ per T.-mi.?

CHAPTER III

LINEAR
PROGRAMMING

1. LINEAR PROGRAMS

In Chapter I, we studied systems of linear inequalities and saw that, in general, such systems do not have unique solutions. Nor is it generally possible to characterize their solutions as simply as those of systems of linear equations. This is not, however, an important consideration, for it is seldom desired to find *all* the solutions to such a system. It is usually only necessary to find *one* solution, i.e., a point that satisfies the constraints (inequalities) of the system. In a few cases, any solution will do, but normally, a special solution is required —the one that is best in a particular sense, say, that maximizes profits or minimizes costs. We can see this best from the following example.

III.1.1 Example. A dietitian must prepare a mixture of foods A, B, and C. Each unit of food A contains 3 oz. protein, 2 oz. carbohydrate, and 6 oz. fat, and costs 60¢. Each unit of food B contains 5 oz. protein, 6 oz. carbohydrate, and 2 oz. fat, and costs $1.25. Each unit of food C contains 3 oz. protein, 4 oz. carbohydrate, and 5 oz. fat, and costs 50¢. The mixture must contain at least 20 oz. protein, 18 oz. carbohydrate, and 30 oz. fat. It is desired to find the mixture of minimal total cost, subject to these constraints.

Letting x, y, and z be the amounts of foods A, B, and C, respectively, and letting w represent the total cost of the mixture, we can summarize the data of the problem in a table (Table III.1.1).

TABLE III.1.1

Food	Amount	Protein	Carbohydrate	Fat	Cost ($)
A	x	3	2	6	0.60
B	y	5	6	2	1.25
C	z	3	4	5	0.50
Total		≥ 20	≥ 18	≥ 30	

We may express the problem in the following form:
Minimize

$$w = 0.60x + 1.25y + 0.50z$$

Subject to

$$3x + 5y + 3z \geq 20$$
$$2x + 6y + 4z \geq 18$$
$$6x + 2y + 5z \geq 30$$
$$x, y, z \geq 0$$

A problem such as this, of maximizing or minimizing a linear function of the variables, subject to linear constraints (equations or, more commonly, inequalities), is called a *linear program*. It is not possible, at this point, to solve the example since we have not yet developed the necessary mathematical tools. We will, therefore, give a strict mathematical analysis of the subject and come back to the example later on.

The science of linear programming is a relatively new branch of mathematics; it was among the many studies that were given impetus by the interest of military planners during World War II. (Prior to this war, only trial-and-error and approximation techniques were available for the solution of linear programs.) The most commonly used method nowadays, the *simplex algorithm*, was developed by G. B. Dantzig during the late 1940's; certain refinements in technique and notation as well as the theory of duality were later introduced by A. W. Tucker.

The standard form of the linear programming problem may be given as

3.1.1 Maximize

$$w = c_1 x_1 + c_2 x_2 + \ldots + c_n x_n$$

Subject to

3.1.2
$$a_{11} x_1 + a_{12} x_2 + \ldots + a_{1n} x_n \leq b_1$$
$$a_{21} x_1 + a_{22} x_2 + \ldots + a_{2n} x_n \leq b_2$$
$$\ldots \ldots \ldots \ldots \ldots \ldots \ldots \ldots \ldots \ldots$$
$$a_{m1} x_1 + a_{m2} x_2 + \ldots + a_{mn} x_n \leq b_m$$

3.1.3 $\qquad$ $x_1, x_2, \ldots, x_n \geq 0$

or, in matrix notation:

3.1.4 Maximize

$$w = \mathbf{c}^t \mathbf{x}$$

Subject to

3.1.5 $\qquad$ $A\mathbf{x} \leq \mathbf{b}$

3.1.6 $\qquad$ $\mathbf{x} \geq 0$

where $A = (a_{ij})$ is the matrix of coefficients, $\mathbf{b}$, and $\mathbf{c}$ are constant column vectors, $\mathbf{x}$ is the column vector of variables, and 0 is the zero column vector. (We add that vector inequalities hold whenever the inequalities hold, component by component.)

We give some definitions. The variable w, given by (3.1.1) or (3.1.4), is called the *objective function*. The inequalities (3.1.2) and (3.1.3) or (3.1.5) and (3.1.6) are the *constraints* of the problem. The set of points satisfying the constraints is called the *constraint set*. Such points are *feasible* points.

Let us suppose that there is a vector, $\mathbf{x}^*$, that satisfies the constraints and such that, if $\mathbf{x}$ is any vector in the constraint set, $\mathbf{c}^t \mathbf{x} \leq \mathbf{c}^t \mathbf{x}^*$. In this case we say that $\mathbf{x}^*$ is the *solution* of the program (3.1.1) through (3.1.3) or (3.1.4) through (3.1.6); while $\mathbf{c}^t \mathbf{x}^*$ is the *value* of the program.

Strictly speaking, the program (3.1.1) through (3.1.3) is not the most general type of linear program. It may be, for example, that the objective function is to be minimized rather than maximized; that some of the constraints have opposite sign, $\geq$ instead of $\leq$; or even that some of the constraints may be equations instead of inequalities. It may also happen that some of the variables are allowed to have negative values. We can always, however, reduce a program to the standard form of (3.1.1) through (3.1.3). In fact, a function can be minimized by maximizing its negative. If an inequality has the form $\geq$, we can reverse it through the simple expedient of multiplying by -1. In turn, an equation $\alpha = \beta$, may be replaced by the two inequalities $\alpha \leq \beta$, and $-\alpha \leq -\beta$. Finally, an unrestricted variable (i.e., one that is allowed to have negative values) may be expressed as the difference of two non-negative variables. The program is, however, as general as we desire; we are concerned, mainly, with programs that have this form, though occasionally problems of other forms may be considered without reduction to the standard form.

Let us analyze the problem by looking at the geometry of the situation. The first thing to be pointed out is that the constraint set is *convex*. Geometrically, a set, S, is convex if, whenever the points P and Q both lie in S, so does the entire line-segment determined by these two points (i.e., every point that lies on the line PQ, *between P and Q*, lies in the set S). Figures III.1.1 (a) and (b) show examples of convex sets. On the other hand Figure III.1.1 (c) shows an example of a set that is not convex.

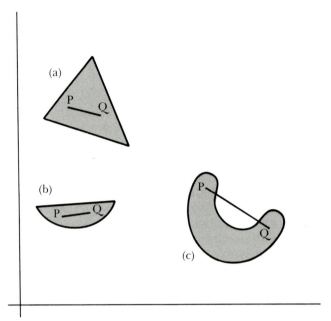

FIGURE III.1.1 (a) and (b) are convex; (c) is not convex.

To prove that the constraint set for a linear program is convex, we recall that each of the constraints (3.1.2) or (3.1.3) is satisfied by the set of points lying on one side of a hyperplane. It is clear that such a set is convex, since, if two points lie above a plane, then so does any point between them. Suppose that P and Q both satisfy all the constraints, and R is any point between them (Figure III.1.2). It is clear that R will lie on the "correct" side of each of the hyperplanes, since P and Q do. But this means that R lies in the constraint set. Thus, the constraint set is convex.

The second thing to be pointed out is that the objective function is linear:

$$w = c_1 x_1 + c_2 x_2 + \ldots + c_n x_n$$

Now, consider any line in n-dimensional space. As we know from previous study, such a line can be expressed by making one of the variables, say x_1, arbitrary and expressing all the other variables in terms of it:

3.1.7 $x_j = \alpha_j x_1 + \beta_j$

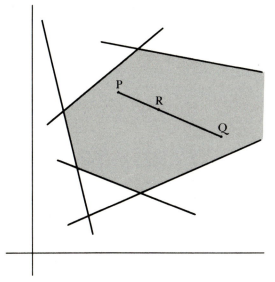

FIGURE III.1.2   The set of points that satisfy a collection of linear inequalities is convex.

where the α_j and β_j are constants, for each value of $j = 2, 3, \ldots, n$. If we substitute (3.1.7) into (3.1.1), we obtain

$$w = \left(c_1 + \sum_{j=2}^{n} c_j\, \alpha_j \right) x_1 + \sum_{j=2}^{n} \beta_j$$

or, more compactly,

3.1.8 $w = \alpha x_1 + \beta$

where α and β depend only on the particular line considered.

Considering Figure III.1.2 once again, we see that, if the value of the coordinate x is greater at Q than at P, then it must increase steadily from P to Q. Now, consider the objective function, w, which, as we have seen, can be expressed *along the line PQ* by (3.1.8). If $\alpha = 0$, then w is constant, equal to β, along this line. If $\alpha > 0$, then w increases whenever x_1 increases, and so it either increases steadily or decreases steadily along PQ. If $\alpha < 0$, then w decreases whenever x_1 increases, and once again, it either decreases or increases steadily along PQ. The important thing is that, along any given line, the function w can do any one of these three things: remain constant, increase steadily, or decrease steadily. It cannot do anything else: it cannot, say, increase from P to R and then decrease from R to Q, or be constant from P to R and then increase from R to Q.

The importance of these observations can best be seen from the following theorem. We already know, from Chapter I, that the constraint set is a polyhedral set; we have seen how its vertices, or extreme points, can be found.

III.1.2 Theorem. The maximum of the objective function (3.1.1) over the constraint set, if such a maximum exists, will be attained at one of the extreme points of the constraint set.

We will not give a strict mathematical proof of this theorem here. The general idea of the proof runs as follows: first, for any point, P, we shall let $w(P)$ denote the value of the objective function, w, at the point P. Now, suppose Q is the point that maximizes the function, i.e., Q lies in the constraint set, and, for any P in the constraint set, $w(P) \leq w(Q)$. If Q is an extreme point, the theorem is proved. If not, then there will be an extreme point, P, of the constraint set, such that the line PQ has points, such as R, which lie "beyond" Q (i.e., Q lies between P and R) but still inside the constraint set (Figure III.1.3).

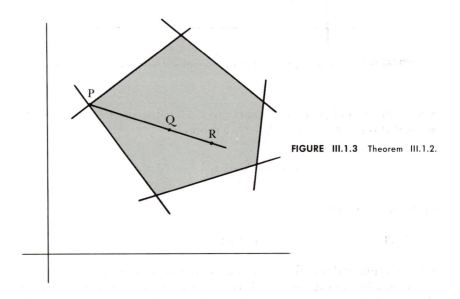

FIGURE III.1.3 Theorem III.1.2.

Suppose, now, that $w(P) < w(Q)$. By the linearity of w, we know that if w increases from P to Q, it must continue to increase from Q to R, and so $w(Q) < w(R)$. But this contradicts the assumption that Q gave the maximum value of w. It follows that we cannot have $w(P) < w(Q)$; we must have $w(P) = w(Q)$. Thus, P also gives the maximum value for w. Since P is extreme, this proves the theorem.

In a sense, Theorem III.1.2 gives us a method for solving the linear program (3.1.1) to (3.1.3). Since the maximum is to be found at an extreme point, all that we have to do is consider all the extreme points, find the value of the objective function at each of them, and take the one that gives the best value. As there are only a finite number of extreme points, it is clear that this method will give us the solution in a finite number of steps.

There are, unfortunately, two difficulties with this. The first is conceptual: we have, generally, no guarantee that the maximum exists, and, if it does not, Theorem III.1.2 will not tell us so. The second difficulty is practical: the number of extreme points, though finite, can be quite large, and finding them all might be beyond the scope of even the largest computers in use today. Nevertheless, we give examples of this method of solution of linear programs. We repeat that it is useful only in very small programs (i.e., those with only a few variables and constraints).

III.1.3 Example

Maximize
$$w = 2x + y + 3z$$
Subject to

$$
\begin{aligned}
x + 2y + \ \ z &\le 25 & (i_1) \\
3x + 2y + 2z &\le 30 & (i_2) \\
x &\ge 0 & (i_3) \\
y &\ge 0 & (i_4) \\
z &\ge 0 & (i_5)
\end{aligned}
$$

We shall, as in Chapter I, look for the extreme points of the constraint set. (It is for this reason that we have numbered the constraints.) As before, we take the constraints, three at a time, and solve them as equations; taking constraints (i_1, i_2, i_3), we obtain the system

$$
\begin{aligned}
x + 2y + \ \ z &= 25 \\
3x + 2y + 2z &= 30 \\
x \qquad\qquad &= 0
\end{aligned}
$$

which has the solution $(0,10,5)$. We check, then, that the remaining constraints are satisfied; since they are, we evaluate the objective function, which has the value of $2 \cdot 0 + 10 + 3 \cdot 5 = 25$ at this point. Continuing in this way with each combination of three inequalities, we obtain Table III.1.2.

TABLE III.1.2

Constraints	Point	Check	w
1,2,3	(0,10,5)	Yes	25
1,2,4	(−20,0,45)	No	
1,2,5	(5/2,45/4,0)	Yes	65/4
1,4,5	(25,0,0)	No	
2,3,4	(0,0,15)	Yes	45
2,3,5	(0,15,0)	No	
1,3,4	(0,0,25)	No	
1,3,5	(0,25/2,0)	Yes	25/2
2,4,5	(10,0,0)	Yes	20
3,4,5	(0,0,0)	Yes	0

We see from the table that, of the six extreme points of the constraint set, the point (0,0,15) gives the best value for the objective function, $w = 45$. We conclude that, if the program has a solution, it is this point.

The program does have a solution. It can be seen, multiplying the first constraint by 3, that $3x + 6y + 3z \leq 75$. Since the variables x,y are non-negative, it follows that $w \leq 75$, i.e., w cannot become arbitrarily large. From this and the fact that the constraint set contains its boundary points, it can be shown that the function w will have a maximum. We have, thus, solved the program: the solution is

$$x = 0$$
$$y = 0$$
$$z = 15$$
$$w = 45$$

III.1.4 Example

Maximize

$$w = x + 2y$$

Subject to

$$-2x + y \leq 10 \qquad (i_1)$$
$$x - 2y \leq 6 \qquad (i_2)$$
$$x \geq 0 \qquad (i_3)$$
$$y \geq 0 \qquad (i_4)$$

Once again, we take the constraints two at a time and solve as equations. We obtain Table III.1.3.

TABLE III.1.3

Constraints	Point	Check	w
1,2	$(-26/3,-22/3)$	No	
1,3	$(0,10)$	Yes	20
1,4	$(-5,0)$	No	
2,3	$(0,-3)$	No	
2,4	$(6,0)$	Yes	6
3,4	$(0,0)$	Yes	0

We see from the table that, if the program has a solution, it must be the point (0,10), which gives a value of 20 for w. Let us, however, look at Figure III.1.4, which shows the constraint set of the program. We see here that the constraint set is not bounded on the upper righthand side; in fact, for any positive value of x, the point (x,x) lies in the constraint set. But this means that $w = x + 2y$ can be made arbitrarily large; the program fails to have a solution. In this case, our method of looking at the extreme points fails us.

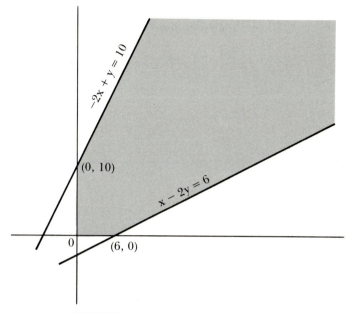

FIGURE III.1.4   Constraint set for Example III.1.4.

PROBLEMS ON LINEAR PROGRAMMING

1. Make a sketch of the constraint set in each of these problems.

(a) Maximize

$$3x + 5y$$

Subject to

$$x + 3y \leq 12$$
$$2x + y \leq 10$$
$$x + y \leq 4$$
$$x \geq 0$$
$$y \geq 0$$

(b) Maximize

$$2x + y$$

Subject to

$$x + 3y \leq 12$$
$$2x + y \leq 10$$
$$x + y \leq 4$$
$$x \geq 0$$
$$y \geq 0$$

(c) Minimize

$$3x + 2y$$

Subject to

$$x + 3y \geq 6$$
$$2x + y \geq 3$$
$$x \geq 0$$
$$y \geq 0$$

(d) Maximize

$$3x - y$$

Subject to

$$x + 2y \geq 4$$
$$x + y \leq 12$$
$$2x + 4y \leq 30$$
$$x \geq 0$$
$$y \geq 0$$

2. A dietitian must prepare a mixture of two foods A and B. Each unit of food A contains 10 gm. of protein and 20 gm. of carbohydrate, and costs 15¢. Each unit of food B contains 20 gm. of protein and 25 gm. of carbohydrate, and costs 20¢. What is the cheapest mixture that can be obtained, subject to the constraint that it must contain at least 500 gm. of protein and 800 gm. of carbohydrate?

3. A nut company has 600 lb. of peanuts and 400 lb. of walnuts. It can sell the peanuts alone at 20¢ per lb.; it can also mix the peanuts and walnuts in a ratio of three parts peanuts and one part walnuts, or in a ratio of one part peanuts and two parts walnuts. The first mixture sells at 35¢ per lb., and the third at 50¢ per lb. How much of each mixture should it produce to maximize sales revenue?

4. A food processing company has 1000 lb. of African coffee, 2000 lb. of Brazilian coffee, and 500 lb. of Colombian coffee. It produces two grades of coffee. Grade A is a mixture of equal parts of African and Brazilian coffees, and sells for 60¢ per lb. Grade B is a mixture of three parts Brazilian and one part Colombian coffee, and sells for 85¢ per lb. How much of each should it produce to maximize sales revenue?

2. THE SIMPLEX ALGORITHM: SLACK VARIABLES

As we have seen from the examples in Section III.1, as a method of solving a linear program, the process of enumerating vertices is both impractical and uncertain. It is uncertain because we generally have no guarantee that the program does, indeed, have a solution; it may just possibly be unbounded, and the method of enumeration will not tell us when this happens. It is impractical because, for a large program, the process of enumerating the extreme points can be extremely long.

To take an example, a program with 6 variables and 10 constraints (including the non-negativity constraints) would require us to solve a total of $\binom{10}{6} = 120$ systems of 6 equations in 6 unknowns.

Even if we could be certain of having the answer when we finished, the amount of work required would necessitate using a medium-sized computer to obtain the solution with reasonable speed. The trouble is that enumeration of the combinations of constraints, while it may be systematic, is not to the point. A need arises, therefore, for a method of enumeration that will be both systematic *and* to the point (i.e., that will discard as many vertices as possible without more than a passing glance, if that much). The *simplex algorithm* is such a method.

To understand more fully the point of the simplex algorithm, let us return to the geometric study of the program. As we have pointed out several times, the constraint set for a linear program will be a convex polyhedral set. We shall say that two vertices (extreme points) of this set are *adjacent extreme points* if they are joined by one of the edges of the polyhedron. In Figure III.2.1, the vertices A and B are adjacent, as are A and C, or C and D. The vertices B and C, however, are not adjacent; neither are D and E, nor C and F.

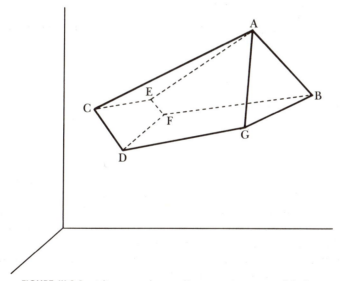

FIGURE III.2.1 Adjacent and non-adjacent vertices in a polyhedron.

The simplex algorithm, as we shall see, is a procedure that allows us to move from one extreme point of the constraint set to an *adjacent* extreme point, in such a way that the value of the objective function is improved at each step, until we find the solution of the program (or until we see that it has no solution). As such, it is a very effective method of implicit enumeration of the extreme points—it automatically discards a great many points before they are even considered.

The mathematical justification for the simplex algorithm is given by the following theorem:

112

III.2.1 Theorem. Let P be an extreme point of the constraint set of the program (3.1.1) through (3.1.3). Then, if P does not maximize the objective function w, there is an edge of the constraint set, starting at P, along which the objective function increases.

Again, we will not give a strict proof. The general idea of the proof, for a two-dimensional set, is as follows. Since P does not maximize w, there is a point in the constraint set, say Q, such that $w(Q) > w(P)$ (see Figure III.2.2). By the convexity of the constraint set, and the linearity of the objective function, we know that if R is on the line-segment PQ, then R is in the constraint set, and $w(R) > w(P)$. Through R, a line may be drawn cutting the two edges that begin at P at the points S and S'.

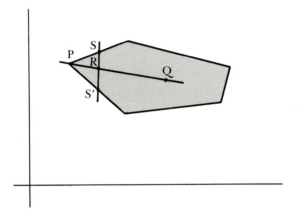

FIGURE III.2.2   Theorem III.2.1.

Now, if $w(S) \geq w(R)$, we have $w(S) > w(P)$, and so w increases along the edge PS. If, on the other hand, $w(S) < w(R)$, then we must (by linearity) have $w(S') > w(R)$, and so $w(S') > w(P)$, so that w increases along the edge PS'. Thus w will increase along one, at least, of the two edges that begin at P.

For higher dimensions, the proof is similar, though somewhat greater care must be observed in the arguments. The theorem is true in any case.

Theorem III.2.1 tells us that the procedure just outlined, considering only extreme points adjacent to those that have already been obtained, will always bring us to the maximum if it exists. At the same time, if the objective function is not bounded, we will always find an edge along which the function increases without bound.

One of the difficulties in dealing with linear programs is that constraints of the type (3.1.2) are generally difficult to handle; this is something that we saw in Chapter I. On the other hand, equations are comparatively easy to handle, as are the non-negativity constraints (3.1.3). It follows that the program will become that much easier to solve if we can replace the constraints (3.1.2) by equations or by constraints of the type (3.1.3). This can be effected by introducing so-called *slack variables*.

Let us consider one of the constraints (3.1.2):

3.2.1 $$a_{i1}x_1 + a_{i2}x_2 + \ldots + a_{in}x_n \leq b_i$$

This can be restated in the form

$$a_{i1}x_1 + a_{i2}x_2 + \ldots + a_{in}x_n - b_i \leq 0$$

Since the left-hand side of this inequality is negative or zero, we can write

3.2.2 $$a_{i1}x_1 + a_{i2}x_2 + \ldots + a_{in}x_n - b_i = -u_i$$

where

3.2.3 $$u_i \geq 0$$

Thus we have replaced the single constraint (3.2.1) by the equation (3.2.2) and the non-negativity constraint (3.2.3). This is what we desired to do. Proceeding in this manner with each of the constraints (3.1.2), we find that the program (3.1.1) through (3.1.3) can be rewritten in the form

3.2.4 Maximize

$$w = c_1 x_1 + \ldots + c_n x_n$$

Subject to

3.2.5
$$
\begin{aligned}
a_{11}x_1 + a_{12}x_2 + \ldots + a_{1n}x_n - b_1 &= -u_1 \\
a_{21}x_1 + a_{22}x_2 + \ldots + a_{2n}x_n - b_2 &= -u_2 \\
&\ldots\ldots\ldots\ldots\ldots\ldots\ldots\ldots \\
a_{m1}x_1 + a_{m2}x_2 + \ldots + a_{mn}x_n - b_m &= -u_m
\end{aligned}
$$

3.2.6
$$x_1, x_2, \ldots, x_n \geq 0$$
$$u_1, u_2, \ldots, u_m \geq 0$$

or, in matrix notation,

3.2.7 Maximize

$$w = c^t x$$

Subject to

3.2.8
$$Ax - b = -u$$
$$x, u \geq 0$$

114   3. THE SIMPLEX TABLEAU

We introduce next, a system of notation that simplifies the computational procedure for the simplex algorithm to a considerable extent. This is the *simplex tableau* or *schema*. In a sense, it is only a shorthand notation for a system of linear equations. The schema

3.3.1

x_1	x_2	$\ldots x_n$	1	
a_{11}	a_{12}	$\ldots a_{1n}$	$-b_1$	$= -u_1$
a_{21}	a_{22}	$\ldots a_{2n}$	$-b_2$	$= -u_2$
. . . .			. . . . .	. . . . .
. . . .			. . . . .	. . . . .
. . . .			. . . . .	. . . . .
a_{m1}	$a_{m2} \ldots$	a_{mn}	$-b_m$	$= -u_m$
c_1	c_2	$\ldots c_n$	0	$= w$

represents the equations (3.2.4) and (3.2.5). Generally speaking, each of the rows inside the box represents a linear equation, namely, the equation obtained by multiplying each entry in the row by the corresponding entry in the top row, and then adding and setting this sum equal to the entry in the right-hand column (outside the box). Thus the first row represents the equation

$$a_{11}x_1 + a_{12}x_2 + \ldots + a_{1n}x_n - b_1 = -u_1$$

which is precisely the first of the equations (3.2.5), and so on for the other rows in the tableau. The bottom row in the tableau is nothing other than the definition of the objective function, w, according to (3.2.4).

Let us, for the present, ignore the non-negativity constraints (3.2.6). The equation constraints (3.2.5) are, of course, a system of m equations in the $m + n$ unknowns x_j, u_i. The system is "solved" for the u_i in terms of the x_j, but, as we know from Chapters I and II, it is normally possible to solve such a system for any m variables in terms of the remaining n (though in practice there may be certain combinations of m variables for which this is not possible). What we are looking for now is a method of solving successively for different combinations of the variables.

To see how this is done, let us suppose that we have, at some time, solved for the variables $r_1, r_2, \ldots, r_m$ in terms of the variables $s_1, s_2, \ldots, s_n$. We have a tableau very similar to (3.3.1):

3.3.2

s_1	s_2	$\ldots$ s_n	1	
a_{11}	a_{12}	$\ldots$ a_{1n}	$-b_1$	$=-r_1$
a_{21}	a_{22}	$\ldots$ a_{2n}	$-b_2$	$=-r_2$
$\cdots$	$\cdots$	$\cdots$	$\cdots$	$\cdot$
$\cdots$	$\cdots$	$\cdots$	$\cdots$	$\cdot$
a_{m1}	a_{m2}	$\ldots$ a_{mn}	$-b_m$	$=-r_m$
c_1	c_2	$\ldots$ c_n	δ	$=\quad w$

(where, of course, the entries are different from those in (3.3.1). For such a tableau, the m variables $r_1, \ldots, r_m$, for which we have solved, are called *basic variables*. The variables $s_1, \ldots, s_n$, in terms of which we have solved, are called *non-basic variables*. Let us suppose, next, that we want to interchange the roles of the basic variable r_i and the non-basic variable s_j, i.e., we wish to solve for $m-1$ of the r's, and s_j, in terms of $n-1$ of the s's, and r_i. Consider, then, the ith row of the tableau (3.3.2):

$$a_{i1}s_1 + a_{i2}s_2 + \ldots + a_{ij}s_j + \ldots + a_{in}s_n - b_i = -r_i$$

We can solve this for s_j, if $a_{ij} \neq 0$, obtaining

3.3.3 $\quad \dfrac{a_{i1}}{a_{ij}}s_1 + \dfrac{a_{i2}}{a_{ij}}s_2 + \ldots + \dfrac{1}{a_{ij}}r_i + \ldots + \dfrac{a_{in}}{a_{ij}}s_n - \dfrac{b_i}{a_{ij}} = -s_j$

We must now substitute this value of s_j in the remaining equations that (3.3.2) represents. Let us take any other equation, say the kth (in which $k \neq i$):

$$a_{k1}s_1 + \ldots + a_{kj}s_j + \ldots + a_{kn}s_n - b_k = -r_k$$

Substituting (3.3.3), we obtain

$$a_{k1}s_1 + \ldots - a_{kj}\left(\frac{a_{i1}}{a_{ij}}s_1 + \ldots + \frac{1}{a_{ij}}r_i + \ldots - \frac{b_i}{a_{ij}}\right) + \ldots + a_{kn}s_n - b_k = -r_k$$

or, collecting terms,

3.3.4 $\quad \left(a_{k1} - \dfrac{a_{kj}a_{i1}}{a_{ij}}\right)s_1 + \ldots - \dfrac{a_{kj}}{a_{ij}}r_i$

$$+ \ldots + \left(a_{kn} - \frac{a_{kj}a_{in}}{a_{ij}}\right)s_n - \left(b_k - \frac{b_i a_{kj}}{a_{ij}}\right) = -r_k$$

Equations (3.3.3) and (3.3.4) tell us, now, how the simplex tableau (3.3.2) is to be changed. We can summarize them as follows:

3.3.5 The variables r_i and s_j are interchanged.

3.3.6 The entry a_{ij}, called the *pivot* of the transformation, is replaced by its reciprocal, $1/a_{ij}$.

3.3.7 The other entries a_{il}, or $-b_i$ in the pivot row, are replaced by a_{il}/a_{ij}, or $-b_i/a_{ij}$, respectively.

3.3.8 The other entries a_{kj}, or c_j in the pivot column, are replaced by $-a_{kj}/a_{ij}$ or $-c_j/a_{ij}$ respectively.

3.3.9 The other entries in the tableau, a_{kl}, $-b_k$, c_l, or δ, are replaced by $a_{kl} - a_{kj}a_{il}/a_{ij}$, $- (b_k - b_i a_{kj}/a_{ij})$, $c_l - c_j a_{il}/a_{ij}$, or $\delta + b_i c_j/a_{ij}$, respectively.

A transformation such as that outlined in (3.3.5) to (3.3.9) is called a *pivot step*, or *pivot transformation*. We give the rules in the schematic form

3.3.10

$$
\begin{array}{cc} p & q \\ r & s \end{array}
\quad \rightarrow \quad
\begin{array}{cc} \dfrac{1}{p} & \dfrac{q}{p} \\ -\dfrac{r}{p} & s - \dfrac{qr}{p} \end{array}
$$

The schema (3.3.10) is to be interpreted as follows: p is the pivot, q is any other entry in the pivot row, r is any other entry in the pivot column, and s is the entry in the row of r and column of q. The arrow shows how the pivot step affects these entries: the pivot is replaced by its reciprocal; other entries in the pivot row are divided by the pivot; other entries in the pivot column are divided by the negative of the pivot. The entry s is replaced by $s - qr/p$, where p is the pivot and q and r are the two entries that form a rectangle with s and p (i.e., q is in the row of p and column of s, while r is in the column of p and row of s).

An example of such a pivot step follows.

III.3.1 Example. Solve the system

$$
\begin{aligned}
r &= x + 2y - 3 \\
s &= 2x + 5y + 1
\end{aligned}
$$

for x and y in terms of r and s.

There are actually two ways of doing this: one is by means of row operations on the matrix of coefficients; the other, by pivot steps. We shall use the latter method.

We form the tableau

$$
\begin{array}{ccc}
x & y & 1 \\
\end{array}
$$

$$
\begin{array}{|ccc|}
\hline
-1^* & -2 & 3 \\
-2 & -5 & -1 \\
\hline
\end{array}
\begin{array}{c}
=-r \\
=-s
\end{array}
$$

We wish to interchange the roles of x and y with those of r and s. This means we must carry out two pivot steps. In the first, we pivot on the starred entry to interchange the variables x and r, thus obtaining

$$
\begin{array}{ccc}
r & y & 1
\end{array}
$$

$$
\begin{array}{|ccc|}
\hline
1/-1 & -2/-1 & 3/-1 \\
-(-2/-1) & -5 - \dfrac{(-2)(-2)}{-1} & -1 - \dfrac{3(-2)}{-1} \\
\hline
\end{array}
\begin{array}{c}
=-x \\
=-s
\end{array}
$$

or, carrying out the operations,

$$
\begin{array}{ccc}
r & y & 1
\end{array}
$$

$$
\begin{array}{|ccc|}
\hline
-1 & 2 & -3 \\
-2 & -1^* & -7 \\
\hline
\end{array}
\begin{array}{c}
=-x \\
=-s
\end{array}
$$

This gives us x and s in terms of r and y. To interchange s and y, we pivot on the starred entry to obtain

$$
\begin{array}{ccc}
r & s & 1
\end{array}
$$

$$
\begin{array}{|ccc|}
\hline
-5 & 2 & -17 \\
2 & -1 & 7 \\
\hline
\end{array}
\begin{array}{c}
=-x \\
=-y
\end{array}
$$

or, in equation form,

$$
\begin{aligned}
x &= 5r - 2s + 17 \\
y &= -2r + s - 7
\end{aligned}
$$

This could also have been done by means of row operations, as in Chapter II: we have the system

$$
\begin{aligned}
-x - 2y + r \quad\;\; &= -3 \\
-2x - 5y \quad\;\; + s &= 1
\end{aligned}
$$

which gives us the augmented matrix

$$
\left(
\begin{array}{cccc|c}
-1 & -2 & 1 & 0 & -3 \\
-2 & -5 & 0 & 1 & 1
\end{array}
\right)
$$

We multiply the first row by -1, and clear the first column:

$$\begin{pmatrix} 1 & 2 & -1 & 0 & | & 3 \\ 0 & -1 & -2 & 1 & | & 7 \end{pmatrix}$$

Now we multiply the second row by -1, and clear the second column, obtaining

$$\begin{pmatrix} 1 & 0 & -5 & 2 & | & 17 \\ 0 & 1 & 2 & -1 & | & -7 \end{pmatrix}$$

which gives us the desired solution. It may be checked that, as the row operations are carried out, the matrices obtained at each step contain an identity matrix formed by two columns. The remaining columns are the same, except for a possible interchange of columns, as in the simplex tableaux obtained previously (the only difference being in the signs of the last column). In fact, the pivot steps are nothing other than a condensed format for the row operations.

III.3.2 *Example.* Interchange the roles of s and y in the tableau

	x	y	1	
	1	4	-1	$=-r$
	5	-9^*	2	$=-s$
	6	-5	-3	$=-t$

In this case, the pivot is the entry -9, in the y-column and s-row. The new tableau will be

	x	s	1	
	$1 - 4 \cdot 5/-9$	$-4/-9$	$-1 - 4 \cdot 2/-9$	$=-r$
	$5/-9$	$1/-9$	$2/-9$	$=-y$
	$6 - 5(-5)/-9$	$-(-5)/-9$	$-3 - 2(-5)/-9$	$=-t$

or

	x	s	1	
	$\dfrac{29}{9}$	$\dfrac{4}{9}$	$\dfrac{-1}{9}$	$=-r$
	$\dfrac{-5}{9}$	$\dfrac{-1}{9}$	$\dfrac{-2}{9}$	$=-y$
	$\dfrac{29}{9}$	$\dfrac{-5}{9}$	$\dfrac{-37}{9}$	$=-t$

1. In each of the following problems, solve the given system of equations for x, y, and z in terms of the other variables.

(a) $3s + 2y - z + 4 = -x$
 $s - 2y + 4z - 2 = -t$
 $4s + 3y - 2z + 1 = -u$

(b) $2x + 3y - 5 = -z$
 $x + 2y - 3 = -s$
 $3x + 7y + 1 = -t$

(c) $x - 2s + 5z - 1 = -y$
 $3x + s - 5z + 3 = -t$
 $-x + 4s + 2z + 1 = -r$

(d) $4x + 3y - 2z - 5 = -r$
 $x + 2y - 3z + 3 = -s$
 $2x + y + 2z + 6 = -t$

(e) $3x + 2y + 6s + 1 = -r$
 $x + y + 2s + 4 = -z$
 $-2x - y + 3s - 5 = -t$

(f) $5r + x - 2y + 4 = -z$
 $3r + x + y - 6 = -s$
 $r - 5x + 2y + 7 = -t$

(g) $x + y + 3z - 2 = -s$
 $2x + y - 2z + 4 = -t$
 $5x + 2y - z + 6 = -u$

4. THE SIMPLEX ALGORITHM: OBJECTIVES

Now that we have seen what a pivot step is, we have to decide what to do with it. We must remember that, essentially, it is nothing other than a method of rewriting the system of equations (3.2.5) to obtain a different but equivalent system. Let us suppose that, after one or several pivot steps, a tableau such as (3.3.2) is reached, which has the property that all the entries in the right-hand column, except possibly the bottom entry, are non-positive; that is, for each value of i, $b_i \geq 0$. Since (3.3.2) is just a "solution" of the equations (3.2.5) for the basic variables $r_1, \ldots, r_m$ in terms of the non-basic variables $s_1, \ldots, s_n$, we may assign values arbitrarily to the non-basic variables. We might let each of the non-basic variables equal zero. This gives us the values

3.4.1 $$s_j = 0 \quad \text{for } j = 1, \ldots, n$$

FINITE MATHEMATICS

3.4.2
$$r_i = b_i \quad \text{for } i = 1, \ldots, m$$

as a particular solution to the system (3.2.5). But we are assuming that $b_i \geq 0$. It follows that these values of the variables satisfy the non-negativity constraints (3.2.6) as well as the equations (3.2.5) and thus give us a point in the constraint set of the program (3.2.4) to (3.2.6).

Let us suppose, further, that each entry in the bottom row, except possibly the right-hand entry, is non-positive, i.e., for each value of j, $c_j \leq 0$. The bottom row of (3.3.2) says simply that

3.4.3
$$w = c_1 s_1 + c_2 s_2 + \ldots + c_n s_n + \delta$$

We know that a feasible point is obtained by letting each s_j equal zero. If we do so, we find that $w = \delta$, i.e., this feasible point gives a value of δ for the objective function. On the other hand, each term on the right side of (3.4.3) except the last is the product of a non-positive constant c_j and a variable s_j that is non-negative for any feasible point. Thus, for any feasible point, the right side of (3.4.3) is the sum of n non-positive terms $c_j s_j$, and δ. It follows that we must have $w \leq \delta$. But the point (3.4.1) and (3.4.2) gives a value of δ, and is thus the *solution* of the program (3.2.4) to (3.2.6); it gives the maximum of the objective function.

We now know what we want to accomplish by means of the pivot steps: we wish to make every entry in the right-hand column (except possibly the bottom entry) non-positive; we wish to make every entry in the bottom row (except possibly the right-hand entry) non-positive. Of course, if we wish to minimize the objective function, we will try to make the entries in the bottom row non-negative; except for this, there is no great difference between maximization and minimization problems. To see how this can best be done, we shall look at the geometric situation. First, however, we give some definitions.

III.4.1   Definition.   A simplex tableau is associated with a point in $(m + n)$-dimensional space, namely, the point obtained by setting the n non-basic variables equal to zero. Such a point is called a *basic point* of the program. If all the entries in the right-hand column (except possibly the bottom entry) are non-positive, then the point is a *basic feasible point* (b.f.p.). Two basic points are said to be *adjacent* if the corresponding simplex tableaux can be obtained from each other by a single pivot step.

Geometrically, the situation is as follows: each of the planes $x_j = 0$ or $u_i = 0$ is one of the bounding hyperplanes of the constraint set. The intersection of n of these determines a basic point of the program. If the remaining constraints are all satisfied, we have (as we saw in Chapter I) a vertex of the constraint set, so that a b.f.p. is simply one of the vertices of the constraint set. An edge of the constraint set is determined by setting $n - 1$ of the variables equal to zero, so that two vertices will lie on a common edge if they have $n-1$ of their non-basic variables in common, i.e., if their tableaux can be obtained from each other by a single pivot step. The idea of adjacent b.f.p.'s corresponds to the geometric idea of adjacent vertices.

5. THE SIMPLEX ALGORITHM: CHOICE OF PIVOTS

The process of solving a linear program through the simplex method generally can be thought of as consisting of two stages. In Stage I, the entries in the right-hand column are made non-negative to obtain a basic feasible point. In Stage II, while care is exercised to maintain a b.f.p. at all times, the entries in the bottom row are made non-positive (or non-negative in case of a minimization problem) to obtain the solution. The two stages have to be considered separately since the objectives in the two cases are distinct. Let us consider Stage II first; the rules for dealing with Stage I can be adapted from those for Stage II.

For Stage II, Figure III.5.1 gives a two-dimensional illustration of what we are trying to do. The vertex A lies at the intersection of $x = 0$ and $r = 0$; i.e., its non-basic variables are r and x. From this point, suppose we see that the objective function increases along the edge AB, i.e., by letting x increase while r remains at zero. We improve the value of the function, then, if we interchange x with one of the basic variables, y,s,t. But which one? Clearly not y, since this would cause

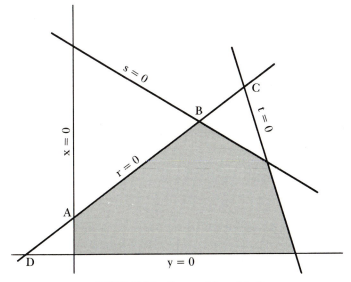

FIGURE III.5.1   Movement from A to B.

x to decrease (giving the point D). Not t either, since this would cause x to increase too much; it would give us the point C, at which s is negative. We conclude that we should interchange x and s, since, of those interchanges that cause x to increase, this is the one that causes the smallest increase. We thus obtain the adjacent vertex, B, and continue from B in the same manner.

Accordingly, we obtain the following rules:

III.5.1 Rules for Choosing a Pivot (Stage II). Let c_j be any positive entry (other than the right-hand entry) in the bottom row of the tableau. This gives us the pivot column. For each *positive* entry a_{ij} in this column, form the quotient b_i/a_{ij} (obtained by dividing a_{ij} into the negative of the corresponding entry in the right-hand column). Let b_k/a_{kj} be the smallest of these quotients. Then a_{kj} will be the pivot. (If there is a tie for the smallest quotient, the pivot may be chosen arbitrarily from among those that tie.)

The rules III.5.1 are for a maximization problem. If we are required to minimize the objective function, the procedure is changed only in that we take c_j to be a negative entry; otherwise, the choice of pivot is the same.

We shall not, at this moment, prove that the rules III.5.1 work in the sense that they will eventually give us the solution of the program if such a solution exists. We give, instead, a numerical example.

III.5.2 Example. A company produces three types of toys, A, B, and C, by using three machines called a forge, a press, and a painter.

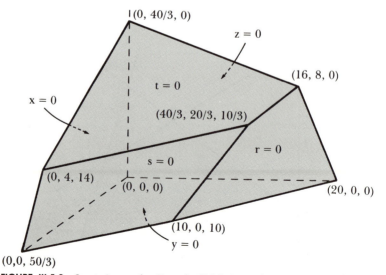

FIGURE III.5.2 Constraint set for Example III.5.2. Note that each bounding plane corresponds to one of the variables.

Each toy of type A must spend 4 hr. in the forge, 2 hr. in the press, and 1 hr. in the painter, and yields a profit of $5. Each toy of type B must spend 2 hr. in the forge, 2 hr. in the press, and 3 hr. in the painter and yields a profit of $3. Each toy of type C spends 4 hr. in the forge, 3 hr. in the press, and 2 hr. in the painter and yields a profit of $4. In turn, the forge may be used up to 80 hr. per week; the press, up to 50 hr., and the painter up to 40 hr. How many of each type of toy should be made in order to maximize the company's profits?

Letting x, y, and z be the amounts produced of toys A, B, and C, respectively, we obtain the linear program:

Maximize

$$5x + 3y + 4z = w$$

Subject to

$$4x + 2y + 4z \leq 80$$
$$2x + 2y + 3z \leq 50$$
$$x + 3y + 2z \leq 40$$
$$x, y, z \geq 0$$

Figure III.5.2 shows the constraint set for this problem.

Introducing the slack variables r, s, and t, we obtain the tableau

x	y	z	1	
4*	2	4	−80	$= -r$
2	2	3	−50	$= -s$
1	3	2	−40	$= -t$
5	3	4	0	$= w$

We see that we have a basic feasible point, since the entries in the right-hand column are negative. On the other hand, there are three positive entries in the bottom row (see also Figure III.5.3). Let us

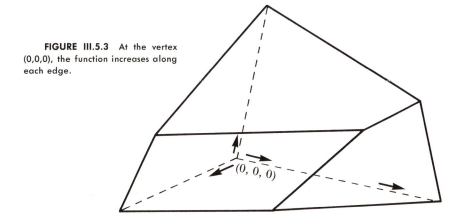

FIGURE III.5.3 At the vertex (0,0,0), the function increases along each edge.

$(0, 0, 0)$

choose the first column, which has one of these positive entries, as the pivot column. According to the rules III.5.1, we divide each of the positive entries in this column into the corresponding b_i. The corresponding quotients b_i/a_{ij} are then:

$$b_1/a_{11} = 80/4 = 20$$
$$b_2/a_{21} = 50/2 = 25$$
$$b_3/a_{31} = 40/1 = 40$$

The first row gives the smallest quotient: the entry 4 (starred) will be the pivot. This gives us the new tableau

r	y	z	1	
1/4	1/2	1	−20	$=-x$
−1/2	1	1	−10	$=-s$
−1/4	5/2*	1	−20	$=-t$
−5/4	1/2	−1	100	$=\ \ w$

This tableau gives an improved b.f.p.: the profit here is $100. It is still not the maximum, since there is a positive entry in the bottom row, second column (see also Figure III.5.4). Once again, we consider the quotients b_i/a_{ij} for each positive a_{ij} in the second column. These quotients are, respectively, $20/(1/2) = 40$, $10/1 = 10$, and $20/(5/2) = 8$. Since the starred entry, 5/2, gives the smallest quotient, we use it as pivot. We then obtain the tableau

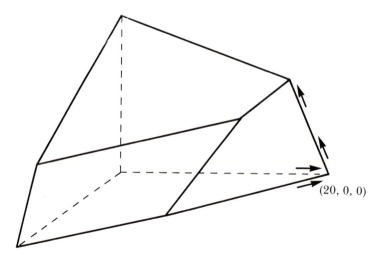

$(20, 0, 0)$

FIGURE III.5.4 At (20,0,0), the function increases along one of the three edges.

r	t	z	1	
$3/10$	$-1/5$	$4/5$	-16	$=-x$
$-2/5$	$-2/5$	$3/5$	-2	$=-s$
$-1/10$	$2/5$	$2/5$	-8	$=-y$
$-6/5$	$-1/5$	$-6/5$	104	$=\ \ w$

(see Figure III.5.5). This gives us the solution: it is obtained by setting the non-basic variables equal to zero, and so

$$\begin{aligned} x &= 16 \\ y &= 8 \\ z &= 0 \\ w &= 104 \end{aligned}$$

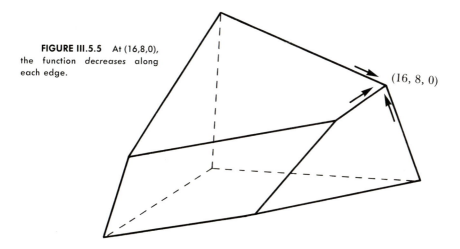

FIGURE III.5.5 At (16,8,0), the function decreases along each edge.

$(16, 8, 0)$

The company should produce 16 type A toys and 8 type B toys every week. Type C toys will not be produced. It may be checked directly that the constraints are satisfied; it is not possible, at least yet, to check that this is indeed the solution. Note that the press is idle 2 hr. a week; this corresponds to the value $s = 2$ for the related slack variable.

III.5.3 Example

Maximize

$$w = 2x + y + 3z$$

Subject to

$$\begin{aligned} x + 2y + z &\leq 25 \\ 3x + 2y + 2z &\leq 30 \\ x, y, z &\geq 0 \end{aligned}$$

126

This is, of course, the same as Example III.1.3. We propose to solve it, now by the simplex algorithm. Introducing slack variables we have the b.f.p.

x	y	z	1	
1	2	1	−25	$=-r$
3	2	2*	−30	$=-s$
2	1	3	0	$= w$

We may choose any one of the columns as the pivot column. Say we choose the third column. The corresponding quotients are $25/1 = 25$ and $30/2 = 15$. Since 15 is the smaller quotient, we choose the starred entry as pivot. We obtain

x	y	s	1	
−1/2	1	−1/2	−10	$=-r$
3/2	1	1/2	−15	$=-z$
−5/2	−2	−3/2	45	$= w$

which gives the solution $x = 0$, $y = 0$, $z = 15$, $w = 45$. This is, of course, the same as was obtained by other means earlier in the chapter.

6. THE SIMPLEX ALGORITHM: RULES FOR STAGE I

In Stage I, the situation is, as we mentioned, somewhat different. The general rule is to take the several entries $-b_i$ in the right-hand column and work on them, one at a time, until all have become non-positive. Some care must be observed, of course, that, while one $-b_i$ becomes non-positive, others do not simultaneously become positive. The method for "working" on a positive entry $-b_i$, then, is as follows: if it can be made non-positive immediately without disturbing the non-positivity of the other entries in the column, we do so. If not, then we treat the $-b_i$ as a sort of "secondary objective": we try to decrease it as much as possible while keeping the other entries in the column non-positive. We have the following rules.

III.6.1 Rules for Choosing a Pivot (Stage I). Let $-b_l$ be the lowest positive entry (other than the bottom entry) in the right-hand column. Let a_{lj} be any negative entry in this row; this gives us the pivot column. Now, for a_{lj}, and for each *positive* entry a_{ij} *below* a_{lj} in this column, form the quotient b_i/a_{ij}. Let b_k/a_{kj} be the smallest of these quotients. Then a_{kj} is the pivot. (Once again, in case of ties, any one of those entries that tie to give the smallest quotient may be chosen as pivot.)

We shall leave the proof that these rules work, i.e., that they eventually give us a b.f.p., until later. We are now in a position to solve Example III.1.1, which follows.

III.6.2 Example

Minimize

$$w = 0.60x + 1.25y + 0.50z$$

Subject to

$$3x + 5y + 3z \geq 20$$
$$2x + 6y + 4z \geq 18$$
$$6x + 2y + 5z \geq 30$$
$$x, y, z \geq 0$$

Introducing slack variables, we have the tableau

x	y	z	1	
-3	-5	-3	20	$= -r$
-2	-6	-4	18	$= -s$
-6	-2	-5^*	30	$= -t$
$3/5$	$5/4$	$1/2$	0	$= w$

which does not represent a b.f.p., since there are several positive entries in the right-hand column. The lowest such entry is in the third row. We take a negative entry in this row, say the -5 in the third column. According to the rules III.6.1, we must consider this entry and any positive entry below it except that in the bottom row. This reduces the choice to no choice at all, of course; and the -5 will be the pivot. We then obtain

x	y	t	1	
$3/5$	$-19/5$	$-3/5^*$	2	$= -r$
$14/5$	$-22/5$	$-4/5$	-6	$= -s$
$6/5$	$2/5$	$-1/5$	-6	$= -z$
0	$21/20$	$1/10$	3	$= w$

which is still not a b.f.p. We now work on the top row: there are two negative entries here, and either one can be used to obtain the pivot column. Let us take the third column. There are no positive entries in this column, other than the bottom entry, so that the starred entry, $-3/5$, will be the pivot. We then obtain

x	y	r	1	
-1	$19/3$	$-5/3$	$-10/3$	$=-t$
2	$2/3$	$-4/3$	$-26/3$	$=-s$
1	$5/3$	$-1/3$	$-20/3$	$=-z$
$1/10$	$5/12$	$1/6$	$10/3$	$=w$

which is not only a b.f.p., but actually the solution, since all the entries in the bottom row are positive (and we want to minimize the objective function). We have then

$$x = 0$$
$$y = 0$$
$$z = 20/3$$
$$w = 10/3$$

and the mixture desired in Example III.1.1 consists of 20/3 units of food of type C, and nothing else, for a total cost of $3.33.

III.6.3 Example

Maximize

$$w = x + 2y + z$$

Subject to

$$x - 10y - 4z \leq -20$$
$$3x + y + z \leq 3$$
$$x, y, z \geq 0$$

We obtain, in this case, the tableau

x	y	z	1	
1	-10	-4	20	$=-r$
3	1	1^*	-3	$=-s$
1	2	1	0	$=w$

There is a positive entry at the top of the right-hand column. We can choose the pivot column to be either the second or third column, since these are the negative entries in the top row. Let us choose the third. The pivot is then chosen from between the negative entry -4 and the positive entry below it (the bottom row does not count, of course). The corresponding quotients are $-20/-4 = 5$ and $3/1 = 3$. Since 3 is smaller, we choose the starred entry as pivot. This gives us

x	y	s	1	
13	−6*	4	8	$= -r$
3	1	1	−3	$= -z$
−2	1	−1	3	$= w$

which is still not a b.f.p. Note how the right-hand entry in the top row has been decreased, while the right-hand entry in the second row remains non-positive.

The only negative entry in the first row is in the second column, which is, therefore, the pivot column. The choice is between the negative entry in the top row and the positive entry below it; the corresponding quotients are $-8/-6 = 4/3$ and $3/1 = 3$, and the pivot is the starred entry. We then get the tableau

x	r	s	1	
−13/6	−1/6	−2/3	−4/3	$= -y$
31/6	1/6*	5/3	−5/3	$= -z$
1/6	1/6	−1/3	13/3	$= w$

which is a b.f.p., but not the solution, since there are two positive entries in the bottom row. Let us choose the second column as the pivot column; the pivot will be the only positive entry in this column (other than the bottom entry) which is starred. We then get the tableau

x	z	s	1	
3	1	1	−3	$= -y$
31	6	10	−10	$= -r$
−5	−1	−2	6	$= w$

which gives us the solution: it is

$$x = 0$$
$$y = 3$$
$$z = 0$$
$$w = 6$$

7. THE SIMPLEX ALGORITHM: PROOF OF CONVERGENCE

Now that we have illustrated the two sets of rules, III.5.1 and III.6.1, for choosing pivots, we shall show that they actually work in the sense of giving us the solution if this exists and telling us that there is no solution if such is the case. We shall make, at first, an assumption of *non-degeneracy:* no more than n of the bounding hyperplanes of the original program (3.1.1) to (3.1.3) pass through a single point. This is the general case and will "almost always" happen, unless the program is of a very special type. In practice, this means that there will be no zeros in the right-hand column (except, possibly, in the bottom row).

III.7.1 Proof for Stage II.

In Stage II, the constant term in the pivot row, $-b_k$, is replaced by

3.7.1
$$-b'_k = -\frac{b_k}{a_{kj}}$$

By hypothesis, $-b_k \leq 0$, and $a_{kj} > 0$. Hence $-b'_k \leq 0$.

The other terms $-b_i$ in the right-hand column are replaced by

3.7.2
$$-b'_i = -b_i + \frac{b_k a_{ij}}{a_{kj}}$$

We consider two possibilities: $a_{ij} \leq 0$ and $a_{ij} > 0$. If $a_{ij} \leq 0$, then, since $b_k \geq 0$ and $a_{kj} > 0$, it will follow that the second term on the right side of (3.7.2) is non-positive. In this case, $-b'_i \leq -b_i$. But, by hypothesis, $-b_i \leq 0$. Hence $-b'_i \leq 0$.

If $a_{ij} > 0$, then (3.7.2) can be rewritten

3.7.3
$$-b'_i = a_{ij} \left(\frac{b_k}{a_{kj}} - \frac{b_i}{a_{ij}} \right)$$

Now, $a_{ij} > 0$, which means that it was among the terms considered as possible pivots. The entry a_{kj} was chosen as pivot, which means (by rules III.5.1) that

$$\frac{b_k}{a_{kj}} \leq \frac{b_i}{a_{ij}}$$

But this means that the parenthesis on the right side of (3.7.3) is non-positive. Since $a_{ij} > 0$, it follows that $-b'_i \leq 0$.

Finally, the term δ, in the lower right-hand corner of the tableau is replaced by

3.7.4
$$\delta' = \delta + \frac{b_i c_j}{a_{ij}}$$

Now, according to the rules, c_j and a_{ij} are both positive. By hypothesis, $b_i \geq 0$; under the non-degeneracy assumption, however, we cannot have $b_i = 0$, so that $b_i > 0$. This means that the second term on the right side of (3.7.4) is positive, and so $\delta' > \delta$.

We have seen, then, that following the rules III.5.1 will give us a new tableau that represents a b.f.p. (since all the $-b_i'$ are non-positive). Moreover, this new b.f.p. will (under the non-degeneracy assumption) give an improved value for the objective function. Thus, every time that we follow the rules, we obtain a new vertex that is an improvement over the previous one. This means that we cannot obtain the same vertex twice. As there are only a finite number of vertices, it follows that we must eventually get a tableau at which the rules III.5.1 cannot be followed.

Suppose, then, that it is impossible to follow the rules. This may be for either of two reasons. One is that there are no positive entries (other than the right-hand entry) in the bottom row. But this means that we have the solution.

The other possibility is that there may be a column (other than the right-hand column) with a positive entry in the bottom row, and no other positive entries, i.e., for some j, $c_j > 0$, but $a_{ij} \leq 0$ for each i. If we let all the non-basic variables, except s_j, vanish, we have

$$r_i = b_i - a_{ij}s_j \quad \text{for each } i$$
$$w = \delta + c_j s_j$$

Now, we can see that, for any positive value of s_j, we will have $r_i \geq b_i$ (since $a_{ij} \leq 0$). But $b_i \geq 0$ (by hypothesis, we have a b.f.p.), and so $r_i \geq 0$. On the other hand, since $c_j > 0$, it follows that we can make w as large as desired simply by making s_j large enough. But this means that the program has no solution: it is unbounded.

We have seen that, starting from any b.f.p., the rules III.5.1 will eventually *either* give us the solution to the program *or* tell us that the program is unbounded. An example of the latter possibility follows.

III.7.2 Example

Maximize

$$w = 2x + y + z$$

Subject to

$$x - 2y - 4z \leq -20$$
$$3x - 5y + z \leq 3$$
$$x, y, z \geq 0$$

We obtain, in this case, the tableau

x	y	z	1	
1	-2	-4	20	$= -r$
3	-5	1^*	-3	$= -s$
1	2	1	0	$= w$

We have a positive entry in the right-hand column. Taking the first column as our pivot column, the rules III.6.1 tell us to pivot on the starred entry, giving us the tableau

x	y	s	1	
13	-22^*	4	8	$=-r$
3	-5	1	-3	$=-z$
-2	7	-1	3	$=w$

which is still not a b.f.p. We pivot on the starred entry (which is the only negative entry in the top row) obtaining

x	r	s	1	
$-13/22$	$-1/22$	$-2/11$	$-4/11$	$=-y$
$1/22^*$	$-5/22$	$1/11$	$-53/11$	$=-z$
$47/22$	$7/22$	$3/11$	$61/11$	$=w$

which is a b.f.p., but not the solution, as there are several positive entries in the bottom row. If we pivot on the starred entry, we obtain the tableau

z	r	s	1	
13	-3	1	-63	$=-y$
22	-5	2	-106	$=-x$
-47	11	-4	232	$=w$

This tableau is not the solution, as there is still a positive entry in the bottom row, second column. There are, however, no other positive entries in the second column, so the rules III.5.1 cannot be followed. We conclude that the program does not have a solution: it is unbounded.

In fact, we may check that, for any positive value of r, the point

$$x = 106 + 5r$$
$$y = 63 + 3r$$
$$z = 0$$
$$w = 232 + 11r$$

satisfies the constraints. By making r large enough, w can be made as large as desired. If, for example, we want w to be greater than 1,000,000, we simply let $r = 100,000$; then

$$x = 500,106$$
$$y = 300,063$$
$$z = 0$$

$$w = 1,100,232$$

is a feasible point of the program, as may be checked directly.

We return to the rules III.6.1. Once again, we make the non-degeneracy assumption: no zeros appear in the right-hand column (except possibly in the bottom row).

III.7.3 Proof for Stage I.

In Stage I, the constant term in the pivot row, $-b_k$, and the other terms in the right-hand column, $-b_i$, are replaced by $-b_k'$ and $-b_i'$, respectively, according to equations (3.7.1) and (3.7.2). It can be shown, exactly as in III.7.1, that, for $i > l, -b_i' \leq 0$ (i.e., the entries in the right-hand column below the lth row remain non-positive.)

As for the entry $-b_l$, it is replaced by $-b_l'$, which will be given by (3.7.1) or (3.7.2), depending on whether the pivot is in the lth row ($k = l$) or in another row ($k \neq l$).

In case the pivot is in the lth row we have

3.7.5
$$-b_l' = -\frac{b_l}{a_{il}}$$

Now, by assumption, b_l and a_{il} are both negative. This means that $-b_l' < 0$.

If the pivot is in another row, we will have

3.7.6
$$-b_l' = -b_l + \frac{b_k a_{il}}{a_{ik}}.$$

By the rules, $a_{il} < 0$ while $a_{ik} > 0$. By the non-degeneracy assumption, $b_k > 0$, since $-b_k$ is a term in the right-hand column, below $-b_l$. It follows that the second term on the right side of (3.7.6) is negative, and so $-b_l' < -b_l$.

We see, thus, that, by following the rules III.6.1, we obtain a new tableau that preserves the non-positivity of the terms in the right-hand column *below* the lth row. Under the non-degeneracy assumption, the term $-b_l$ (the lowest positive entry in the right-hand column) will be decreased. Using once again the fact that any program can give only a finite number of distinct tableaux, we conclude that, so long as we follow the rules III.6.1, the term $-b_l$ will eventually become negative. We then work on the lowest remaining positive term in the right-hand column, and it follows that, eventually, *either* a b.f.p. will be reached *or* a tableau will appear at which it is impossible to follow rules III.6.1.

Suppose then, that we find it impossible to follow the rules III.6.1, although we do not yet have a b.f.p. This will happen if a row has a positive entry in the right-hand column, and only non-negative entries otherwise, i.e., if for an i, $-b_i > 0$, and $a_{ij} \geq 0$ for all j. The ith row then represents the equation

3.7.7 $$r_i = b_i - (a_{i1}s_1 + a_{i2}s_2 + \ldots + a_{in}s_n)$$

Now, each term inside the parenthesis in the right side of (3.7.7) is the product of a non-negative constant a_{ij} and a non-negative variable s_j. It follows that the expression inside the parenthesis is non-negative, and so $r_i \leq b_i$. But $b_i < 0$, and so $r_i < 0$. But r_i must be non-negative. We have, thus, a contradiction. It means that the program is *infeasible*: the constraints cannot be satisfied.

We see, then, that the rules for Stage I will either lead us to a b.f.p. or tell us that the program is infeasible. An example of this possibility is:

III.7.4 Example

Maximize

$$w = x - 4y$$

Subject to

$$x + 3y \geq 2$$
$$2x + 5y \leq 1$$
$$x + y \leq 4$$
$$x, y \geq 0$$

We have here the tableau

	x	y	1	
	-1	-3	2	$= -r$
	2	5^*	-1	$= -s$
	1	1	-4	$= -t$
	1	-4	0	$= w$

which has a positive entry in the right-hand column, top row. There are two negative entries in this row: let us choose the second column as the pivot column. The negative entry in the top row and both positive entries below it have to be considered in choosing a pivot. The corresponding quotients are 2/3, 1/5, and 4. The smallest corresponds to the second row; we pivot, therefore, on the starred entry, to obtain the tableau

	x	s	1	
	1/5	3/5	7/5	$= -r$
	2/5	1/5	$-1/5$	$= -y$
	3/5	$-1/5$	$-19/5$	$= -t$
	13/5	4/5	$-4/5$	$= w$

This is not a b.f.p., since there is still a positive entry (albeit smaller) in the right-hand column. In fact, however, all the entries in the top row are positive. We conclude that the program is infeasible.

8. EQUATION CONSTRAINTS

In some cases, some of the constraints in a program may be equations instead of inequalities. As we mentioned earlier, an equation may be replaced by two opposite inequalities, so that it is always possible to reduce the program to standard form. On the other hand, it should be clear that if we replace an equation by two inequalities, and then introduce two slack variables to make each of these inequalities into an equation, we are merely multiplying work for ourselves. It is better to leave the equations as they are; there is then, of course, no slack variable corresponding to the constraint. The technique is slightly different, though the same general rules apply. The following example shows this.

III.8.1 Example. A steel company has two warehouses, W_1 and W_2, and three distributors, D_1, D_2, and D_3. The two warehouses hold, respectively, 50,000 and 60,000 tons of steel. Of these, 30,000 tons must be shipped to D_1, 40,000 tons to D_2, and 40,000 tons to D_3. The transportation costs, in dollars per ton, between warehouses and distributors, are given in Table III.8.1.

TABLE III.8.1

From Warehouse	To Distributor		
	D_1	D_2	D_3
W_1	1	4	1
W_2	5	9	2

The required amounts must be transported in such a way as to minimize the total costs.

Letting x_{ij} represent the amount shipped from warehouse W_i to distributor D_j, we have the program

Minimize

$$w = x_{11} + 4x_{12} + x_{13} + 5x_{21} + 9x_{22} + 2x_{23}$$

FINITE MATHEMATICS

Subject to

$$x_{11} + x_{12} + x_{13} \qquad\qquad\qquad = 50$$
$$x_{21} + x_{22} + x_{23} = 60$$
$$x_{11} \qquad\qquad + x_{21} \qquad\qquad = 30$$
$$x_{12} \qquad\qquad + x_{22} \qquad = 40$$
$$x_{13} \qquad\qquad + x_{23} = 40$$
$$x_{ij} \geq 0$$

where, for example, the first constraint means that the amounts shipped from W_1 to the three distributors must add up to 50, and so on for the other constraints (we are measuring in units of a thousand). Of the five equation constraints, one can be dispensed with: the fifth constraint is equal to the sum of the first two, minus the sum of the third and fourth equations. (This is, of course, because the total amounts available at the warehouses are equal to the total requirements by the distributors. If D_1 and D_2 receive what they require, then the amount left for D_3 is eactly what he requires.) Discarding the fifth equation, we get the tableau

x_{11}	x_{12}	x_{13}	x_{21}	x_{22}	x_{23}	1	
1	1	1	0	0	0	−50	= 0
0	0	0	1	1	1*	−60	= 0
1	0	0	1	0	0	−30	= 0
0	1	0	0	1	0	−40	= 0
1	4	1	5	9	2	0	= w

It is clear that zeros have no business in the space reserved for basic variables. To remove them, we pivot in the usual manner, say, on the starred entry. This will bring one of the zeros into the row of non-basic variables:

x_{11}	x_{12}	x_{13}	x_{21}	x_{22}	0	1	
1	1	1	0	0	0	−50	= 0
0	0	0	1	1	1	−60	= −x_{23}
1	0	0	1	0	0	−30	= 0
0	1	0	0	1*	0	−40	= 0
1	4	1	3	7	−2	120	= w

We cross out the column corresponding to the zero, since the elements in that column would merely be multiplied by zero. Pivoting on the next starred entry, we get

x_{11}	x_{12}	x_{13}	x_{21}	0	1	
1	1	1	0	0	−50	= 0
0	−1	0	1	−1	−20	= −x_{23}
1	0	0	1*	0	−30	= 0
0	1	0	0	1	−40	= −x_{22}
1	−3	1	3	−7	400	= w

Again, we cross out the column corresponding to zero, and pivot on the starred entry

x_{11}	x_{12}	x_{13}	0	1	
1	1	1*	0	−50	= 0
−1	−1	0	−1	10	= −x_{23}
1	0	0	1	−30	= −x_{21}
0	1	0	0	−40	= −x_{22}
−2	−3	1	−3	490	= w

We repeat the procedure:

x_{11}	x_{12}	0	1	
1	1	1	−50	= −x_{13}
−1	−1*	0	10	= −x_{23}
1	0	0	−30	= −x_{21}
0	1	0	−40	= −x_{22}
−3	−4	−1	540	= w

Having reduced the tableau to the desired form, we now proceed to solve the problem by applying rules III.5.1 and III.6.1. We show the pivots, in each case, by asterisks.

x_{11}	x_{23}	1	
0	1	−40	= −x_{13}
1	−1	−10	= −x_{12}
1	0	−30	= −x_{21}
−1	1*	−30	= −x_{22}
1	−4	500	= w

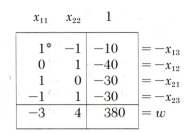

x_{11}	x_{22}	1	
1^*	-1	-10	$= -x_{13}$
0	1	-40	$= -x_{12}$
1	0	-30	$= -x_{21}$
-1	1	-30	$= -x_{23}$
-3	4	380	$= w$

x_{13}	x_{22}	1	
1	-1	-10	$= -x_{11}$
0	1	-40	$= -x_{12}$
-1	1	-20	$= -x_{21}$
1	0	-40	$= -x_{23}$
3	1	350	$= w$

This last tableau gives us the minimum cost solution: it is

$$x_{11} = 10 \qquad x_{12} = 40 \qquad x_{13} = 0$$
$$x_{21} = 20 \qquad x_{22} = 0 \qquad x_{23} = 40$$

with a total cost of 350. (Measured in thousands, this is really $350,000.)

Such problems, called *transportation problems,* can be solved by a considerably simpler method; we shall see this in a later section of this chapter.

9. DEGENERACY PROCEDURES

In case of degeneracy, i.e., assuming that some zeros appear in the right-hand column of the simplex tableau, the foregoing arguments do not hold. The main difficulty is that, in Stage II, for instance, the new tableau obtained need not give a better value for the objective function. But it is precisely the improvement in the objective function that guarantees that the process will terminate; in a case of degeneracy we might, with bad luck, find that the process "cycles"—it brings us back to a previous position. Similar difficulties arise in Stage I.

From the geometric point of view, degeneracy arises when more than n of the bounding planes of the constraint set pass through a point (n, here, is the dimension of the space). Figure III.9.1 illustrates this: it shows the set satisfying

$$\begin{aligned} x \quad + z &\leq 1 \\ y + z &\leq 1 \\ x, y, z &\geq 0 \end{aligned}$$

3.9.1

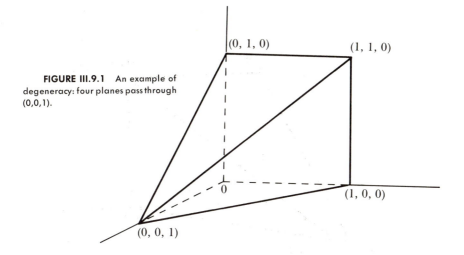

FIGURE III.9.1 An example of degeneracy: four planes pass through (0,0,1).

As may be seen, the planes $x + z = 1$, $y + z = 1$, $x = 0$, and $y = 0$ all pass through the point (0,0,1). This is not normal: in three-dimensional space, four planes will usually "meet," three at a time, at four distinct points.

Precisely because this situation is not normal, we can change it by what is called a *perturbation*. Let us consider the slightly different set

3.9.2
$$\begin{aligned} x \quad + z &\leq 11/10 \\ y + z &\leq 1 \\ x, y, z &\geq 0 \end{aligned}$$

which differs from the previous system in that the first inequality has been slightly "perturbed," i.e., the right side has been changed by 1/10. Figure III.9.2 shows the new constraint set: the vertex at (0,0,1) has been split into two vertices, one of which is now at (1/10,0,1). Two of the other vertices have been slightly perturbed, the others not at all.

Suppose, now, that we are given a program with (3.9.1) as constraint set. We replace the constraint set by (3.9.2) and solve. The solution of the new program will be at one of the vertices of Figure III.9.2. Each of these vertices corresponds to a *unique* vertex of Figure III.9.1: for instance, (0,1,0) corresponds to (0,1,0), while (11/10,0,0) corresponds to (1,0,0). (The converse is not true: a vertex of Figure III.9.1 may correspond to more than one vertex of Figure III.9.2.) That vertex of Figure III.9.1 that corresponds to the solution of the program (3.9.2) will *probably* be the solution of the program (3.9.1); if (3.9.2) has its solution at (1/10,0,1), then (3.9.1) will *probably* have its solution at (0,0,1). We say probably, because it may be that if the perturbation has been too large the solution of the perturbed problem (3.9.2) does not correspond to the solution of (3.9.1). In that

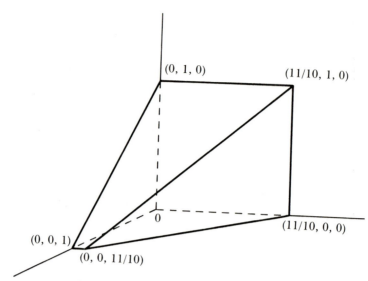

(0, 1, 0)

(11/10, 1, 0)

(0, 0, 1)

(0, 0, 11/10)

0

(11/10, 0, 0)

FIGURE III.9.2 A slight perturbation of Figure III.9.1.

case, the perturbation should have been smaller: the first constraint may be changed, say, to $x + z \leq 101/100$.

This, then, is the usual technique for perturbation. Each time a zero appears in the right-hand column, we perturb slightly, replacing the zero by a small number. Since we do not know how small is small enough to preserve the position of the solution, we generally use the letter $-\epsilon$, representing a very small but negative amount. The second time that we obtain a zero in the right-hand column, we replace it by $-\epsilon^2$, and so on. The solution, when obtained, will be in terms of ϵ. We then obtain the solution to the unperturbed problem by letting ϵ be equal to zero.

In a sense, this technique is needlessly complicated; careful selection of the pivot will generally make it unnecessary. The procedure is described here only because it helps to complete the proof of convergence for the simplex algorithm, using rules III.5.1 and III.6.1. Normally, "cycling" occurs only when there is a very systematic choice of the pivot. A recommended procedure when using computers for the solution of linear programs is to instruct the computer to choose its pivots at random whenever the rules allow it latitude for such choice. This makes certain that the pivot is not chosen systematically, and cycling is avoided.

Although the perturbation method is not generally recommended, we give the following example.

III.9.1 Example

Maximize

$$w = x + y + 3z$$

Subject to

$$x \quad + z \leq 1$$
$$y + z \leq 1$$
$$x, y, z \geq 0$$

We have here the tableau

x	y	z	1	
1	0	1	-1	$= -r$
0	1	1^*	-1	$= -s$
1	1	3	0	$= w$

Let us assume we take our pivot from the third column; the two quotients are equal, and so we may pivot on either one of the entries in this column. *It is this situation that gives rise to degeneracy.* Pivoting on the starred entry, we get the tableau

x	y	s	1	
1	-1	-1	0	$= -r$
0	1	1	-1	$= -z$
1	-2	-3	3	$= w$

This shows degeneracy: we therefore perturb the tableau to get

x	y	s	1	
1^*	-1	-1	$-\epsilon$	$= -r$
0	1	-1	-1	$= -z$
1	-2	-3	3	$= w$

Pivoting on the starred entry, we then have

r	y	s	1	
1	-1	-1	$-\epsilon$	$= -x$
0	1	1	-1	$= -z$
-1	-1	-2	$3+\epsilon$	$= w$

Thus the solution to the perturbed problem is $x = \epsilon$, $y = 0$, $z = 1$, with a value $w = 3 + \epsilon$ for the objective function. We conclude that the unperturbed problem has the solution $x = 0$, $y = 0$, $z = 1$, and $w = 3$.

142   10. SOME PRACTICAL COMMENTS

The number of pivot steps necessary to solve an $m \times n$ program (i.e., a program with m inequalities in n variables, not counting non-negativity constraints) is generally of the order of magnitude of $m + n$. This means that, in most cases, it will be between, say, $(m + n)/2$ and $2(m + n)$. This is certainly a tremendous gain over the "brute force" method described in Section III.1. For instance, if $m = n = 10$, we expect that the simplex algorithm will use between 10 and 40 pivot steps, whereas inspecting all the intersections of 10 of the planes would imply solving $\binom{20}{10} = 184{,}756$ systems of 10 equations in 10 unknowns! Nevertheless, for large programs, it is always desirable to decrease even the number of steps that the simplex method uses. This can be done by judicious choice of the pivot. In fact, rules III.5.1 and III.6.1 often give us great latitude for choice, for instance, there may be many positive entries in the bottom row. Much work has been done in perfecting the choice of pivot so as to cut the number of steps considerably, in some cases by as much as one half, over indiscriminate choices. Unfortunately, we cannot, within the scope of this book, study all these systems. Interested students should read more advanced treatises on linear programming.

PROBLEMS ON THE SIMPLEX ALGORITHM

1. Maximize

$$w = 2x + 3y + z$$

Subject to

$$
\begin{aligned}
x + 4y - 2z &\leq 10 \\
3y + z &\leq 7 \\
2x - 3y + 4z &\leq 12 \\
x &\geq 0 \\
y &\geq 0 \\
z &\geq 0
\end{aligned}
$$

2. Minimize

$$w = x + 2y + z$$

Subject to

$$
\begin{aligned}
3x + y + 2z &\geq 20 \\
x - 3y + 4z &\geq 15 \\
-x + y + z &\geq 8 \\
x &\geq 0 \\
y &\geq 0 \\
z &\geq 0
\end{aligned}
$$

3. Maximize

$$w = x - 2y + 4z$$

Subject to

$$
\begin{aligned}
x + y + 2z &\le 20 \\
x - y + z &\le 5 \\
2x + 3y + 2z &\le 35 \\
x &\ge 0 \\
y &\ge 0 \\
z &\ge 0
\end{aligned}
$$

4. Minimize

$$w = 5x + 2y + 3z$$

Subject to

$$
\begin{aligned}
3x - 2y + 6z &\ge 25 \\
x + y + 2z &\ge 15 \\
3x + y - 3z &\ge 10 \\
x &\ge 0 \\
y &\ge 0 \\
z &\ge 0
\end{aligned}
$$

5. A dietitian is given three foods. Each unit of food A contains 10 oz. protein, 4 oz. fat, and 6 oz. carbohydrate, and costs $1. A unit of food B has 4 oz. protein, 6 oz. fat, and 8 oz. carbohydrate, and costs 50¢. A unit of food C contains 2 oz. protein, 12 oz. fat, and 8 oz. carbohydrate, and costs 30¢. A mixture of these three foods must be made, containing at least 50 oz. protein, 60 oz. fat, and 65 oz. carbohydrate. What is the cheapest such mixture possible?

6. A toy company makes three types of toy, which must be processed through three machines called a twister, a bender, and a painter. Toy A requires 1 hr. in the twister, 2 hr. in the bender, and 1 hr. in the painter, and sells for $4. Toy B requires 2 hr. in the twister, 1 hr. in the bender, and 3 hr. in the painter, and sells for $5. Toy C requires 3 hr. in the twister, 2 hr. in the bender, and 1 hr. in the painter, and sells for $9. The twister can work at most 50 hr. per week, the bender, at most 40 hr.; and the painter, at most 60 hr. How many toys of each type should the company produce to maximize revenues?

7. A furniture firm manufactures chairs, tables, and beds, and has three workers. The time in hours that each employee must spend to produce one of these, and the corresponding profit, are given in the table:

	A	B	C	Profit ($)
Chair	1	3	2	10
Table	3	5	4	25
Bed	5	4	8	35

Employee A can work only 20 hr. per week, while B and C can work 50 hr. each. How many chairs, beds, and tables should be made so as to maximize profits?

8. A food processing company has 10,000 lb. of African coffee, 12,000 lb. of Brazilian coffee, and 7000 lb. of Colombian coffee. It sells four blends of coffee, whose contents (in ounces of the coffee per pound of blend) and prices are given in the table:

	African	Brazilian	Colombian	Price (¢)
A	0	0	16	90
B	0	12	4	70
C	6	8	2	60
D	10	6	0	50

How many pounds of each blend should be produced to maximize revenues?

9. A candy manufacturer has 500 lb. of chocolate, 100 lb. of nuts, and 50 lb. of fruit in inventory. He produces three types of candy. A box of type A uses 3 lb. chocolate, 1 lb. nuts, and 1 lb. fruit, and sells for $10. A box of type B uses 4 lb. chocolate and 1/2 lb. nuts, and sells for $6. A box of type C contains 5 lb. chocolate, and sells for $4. How many boxes of each type should be made to maximize revenues?

10. A metal company needs 50 tons of iron, 5 tons of copper, and 1 ton of nickel to fulfill a contract. It is offered three types of ore. Type A contains 10 per cent iron and 1 per cent copper, and costs $5 per ton. Type B contains 5 per cent copper and 2 per cent nickel, and sells for $15 per ton. Type C contains 3 per cent iron, 1 per cent copper, and 2 per cent nickel, and sells for $6 per ton. How many tons of each ore should be bought to minimize costs?

11. DUALITY

Let us consider the two linear programs

3.11.1 Maximize

$$w = c_1 x_1 + c_2 x_2 + \ldots + c_n x_n$$

Subject to

3.11.2
$$\begin{cases} a_{11}x_1 + a_{12}x_2 + \ldots + a_{1n}x_n \leq b_1 \\ a_{21}x_1 + a_{22}x_2 + \ldots + a_{2n}x_n \leq b_2 \\ \cdots\cdots\cdots\cdots\cdots\cdots\cdots\cdots\cdots \\ \cdots\cdots\cdots\cdots\cdots\cdots\cdots\cdots\cdots \\ a_{m1}x_1 + a_{m2}x_2 + \ldots + a_{mn}x_n \leq b_m \end{cases}$$

3.11.3
$$x_1, x_2, \ldots, x_n \geq 0$$

and

3.11.4 Minimize

$$z = b_1 y_1 + b_2 y_2 + \ldots + b_m y_m$$

Subject to

3.11.5
$$\begin{cases} a_{11}y_1 + a_{21}y_2 + \ldots + a_{m1}y_m \geq c_1 \\ a_{12}y_1 + a_{22}y_2 + \ldots + a_{m2}y_m \geq c_2 \\ \cdots\cdots\cdots\cdots\cdots\cdots\cdots\cdots\cdots \\ \cdots\cdots\cdots\cdots\cdots\cdots\cdots\cdots\cdots \\ a_{1n}y_1 + a_{2n}y_n + \ldots + a_{mn}y_m \geq c_m \end{cases}$$

3.11.6
$$y_1, y_2, \ldots, y_m \geq 0$$

In matrix notation, these programs are

3.11.7 Maximize

$$w = \mathbf{c}^t \mathbf{x}$$

Subject to

3.11.8
$$A\mathbf{x} \leq \mathbf{b}$$

3.11.9
$$\mathbf{x} \geq 0$$

and

3.11.10 Minimize

$$z = \mathbf{b}^t \mathbf{y}$$

Subject to

3.11.11 $$A^t y \geq c$$

3.11.12 $$y \geq 0$$

where the matrix A and the vectors b and c are the same for both programs. We shall call the program (3.11.10) to (3.11.12) — or (3.11.4) to (3.11.6) — the *derived program* of program (3.11.7) to (3.11.9) — or (3.11.1) to (3.11.3).

Let us see whether we can find the derived program of (3.11.10) to (3.11.12). We notice, of course, that (3.11.10) to (3.11.12) is not the same type of program as (3.11.7) to (3.11.9), since it is a minimization problem, and, moreover, the inequalities (3.11.11) go the wrong way. We dispose of these difficulties by multiplying both the objective function and the constraints (3.11.11) by -1; this gives us a restatement of (3.11.10) to (3.11.12) in the form

Maximize

$$-z = (-b)^t y$$

Subject to

$$(-A^t) y \leq -c$$
$$y \geq 0$$

We can now form the derived program of this; it is done by taking the transpose of the coefficient matrix and interchanging the two vectors $-b$ and $-c$. We then obtain

3.11.13 Minimize

$$r = (-c)^t s$$

Subject to

3.11.14 $$(-A^t)^t s \geq -b$$

3.11.15 $$s \geq 0$$

We can, however, do the same to this program as we just did to (3.11.10) to (3.11.12). Using the fact that $A^{tt} = A$, and multiplying through by -1, we have the equivalent program

3.11.16 Maximize

$$-r = c^t s$$

Subject to

3.11.17 $$As \leq b$$

3.11.18 $$s \geq 0$$

It may be seen that this program (3.11.16) to (3.11.18) is the same as (3.11.7) to (3.11.9): in fact, the only difference between the two lies in the fact that the variable vector is **x** in one case, and **s** in the other; in other words, the only difference lies in the labeling of the variables. But this is no difference at all. We find, thus, that the derived of the derived program (3.11.10) to (3.11.12) is the original program (3.11.7) to (3.11.9). When a relation is of this sort (in which each of two problems may be obtained from the other by the same set of rules) it is generally called a *duality* relation: the two programs, in this case, are said to be *dual* to each other.

III.11.1 Definition. The two programs (3.11.1) to (3.11.3) and (3.11.4) to (3.11.6) are *dual* programs. Each of these two programs is the *dual* of the other.

From a purely theoretical point of view, each of the two programs (3.11.1) to (3.11.3) and (3.11.4) to (3.11.6) is the dual of the other; hence neither enjoys any primacy over the other. From the practical point of view, of course, this is not so. The decision-maker is given a practical problem, which he expresses (if possible) in the form of a linear program. The dual program has, in general, no practical interpretation (though we shall see that it may be given one in terms of so-called *shadow prices*). Thus, the original practical program does enjoy a definite primacy and will, therefore, be called the *primal*. The other problem will be called the *dual*.

The relation between two dual linear programs is extremely strong, as is evidenced by the theorems that follow.

III.11.2 Theorem. Let the vectors $\mathbf{x} = (x_1, \ldots, x_n)^t$ and $\mathbf{y} = (y_1, \ldots, y_m)^t$ satisfy the constraints (3.11.8) and (3.11.9) and (3.11.11) and (3.11.12), respectively. Then

$$\mathbf{c}^t \mathbf{x} \leq \mathbf{b}^t \mathbf{y}$$

Proof. We prove this theorem by the following sequence of relations:

By the commutative law for vector multiplication,

$$\mathbf{c}^t \mathbf{x} = \mathbf{x}^t \mathbf{c}$$

Now, $\mathbf{c} \leq A^t \mathbf{y}$, and $\mathbf{x}^t \geq 0$, and so

$$\mathbf{x}^t \mathbf{c} \leq \mathbf{x}^t (A^t \mathbf{y})$$

By the associative law for matrix multiplication,

$$\mathbf{x}^t (A^t \mathbf{y}) = (\mathbf{x}^t A^t) \mathbf{y}$$

Now, $Ax \leq b$, so $x^t A^t \leq b^t$; since $y \geq 0$, then

$$(x^t A^t)y \leq b^t y$$

This gives us the chain of relations

$$c^t x = x^t c \leq x^t (A^t y) = (x^t A^t) y \leq b^t y$$

which proves the theorem.

III.11.3 Corollary. Suppose x^* and y^* satisfy the constraints (3.11.8) to (3.11.9) and (3.11.11) to (3.11.12) respectively, and, moreover,

$$c^t x^* = b^t y^*$$

Then x^* and y^* are the solutions of the programs (3.11.7) to (3.11.9) and (3.11.10) to (3.11.12) respectively.

Proof. This is clear since, if x is any vector satisfying (3.11.8) to (3.11.9), then, by Theorem III.11.2,

$$c^t x \leq b^t y^* = c^t x^*$$

Hence x^* solves (3.11.7) to (3.11.9), and, similarly, y^* solves (3.11.10) to (3.11.12).

Corollary III.11.3 gives us a method of checking that a given pair of vectors, x^* and y^*, solve the two dual programs (3.11.7) to (3.11.9) and (3.11.10) to (3.11.12), respectively. In fact, if they satisfy the constraints of the programs, and give the same values for the objective functions, then III.11.3 confirms that they are solutions. The converse of III.11.3 is also true; if x^* and y^* are solutions to the two mutually dual programs (3.11.7) to (3.11.9) and (3.11.10) to (3.11.12), then $c^t x^* = b^t y^*$. This equality of values is both necessary and sufficient for the vectors x^* and y^* to solve the two programs. (We shall prove this statement later.) We give an example of this property now.

III.11.4 Example. Consider the two programs

Maximize

$$w = 3x_1 + 2x_2 + 4x_3$$

Subject to

$$x_1 + 3x_2 + 2x_3 \leq 10$$
$$2x_1 + x_2 + x_3 \leq 8$$
$$x_1, x_2, x_3 \geq 0$$

and

Minimize

$$z = 10y_1 + 8y_2$$

Subject to

$$y_1 + 2y_2 \geq 3$$
$$3y_1 + y_2 \geq 2$$
$$2y_1 + y_2 \geq 4$$
$$y_1, y_2 \geq 0$$

The maximizing problem gives us the tableau

x_1	x_2	x_3	1	
1	3	2*	−10	$= -u_1$
2	1	1	−8	$= -u_2$
3	2	4	0	$= w$

Pivoting on the starred entries, we obtain the tableaux

x_1	x_2	u_1	1	
1/2	3/2	1/2	−5	$= -x_3$
3/2*	−1/2	−1/2	−3	$= -u_2$
1	−4	−2	20	$= w$

and

u_2	x_2	u_1	1	
−1/3	5/3	2/3	−4	$= -x_3$
2/3	−1/3	−1/3	−2	$= -x_1$
−2/3	−11/3	−5/3	22	$= w$

which gives us the solution; it is $x_1 = 2$, $x_2 = 0$, $x_3 = 4$, and gives a value $w = 22$.

Consider, now, the minimizing problem. Its tableau is

y_1	y_2	1	
−1	−2	3	$= -v_1$
−3	−1	2	$= -v_2$
−2*	−1	4	$= -v_3$
10	8	0	$= z$

150   We pivot on the starred entries to obtain

	v_3	y_2	1	
	$-1/2$	$-3/2^*$	1	$=-v_1$
	$-3/2$	$1/2$	-4	$=-v_2$
	$-1/2$	$1/2$	-2	$=-y_1$
	5	3	20	$=z$

and

	v_3	v_1	1	
	$1/3$	$-2/3$	$-2/3$	$=-y_2$
	$-5/3$	$1/3$	$-11/3$	$=-v_2$
	$-2/3$	$1/3$	$-5/3$	$=-y_1$
	4	2	22	$=z$

which gives the solution: $y_1 = 5/3$, $y_2 = 2/3$ with a value $z = 22$. It is now possible to check that the given vectors are, indeed, the solutions to the two programs. In fact, they satisfy the constraints, and give the same value for the objective functions.

Looking over the solutions of the pair of dual programs, we can see that the implications of duality are far greater than the simple fact that the values of the two programs (i.e., the maximum and minimum values, respectively, of their objective functions) are equal. In fact, it is not only the objective functions that give the same value: the corresponding tableaux for the two programs are, at each step, dually related. Each tableau is "almost" the negative transpose of the corresponding tableau for the other problem. That this is so for any pair of dual programs is immensely important.

Let us, by the introduction of slack variables, rewrite (3.11.7) to (3.11.9) in the form

3.11.19   Maximize

$$w = \mathbf{c}^t \mathbf{x}$$

Subject to

3.11.20 $\mathbf{Ax} - \mathbf{b} = -\mathbf{u}$

3.11.21 $\mathbf{x}, \mathbf{u} \geq 0$

Similarly, we may rewrite (3.11.10) to (3.11.12) in the form

3.11.22   Minimize

$$z = \mathbf{b}^t \mathbf{y}$$

Subject to

3.11.23 $$A^t y - c = v$$

3.11.24 $$y, v \geq 0$$

We may write these two programs simultaneously in the tableau

3.4.25

	x_1	x_2	$\ldots$	x_n	1	
y_1	a_{11}	a_{12}	$\ldots$	a_{1n}	$-b_1$	$= -u_1$
y_2	a_{21}	a_{22}	$\ldots$	a_{2n}	$-b_2$	$= -u_2$
$\ldots$						
y_m	a_{m1}	a_{m2}	$\ldots$	a_{mn}	$-b_m$	$= -u_m$
-1	c_1	c_2	$\ldots$	c_n	0	$= w$

$$= v_1 \quad = v_2 \quad \ldots \quad v_n \quad = -z$$

The rows of (3.11.25) represent, as usual, the equations (3.11.19) and (3.11.20); the columns, now, represent the equations (3.11.22) and (3.11.23). Thus, the single tableau does double duty as a representation of the pair of problems simultaneously. The pivot steps of the simplex algorithm were devised precisely to maintain the validity of the row equations. What we will show is that they have the property of preserving the column equations as well. Let us, then, consider the column system of a tableau:

3.11.26

	a_{11}	a_{12}	$\ldots$	a_{1n}	$-b_1$
q_1	a_{11}	a_{12}	$\ldots$	a_{1n}	$-b_1$
q_2	a_{21}	a_{22}	$\ldots$	a_{2n}	$-b_2$
$\ldots$					
q_m	a_{m1}	a_{m2}	$\ldots$	a_{mn}	$-b_m$
-1	c_1	c_2	$\ldots$	c_n	δ

$$= p_1 \quad = p_2 \quad \ldots \quad = p_n \quad = -z$$

In (3.11.26), the p_j are the basic variables; the q_i are non-basic. Suppose that we wish to interchange p_j and q_i. The jth column represents the equation

3.11.27 $$a_{ij}q_1 + a_{2j}q_2 + \ldots + a_{ij}q_i + \ldots - c_j = p_j$$

We can solve (3.11.27) for q_i, obtaining

3.11.28 $$-\frac{a_{1j}}{a_{ij}}q_1 - \frac{a_{2j}}{a_{ij}}q_2 - \ldots + \frac{1}{a_{ij}}p_j - \ldots + \frac{c_j}{a_{ij}} = q_i$$

In turn, the lth column represents

3.11.29 $\qquad a_{1l}q_1 + a_{2l}q_2 + \ldots + a_{il}q_i + \ldots - c_l = p_l$

Let us, now, substitute (3.11.28) in (3.11.29). This gives us

3.11.30 $\quad \left(a_{1l} - \dfrac{a_{il}a_{1j}}{a_{ij}}\right)q_1 + \left(a_{2l} - \dfrac{a_{il}a_{2j}}{a_{ij}}\right)q_2 + \ldots + \dfrac{a_{il}}{a_{ij}}\,p_j$

$$+ \ldots - \left(c_l - \dfrac{a_{il}c_j}{a_{ij}}\right) = p_l$$

The changes in the coefficients can be represented by the schema

3.11.31

$$
\begin{bmatrix} p & q \\ \\ r & s \end{bmatrix}
\rightarrow
\begin{array}{|cc|}
\hline
\dfrac{1}{p} & \dfrac{q}{p} \\ \\
-\dfrac{r}{p} & s - \dfrac{qr}{p} \\
\hline
\end{array}
$$

in which, as in Section III.2, p represents the pivot, q an entry in the pivot row, r an entry in the pivot column, and s the entry in the row of r and column of q. But (3.11.31) is exactly the same as (3.3.10). Thus, *the pivot transformations that preserve the row equations also preserve the column equations.*

Let us suppose that the program (3.11.1) to (3.11.3) has a solution. This solution is obtained, as we know, from a tableau in which all the entries in the right-hand column and in the bottom row (except possibly the bottom right-hand entry) are non-positive. Consider then, the column equations in this tableau:

3.11.32

q_1	a_{11}	a_{12}	$\ldots$	a_{1n}	$-b_1$
q_2	a_{21}	a_{22}	$\ldots$	a_{2n}	$-b_2$
$\ldots$					
q_m	a_{m1}	a_{m2}	$\ldots$	a_{mn}	$-b_m$
-1	c_1	c_2	$\ldots$	c_n	δ
	$=p_1$	$=p_2$	$\ldots$	$=p_n$	$=-z$

It may be seen from (3.11.32) that, if we set the non-basic variables $q_1, \ldots, q_m$ equal to zero, each of the basic variables, $p_1, \ldots, p_n$ will be equal to the negative of the corresponding entry in the bottom row: $p_j = -c_j$. But, by hypothesis, $c_j \leq 0$. Hence $p_j \geq 0$. Thus the tableau (3.11.32) gives a b.f.p. for the dual program (3.11.4) to (3.11.6).

Consider, next, the right-hand column of (3.11.32). It represents the equation

3.11.33 $$z = b_1 q_1 + b_2 q_2 + \ldots + b_m q_m + \delta$$

Now, by hypothesis, each of the terms $b_i q_i$ is non-negative (being the product of a non-negative constant b_i and a non-negative variable q_i). It follows that, for every point in the constraint set, $z \geq \delta$. We have here, however, a point that gives a value $z = \delta$ for the objective function. It follows that the b.f.p. of (3.11.32) is the solution of (3.11.4) to (3.11.6); *the solutions to both programs (3.11.1) to (3.11.3) and (3.11.4) to (3.11.6) are given by the same tableau (3.11.32).* Therefore, the optimal values of the objective functions must be the same for both programs, i.e., δ. We obtain the following theorem, known as the *fundamental theorem of linear programming.*

III.11.5 Theorem. If either one of the two programs (3.11.1) to (3.11.3) and (3.11.4) to (3.11.6) has a solution, then so does the other, and the solutions give the same value to the objective functions. If both programs are feasible, then both have solutions. If one of the two programs is feasible but unbounded, then the other is infeasible. Finally, if one of the two programs is infeasible, then the other is either infeasible or unbounded.

Proof. The first statement of this theorem has just been proved. The other statements depend on the fact that a program has a solution if it is bounded and feasible (seen in our development of the simplex algorithm), as well as on Theorem III.11.2.

That the same tableau solves both of a pair of dual problems is additionally advantageous because very often the first tableau obtained is a b.f.p., not for the primal problem, but for the dual. By attacking the dual, rather than the primal, we are able to go into Stage II immediately, avoiding perhaps half of the pivot steps normally necessary.

III.11.6 *Example.* A mixture is to be made of foods, A, B, and C. Each unit of food A contains 3 oz. protein and 4 oz. carbohydrate, and costs \$1. Each unit of food B contains 5 oz. protein and 3 oz. carbohydrate, and costs \$3. Each unit of food C contains 1 oz. protein and 4 oz. carbohydrate, and costs 50c. The mixture is to contain at least 15 oz. protein and 10 oz. carbohydrate. Find the mixture of minimal cost.

The program here is:

Minimize

$$z = y_1 + 3y_2 + \frac{1}{2} y_3$$

154

Subject to

$$3y_1 + 5y_2 + y_3 \geq 15$$
$$4y_1 + 3y_2 + 4y_3 \geq 10$$
$$y_1, y_2, y_3 \geq 0$$

The first tableau obtained for this program will not be a b.f.p.; $y = 0$ is not a feasible vector. On the other hand, we may write the tableau that represents this program as the column system: the dual will then be the row system:

	x_1	x_2	1	
y_1	3^*	4	-1	$= -u_1$
y_2	5	3	-3	$= -u_2$
y_3	1	4	$-\dfrac{1}{2}$	$= -u_3$
-1	15	10	0	$= w$
	$= v_1$	$= v_2$	$= -z$	

It may be seen that this is a b.f.p. for the dual problem. We pivot on the starred element, obtaining

	u_1	x_2	1	
v_1	$1/3$	$4/3$	$-1/3$	$= -x_1$
y_2	$-5/3$	$-11/3$	$-4/3$	$= -u_2$
y_3	$-1/3$	$8/3$	$-1/6$	$= -u_3$
-1	-5	-10	5	$= w$
	$= y_1$	$= v_2$	$= -z$	

which gives the solution: $y_1 = 5$, $y_2 = 0$, $y_3 = 0$ with a value $z = 5$. Note how Stage I was avoided by solving the dual program.

In a pair of mutually dual programs, generally one of the pair has a clear practical application: it has been formulated, precisely, to solve a practical problem. Yet, what about the dual? Is it possible to obtain a "practical" interpretation of the dual program?

Let us consider the dual of Example III.11.6:

Maximize

$$15x_1 + 10x_2$$

Subject to

$$3x_1 + 4x_2 \leq 1$$
$$5x_1 + 3x_2 \leq 3$$
$$x_1 + 4x_2 \leq \frac{1}{2}$$
$$x_1, x_2 \geq 0$$

and see whether some interpretation may be obtained. Each of the two variables, x_1 and x_2, corresponds to one of the constraints. But these constraints correspond to the two nutrients, protein and carbohydrate. Thus, x_1 and x_2 are numbers assigned, respectively, to protein and carbohydrate; the type of number can be seen from the fact that the constants in the constraints correspond to costs—the costs of the different foods, A, B, and C. It follows that we may consider x_1 and x_2 as representing some sort of value, in dollars per ounce, for the two nutrients. We are assigning such values, known as *shadow prices*, to these nutrients. The constraints represent, in a sense, a "profitless" system: each of three foods costs at least as much as the values of the nutrients it contains. The program seeks to maximize the total shadow prices of the required diet.

The solution of the dual problem is given by the tableau: $x_1 = 1/3$, $x_2 = 0$, with a value $w = 5$. Note that this seems to make the carbohydrate valueless. This is not to say that the carbohydrate is useless; rather, it represents the well-known economic fact that surplus goods lose all commercial value, and in this case the solution to the primal problem gives a surplus of carbohydrate. Note also that the solution of the primal does not use any of foods B and C, coincident with the fact that the costs of these foods are higher than the shadow prices of the nutrients contained (according to the dual solution).

Generally speaking, the dual may be interpreted in terms of shadow prices whenever the primal problem deals with costs or profits.

III.11.7 Example. A company makes three products, A, B, and C, which must be processed by three employees, D, E, and F. Each unit of product A must be processed 1 hr. by employee D, 2 hr. by E, and 3 hr. by F, and yields a profit of $50. Each unit of B must be processed 4 hr. by D, 1 hr. by E, and 1 hr. by F, and yields a profit of $30. Each unit of C must be processed 1 hr. by D, 3 hr. by E, and 2 hr. by F, and yields a profit of $20. Employee D can work at most 40 hr. per week; E, at most 50 hr., and F, at most 30 hr. It is required to maximize the profit.

The program here is:

Maximize

$$w = 50x_1 + 30x_2 + 20x_3$$

Subject to

$$x_1 + 4x_2 + x_3 \leq 40$$
$$2x_1 + x_2 + 3x_3 \leq 50$$
$$3x_1 + x_2 + 2x_3 \leq 30$$
$$x_1, x_2, x_3 \geq 0$$

156

The *dual* program is:

Minimize

$$z = 40y_1 + 50y_2 + 30y_3$$

Subject to

$$y_1 + 2y_2 + 3y_3 \geq 50$$
$$4y_1 + y_2 + y_3 \geq 30$$
$$y_1 + 3y_2 + 2y_3 \geq 20$$
$$y_1, y_2, y_3 \geq 0$$

which may be interpreted, in one sense, as the assignation of "natural" wages to the three employees. It may be checked that the solutions to the two problems are

$$x_1 = \frac{80}{11}, \; x_2 = \frac{90}{11}, \; x_3 = 0, \; w = \frac{6700}{11}$$

for the primal, and

$$y_1 = \frac{40}{11}, \; y_2 = 0, \; y_3 = \frac{170}{11}, \; z = \frac{6700}{11}$$

for the dual. This should not be taken to mean that employee E should receive no wages; it does mean, however, that he should be encouraged to work a shorter week, since he is clearly idle more than half the time. Conversely, this does not suggest that employee F should be paid over $15 per hour; it does say that his week should be lengthened if extra hours for him will cost less than $15.45 per hour. The "shadow wages" tend to represent the *marginal productivity* of the three employees concerned. (The classic economist would explain that employee F's wages would seek the high level causing him to work longer hours until an equilibrium was found.)

PROBLEMS ON DUALITY

1. Give the duals of the following linear programs. Obtain solutions to each pair of programs.

(a) Maximize

$$2x + 4y + 7z = w$$

Subject to

$$3x + y + 2z \leq 25$$
$$x + 3y - z \leq 10$$
$$4y + z \leq 30$$
$$x \geq 0$$
$$y \geq 0$$
$$z \geq 0$$

(b) Minimize

$$6x + 2y + 3z = w$$

Subject to

$$x + 4y - 3z \geq 15$$
$$2x + 3y + z \geq 25$$
$$-x + 4y + 2z \geq 12$$
$$x \geq 0$$
$$y \geq 0$$
$$z \geq 0$$

(c) Minimize

$$2x + y + 2z = w$$

Subject to

$$x + 3y + z \geq 30$$
$$x - y \geq -10$$
$$3x + y + z \geq 40$$
$$x \geq 0$$
$$y \geq 0$$
$$z \geq 0$$

(d) Maximize

$$3x + 2y + z = w$$

Subject to

$$x + y + 3z \leq 25$$
$$x - y \leq -10$$
$$y + 2z \leq 15$$
$$x \geq 0$$
$$y \geq 0$$
$$z \geq 0$$

2. Give the duals of problems 5 to 10 on pages 143-144. Solve these duals and give a heuristic interpretation (shadow prices).

3. Show that, to an equation constraint in the primal problem, there corresponds an unrestricted (i.e., not necessarily non-negative) variable in the dual, and vice versa.

158   12. TRANSPORTATION PROBLEMS

The simplex algorithm is probably the most efficient method available for the solution of linear programs in general. This does not mean, however, that it will be *uniformly* the most efficient method. The reason is that there are often special types of problem that, precisely because of their special type, can be handled more efficiently by a method other than the simplex algorithm. Much of the present research in linear programming consists precisely in the search for new methods of solving special types of linear programs — generally those that the researcher (an applied mathematician or operations analyst) has encountered in practical experience. Some of these new methods are tremendously efficient — for the particular type of problem concerned, generally considerably larger than could be solved with the simplex algorithm.

We consider here a special type of linear program: the *network* program. Such a program was seen earlier as Example III.8.1. Its characteristic property is that, as may be seen from inspection of the several tableaux in III.8.1, the entries in the inner part of the tableau (i.e., other than on the bottom row or right-hand column) are always 0, 1, or −1. The importance of this fact is that the pivot is always 1 or −1; in effect, this means that no multiplications or divisions will ever be necessary — the simpler operations of addition and subtraction will always be sufficient for solution.

Example III.8.1 is a typical *transportation problem*. In the general case, we might picture a company with a total of m warehouses, each containing a certain amount of the company's product, and n distributors, each having a demand for a certain amount of the product. The problem is to transport the desired amounts from the warehouses to the distributors in such a way as to minimize shipping costs.

More precisely, let a_i be the availability at the ith warehouse; let b_j be the requirement at the jth distributor. We make the assumption:

3.12.1
$$\sum_{i=1}^{m} a_i = \sum_{j=1}^{n} b_j$$

which means that the total availabilities at the m warehouses are exactly equal to the total requirements at the n distributors. Let c_{ij} be the cost of transporting a unit from the ith warehouse to the jth distributor. The problem can then be expressed as a linear program:

3.12.2 Minimize

$$\sum_{i=1}^{m} \sum_{j=1}^{n} c_{ij} x_{ij}$$

Subject to

3.12.3
$$\sum_{j=1}^{n} x_{ij} = a_i \text{ for each } i$$

3.12.4
$$\sum_{i=1}^{m} x_{ij} = b_j \text{ for each } j$$

3.12.5
$$x_{ij} \geq 0 \text{ for each } i,j$$

The first thing to notice is that there is no need for slack variables, since all the constraints (3.12.3) and (3.12.4) are equations rather than inequalities. This means that the "flows" x_{ij} are the only variables. Since each b.f.p. of a program is determined by the non-basic variables, to determine a b.f.p. of this program, it is sufficient to state which of the variables x_{ij} are equal to zero. This being so, it follows that we can represent each b.f.p. by a schematic graph.

Technically speaking, a graph is a collection of points called *nodes* and segments of curved lines called *arcs* that connect some of the nodes. (This is a *topological* graph, as distinguished from the graph of a function, which we saw in Chapter I.) To each b.f.p. we assign a graph as follows: there is a node W_i for each of the warehouses and a node D_j for each of the distributors; there will be an arc joining W_i to D_j if x_{ij} is positive, and there will be no other arcs. Thus, the tableau for the first b.f.p. reached in the solution of Example III.8.1 can be represented either by a graph or by a matrix:

$$\begin{array}{c} & D_1 & D_2 & D_3 \\ W_1 & \left(\begin{array}{ccc} 0 & 10 & 40 \\ 30 & 30 & 0 \end{array} \right. \\ W_2 & \end{array}$$

We repeat that the graph determines the b.f.p. entirely: in fact, the "missing" arcs, i.e., from W_1 to D_1 and from W_2 to D_3 in this example, correspond to the non-basic variables. These determine the values of the basic variables entirely.

The following theorem is a fundamental part of the method of solution of transportation problems. We say a graph is connected if any two of the nodes can be joined by a sequence of arcs from the graph.

III.12.1 Theorem. The graph corresponding to a b.f.p. will not contain any closed loops. Moreover, assuming non-degeneracy of the program, the graph will be connected, and so the addition of any arc will close a loop.

Proof. Let us suppose that the graph corresponding to a b.f.p. does contain a closed loop. This loop must have an even number of arcs, since each arc connects a warehouse to a distributor. Let α be the smallest flow along any of the arcs in the loop. If we number these arcs consecutively, it may be seen that we can increase the flow along the odd-numbered arcs by $\alpha/2$ and decrease it along the even-numbered arcs by the same amount without contradicting any of the constraints. This gives us a different point that has the same arcs "missing" from its graph, which contradicts the fact that the non-basic variables entirely determine a b.f.p. The contradiction proves the first statement of the theorem.

To prove the second statement, we point out that the graph can be disconnected only if the availabilities at a subcollection of the warehouses are equal to the requirements at a subcollection of the distributors. This is, precisely, degeneracy. Assuming that the graph is connected, any new arc will close a loop, since it will connect two nodes that were already connected in some other manner.

Now that we have replaced the simplex tableau by a graph, we must see what form the pivot-step takes. We know that a pivot-step has the property of increasing the value of one of the non-basic variables, and it will, thus, correspond to adding one of the missing arcs to the graph. This will cause one of the arcs presently on the graph to be removed. Let us see which one.

In the preceding graph, there are two missing arcs: $W_1 D_1$ and $W_2 D_3$ (corresponding to $x_{11} = x_{23} = 0$). We can make x_{11} positive by adding the arc $W_1 D_1$ to the graph: this closes a loop, which we denote as $W_1 D_1 W_2 D_2 W_1$. Suppose, that we do increase the variable x_{11}: this means that because a smaller quantity of the product is left at W_1, we must decrease either x_{12} or x_{13} (the flows from W_1 to the other two D_j). We cannot decrease x_{13}, as this would entail an increase in x_{23}, and we want to keep x_{23} fixed at zero (the fundamental idea of pivot-steps). Instead we decrease x_{12}. Now the requirement at D_2 is not being met, and this deficiency must be made up by increasing x_{22}. In turn, this causes x_{21} to decrease. The decrease in x_{21} will be exactly enough to meet the increase in x_{11}, keeping the required quantity at D_1. Thus, an increase in x_{11} will cause an equal increase in x_{22}, and corresponding equal decreases in x_{12} and x_{21}. We increase x_{11}, then, until either x_{12} or x_{21} has decreased to zero. Since $x_{12} = 10$ and $x_{21} = 30$, it is clear that x_{12} will be the first to vanish: adding the arc $W_1 D_1$ to the graph has caused us to remove the arc $W_1 D_2$. We obtain the new b.f.p.

$$
\begin{array}{c}
\begin{array}{ccc}
 & D_1 & D_2 & D_3 \\
W_1 & 10 & 0 & 40 \\
W_2 & 20 & 40 & 0
\end{array}
\end{array}
$$

It remains to be seen whether this new b.f.p. is an improvement over the previous one. This can actually be calculated directly; if we do, we see that the total costs for this b.f.p. are 510, whereas, for the previous b.f.p., they were only 500. Since we wish to minimize costs, this is not an improvement. On the other hand, we could also have seen that the change would increase the costs in the following manner: each unit increase in x_{11} causes an equal increase in x_{22}, and equal decreases in x_{21} and x_{12}. The total change in costs for a unit change in x_{11} is

$$c_{11} - c_{12} + c_{22} - c_{21} = 1 - 4 + 9 - 5 = 1$$

Since this is positive, we would conclude that the change would cause an increase in total costs, and therefore reject it.

This describes, then, the nature of the fundamental step for the transportation problem. Under the assumption of non-degeneracy, Theorem III.12.1 tells us that each missing arc will, when added to the graph, close a loop. If we number the arcs in this loop consecutively, starting with the newly included arc, we see that an increase along the first arc will cause equal increases along the odd-numbered arcs, and decreases along the even-numbered arcs. To see whether the change will be an improvement, we add the unit costs along the odd arcs, and subtract from these the unit costs along the even arcs. If the result is negative the change is an improvement (it will decrease costs). The new arc is then added to the diagram, while one of the even arcs in the loop (the one with the lowest flow) is removed.

With this method explained, we proceed to solve the problem of Example III.8.1.

III.12.2 *Example.* Solve the 2×3 transportation problem with availabilities

$$a_1 = 50 \qquad a_2 = 60$$

requirements

$$b_1 = 30 \qquad b_2 = 40 \qquad b_3 = 40$$

and transportation costs given by the table:

TABLE III.12.1

From Warehouse To Distributor		
	D_1	D_2	D_3
W_1	1	4	1
W_2	5	9	2

We already have obtained a b.f.p., given by

$$W_1 \begin{pmatrix} D_1 & D_2 & D_3 \\ 0 & 10 & 40 \\ 30 & 30 & 0 \end{pmatrix}$$

of the two missing arcs, we have already seen that $W_1 D_1$ will not cause a favorable change. Consider, then, the arc $W_2 D_3$. This will close the loop $W_2 D_3 W_1 D_2 W_2$. The change in cost per unit is given by

$$c_{23} - c_{13} + c_{12} - c_{22} = 2 - 1 + 4 - 9 = -4$$

Since this is negative, the change is a favorable one. We therefore add the arc $W_2 D_3$ to the graph. Of the even arcs in the loop, $W_1 D_3$ and $W_2 D_2$, we see that $W_2 D_2$ has the smaller flow: $x_{13} = 40$ and $x_{22} = 30$. Hence, we remove $W_2 D_2$ from the graph; x_{23} and x_{12} increase by 30, while x_{13} and x_{22} decrease by 30.

$$W_1 \begin{pmatrix} D_1 & D_2 & D_3 \\ 0 & 40 & 10 \\ 30 & 0 & 30 \end{pmatrix}$$

We repeat the procedure. Of the two arcs missing in the new graph, $W_2 D_2$ will clearly not give any improvement; we would simply be replacing the arc we removed in the previous step. Consider, then, the arc $W_1 D_1$. This closes the loop $W_1 D_1 W_2 D_3 W_1$. We calculate

$$c_{11} - c_{21} + c_{23} - c_{13} = 1 - 5 + 2 - 1 = -3$$

This also is negative, so $W_1 D_1$ is added to the graph. We see, next, that $x_{21} = 30$, while $x_{13} = 10$. The arc $W_1 D_3$ should be removed. We increase x_{11} and x_{23} by 10, while decreasing x_{21} and x_{13} by the same amount. This gives us

$$W_1 \begin{pmatrix} D_1 & D_2 & D_3 \\ 10 & 40 & 0 \\ 20 & 0 & 40 \end{pmatrix}$$

This new scheme is the solution. Of the two missing arcs, $W_1 D_3$, which was just removed, can give no improvement; as for $W_2 D_2$, it closes the loop $W_2 D_2 W_1 D_1 W_2$. We calculate

$$c_{22} - c_{12} + c_{11} - c_{21} = 9 - 4 + 1 - 5 = 1$$

Since this is also positive, it follows that this arc gives no improvement. There are no other arcs. We conclude that this is the solution.

We have seen the simplification of Stage II of the simplex algorithm for the transportation problem. Let us consider next, Stage I: obtaining a b.f.p. A very easy method of obtaining a b.f.p.

for these problems is called the *northwest corner* rule; it can be illustrated by referring once again to Example III.8.2.

We wish to obtain a b.f.p. for the problem with availabilities

$$a_1 = 50 \qquad a_2 = 60$$

and requirements

$$b_1 = 30 \qquad b_2 = 40 \qquad b_3 = 40$$

We start at the upper left-hand (i.e., northwest) corner. We have to get 30 units to D_1. Since W_1, has all this (and more) available, we let $x_{11} = 30$. There are still 20 units at W_1. Now, D_2 requires 40 units. We give it the 20 remaining units at W_1, and 20 others from W_2, obtaining $x_{12} = 20$, $x_{22} = 20$. There are still 40 units left at W_2; these are needed at D_3, so we let $x_{23} = 40$. We obtain the scheme

$$
\begin{array}{c}
\quad\ D_1\ \ D_2\ \ D_3 \\
\begin{array}{c} W_1 \\ W_2 \end{array}
\left(\begin{array}{ccc} 30 & 20 & 0 \\ 0 & 20 & 40 \end{array}\right)
\end{array}
$$

which may be seen to be a b.f.p.

This, then, is the general idea behind the northwest corner rule: as many of the D_j as possible are serviced (on a first-come, first-served basis) from W_1; the remainder are serviced, in turn, from W_2, W_3, and so on. Mathematically, the description is as follows: let x_{11} be the smaller of a_1 and b_1. If $x_{11} < a_1$, then let x_{12} be the smaller of $a_1 - x_{11}$ and b_2. If, on the other hand, $x_{11} < b_1$, then let x_{21} be the smaller of a_2 and $b_1 - x_{11}$. Note that each step, in a sense, reduces the problem to that of finding a b.f.p. for a smaller transportation problem. The same process can be continued until all the availabilities and requirements have been exhausted. It is not obvious, but this scheme will actually give a b.f.p. (that is, a vertex of the constraint set, rather than an arbitrary point). The reason is that each of the variables x_{ij} has been found by solving an equation: thus, the point is an intersection of sufficiently many of the bounding planes of the constraint set.

We illustrate this procedure with an example:

III.12.3 Example. Find a b.f.p. for the 6 × 10 transportation program with availabilities

$$a = (40, 30, 50, 20, 80, 30)$$

and requirements

$$b = (30, 20, 10, 40, 20, 20, 30, 10, 10, 60)$$

164

FINITE MATHEMATICS

We check first that $\Sigma a_i = \Sigma b_j = 250$. Using the first warehouse, we service as many of the distributors as possible, starting with D_1. Thus, $a_1 = 40$, and $b_1 = 30$, so we let $x_{11} = 30$. This leaves 10 units at W_1. Since $b_2 = 20$, we set $x_{12} = 10$, and continue to service D_2 from W_2. We get $x_{22} = 10$. There are still 20 units left at W_2; we service D_3 by letting $x_{23} = 10$. The remaining 10 units at W_2 go to D_4, and $x_{24} = 10$. Then we set $x_{34} = 30$ to complete the requirements at D_4. The remaining 20 units at W_3 will go to D_5. We continue in this manner until we obtain the scheme

$$
\begin{array}{c}
 & D_1 & D_2 & D_3 & D_4 & D_5 & D_6 & D_7 & D_8 & D_9 & D_{10} \\
W_1 & 30 & 10 & 0 & 0 & 0 & 0 & 0 & 0 & 0 & 0 \\
W_2 & 0 & 10 & 10 & 10 & 0 & 0 & 0 & 0 & 0 & 0 \\
W_3 & 0 & 0 & 0 & 30 & 20 & 0 & 0 & 0 & 0 & 0 \\
W_4 & 0 & 0 & 0 & 0 & 0 & 20 & 0 & 0 & 0 & 0 \\
W_5 & 0 & 0 & 0 & 0 & 0 & 0 & 30 & 10 & 10 & 30 \\
W_6 & 0 & 0 & 0 & 0 & 0 & 0 & 0 & 0 & 0 & 30 \\
\end{array}
$$

It may be checked directly that this scheme satisfies the constraints of the example.

III.12.4 Example. Solve the 3×5 transportation problem with availabilities

$$a = (50, 45, 65)$$

requirements

$$b = (20, 40, 40, 35, 25)$$

and unit costs given by the table:

TABLE III.12.2

From Warehouse	To Distributor				
	D_1	D_2	D_3	D_4	D_5
W_1	2	6	5	3	5
W_2	8	9	7	9	3
W_3	2	3	8	4	6

First of all, we obtain a b.f.p. by the northwest corner method. Thus, W_1 supplies all 20 units for D_1 and 30 units for D_2. W_2 supplies the remaining 10 units for D_2 and 35 units for D_3. W_3 supplies the remainder. We obtain the scheme

	D_1	D_2	D_3	D_4	D_5
W_1	20	30	0	0	0
W_2	0	10	35	0	0
W_3	0	0	5	35	25

We consider, now, the missing arcs; there are eight of these. The arc W_1D_3 closes the loop $W_1D_3W_2D_2W_1$; accordingly, we calculate

$$c_{13} - c_{23} + c_{22} - c_{12} = 5 - 7 + 9 - 6 = 1$$

Since this is positive, we conclude that the change would not be profitable. We consider other arcs: the arc W_2D_5, finally, is seen to bring a favorable change. In fact, W_2D_5 closes the loop $W_2D_5W_3D_3W_2$. We calculate

$$c_{25} - c_{35} + c_{33} - c_{23} = 3 - 6 + 8 - 7 = -2$$

which is negative. Now, $x_{35} = 25$, and $x_{23} = 35$. As $x_{35} < x_{23}$, we remove the arc W_3D_5; we obtain

	D_1	D_2	D_3	D_4	D_5
W_1	20	30	0	0	0
W_2	0	10	10	0	25
W_3	0	0	30	35	0

For this new scheme, we consider the missing arcs. Arc W_3D_1, will close the loop $W_3D_1W_1D_2W_2D_3W_3$. We calculate

$$c_{31} - c_{11} + c_{12} - c_{22} + c_{23} - c_{33} = 2 - 2 + 6 - 9 + 7 - 8 = -4$$

which is negative. We have $x_{11} = 20$, $x_{22} = 10$, and $x_{33} = 30$. As x_{22} is the smallest of these, we shall add the arc W_3D_1 and remove W_2D_2. We now have

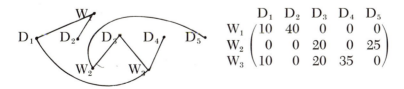

	D_1	D_2	D_3	D_4	D_5
W_1	10	40	0	0	0
W_2	0	0	20	0	25
W_3	10	0	20	35	0

For this scheme, the arc W_1D_3 closes the loop $W_1D_3W_3D_1W_1$. We have

$$c_{13} - c_{33} + c_{31} - c_{11} = 5 - 8 + 2 - 2 = -3$$

We have $x_{33} = 20$ and $x_{11} = 10$. Accordingly, we adjoin $W_1 D_3$ and remove $W_1 D_1$, obtaining

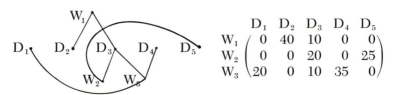

$$\begin{array}{c|ccccc} & D_1 & D_2 & D_3 & D_4 & D_5 \\ W_1 & 0 & 40 & 10 & 0 & 0 \\ W_2 & 0 & 0 & 20 & 0 & 25 \\ W_3 & 20 & 0 & 10 & 35 & 0 \end{array}$$

The arc $W_3 D_2$ closes the loop $W_3 D_2 W_1 D_3 W_3$. We have

$$c_{32} - c_{12} + c_{13} - c_{33} = 3 - 6 + 5 - 8 = -6$$

We have, also, $x_{12} = 40$ and $x_{33} = 10$. Therefore, we add the arc $W_3 D_2$ and remove $W_3 D_3$. This gives us

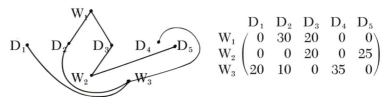

$$\begin{array}{c|ccccc} & D_1 & D_2 & D_3 & D_4 & D_5 \\ W_1 & 0 & 30 & 20 & 0 & 0 \\ W_2 & 0 & 0 & 20 & 0 & 25 \\ W_3 & 20 & 10 & 0 & 35 & 0 \end{array}$$

For this scheme, $W_1 D_1$ closes the loop $W_1 D_1 W_3 D_2 W_1$. We calculate

$$c_{11} - c_{31} + c_{32} - c_{12} = 2 - 2 + 3 - 6 = -3$$

We have $x_{31} = 20$ and $x_{12} = 30$, so we add the arc $W_1 D_1$ and remove $W_3 D_1$. We have, now,

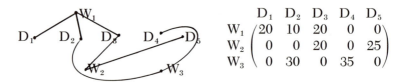

$$\begin{array}{c|ccccc} & D_1 & D_2 & D_3 & D_4 & D_5 \\ W_1 & 20 & 10 & 20 & 0 & 0 \\ W_2 & 0 & 0 & 20 & 0 & 25 \\ W_3 & 0 & 30 & 0 & 35 & 0 \end{array}$$

Here, $W_1 D_4$ closes the loop $W_1 D_4 W_3 D_2 W_1$. We compute

$$c_{14} - c_{34} + c_{32} - c_{12} = 3 - 4 + 3 - 6 = -4$$

We have $x_{34} = 35$ and $x_{12} = 10$. We therefore add $W_1 D_4$ and delete $W_1 D_2$, which gives us

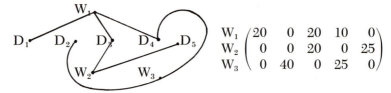

$$\begin{array}{c|ccccc} & D_1 & D_2 & D_3 & D_4 & D_5 \\ W_1 & 20 & 0 & 20 & 10 & 0 \\ W_2 & 0 & 0 & 20 & 0 & 25 \\ W_3 & 0 & 40 & 0 & 25 & 0 \end{array}$$

Now, the arc W_3D_1 closes the loop $W_3D_1W_1D_4W_3$. We calculate

$$c_{31} - c_{11} + c_{14} - c_{34} = 2 - 2 + 3 - 4 = -1$$

We have $x_{11} = 20$, $x_{34} = 25$. Adding W_3D_1 and deleting W_1D_1, we obtain

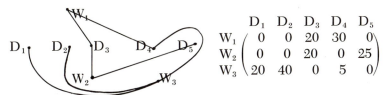

	D_1	D_2	D_3	D_4	D_5
W_1	0	0	20	30	0
W_2	0	0	20	0	25
W_3	20	40	0	5	0

This scheme is optimal: none of the missing arcs can give an improvement. The total cost, as may be verified directly, is 585, which compares most favorably with the cost of 885, given by the first b.f.p. obtained.

13. ASSIGNMENT PROBLEMS

Example III.12.4 shows the strength of the flow-type algorithm developed by T. C. Koopmans. It should be pointed out that this example has 15 variables: using the simplex algorithm, its tableaux would have eight rows and nine columns each. Clearly, there is a very real economy of time and effort in using this method. In fact, transportation problems involving some 20 sources and 1000 destinations (and therefore some 20,000 variables) are easily handled in this way with the aid of medium-sized computers.

In case of degeneracy, of course, the Koopmans algorithm fails because, for degenerate programs, adding a new arc to the graph of a b.f.p. will not necessarily close a loop. If the program is not greatly degenerate, a perturbation of the program is usually sufficient to solve it; it is enough to put a small flow—an amount ϵ—on a missing arc that will not close a loop. (Generally, the arc chosen is the one with smallest unit costs among those that do not close a loop.) If, on the other hand, the program is very degenerate—and there are some problems that by their very nature give rise to very degenerate programs—then substantially different techniques are necessary. The *assignment problem* is one such.

Let us assume that a company has several positions to be filled and an equal number of job applicants; the only problem is to determine which of the applicants should be given which position. Thanks to aptitude tests, or, perhaps, to some study of the applicants' backgrounds, it is estimated that (for each value of i and j) the company will derive a "utility" equal to a_{ij} if the ith applicant is assigned to the jth job. The assignment of applicants to jobs is to be made, then, in such a way as to maximize the sum of these expected utilities.

To describe this problem mathematically, we give a definition.

III.13.1 Definition. An $n \times n$ matrix $X = (x_{ij})$ is said to be a *permutation matrix* if all its entries are either 0 or 1, and, moreover, there is exactly one 1 in each row and one 1 in each column.

Thus, the following 5×5 matrices are permutation matrices:

(a)
$$\begin{pmatrix} 1 & 0 & 0 & 0 & 0 \\ 0 & 1 & 0 & 0 & 0 \\ 0 & 0 & 1 & 0 & 0 \\ 0 & 0 & 0 & 1 & 0 \\ 0 & 0 & 0 & 0 & 1 \end{pmatrix}$$

(b)
$$\begin{pmatrix} 1 & 0 & 0 & 0 & 0 \\ 0 & 0 & 0 & 1 & 0 \\ 0 & 1 & 0 & 0 & 0 \\ 0 & 0 & 0 & 0 & 1 \\ 0 & 0 & 1 & 0 & 0 \end{pmatrix}$$

(c)
$$\begin{pmatrix} 0 & 1 & 0 & 0 & 0 \\ 1 & 0 & 0 & 0 & 0 \\ 0 & 0 & 0 & 1 & 0 \\ 0 & 0 & 0 & 0 & 1 \\ 0 & 0 & 1 & 0 & 0 \end{pmatrix}$$

It is clear that we can think of each permutation matrix X as an assignment: the ith applicant is assigned to the jth job if $x_{ij} = 1$. Definition III.13.1 guarantees that no person is assigned to two jobs and that no two persons are assigned to the same job.

We can now express the assignment problem, mathematically, in the form

3.13.1   Maximize

$$\sum_{j=1}^{n} \sum_{i=1}^{n} a_{ij} x_{ij}$$

Subject to

3.13.2 X is a permutation matrix.

Now, this is not, strictly speaking, a linear programming problem. In fact, the constraint set is finite rather than convex; this type of problem is called a *combinatorial* problem. It would not seem, at first glance, that the assignment problem could be solved by linear programming methods. On the other hand, we should remember that the solution to a linear program (if it exists) can always be found by considering a finite set of points, the extreme points of the constraint set. Hence the assignment problem reduces to a linear program if we can only choose a constraint set whose extreme points are precisely the permutation matrices.

III.13.2   Definition.   An $n \times n$ matrix $X = (x_{ij})$ is said to be *doubly stochastic* if $x_{ij} \geq 0$ and, moreover,

$$\sum_{j=1}^{n} x_{ij} = 1 \qquad \text{for each } i$$

and

$$\sum_{i=1}^{n} x_{ij} = 1 \qquad \text{for each } j$$

In essence, a doubly stochastic matrix is a non-negative matrix whose row sums and column sums are all equal to 1. It is clear that an $n \times n$ matrix can be thought of as representing a point in n^2-dimensional space. If so, the set of doubly stochastic matrices forms a convex set, since it is defined solely by linear equations and inequalities. It is easy to see that a permutation matrix is doubly stochastic; the relation between doubly stochastic matrices and permutation matrices is given by the following theorem.

III.13.3   Theorem.   The extreme points of the set of doubly stochastic matrices are precisely the permutation matrices.

We will not give a proof of Theorem III.13.3; it may be obtained by an adaptation of the proof of III.12.1. We can now restate the assignment problem in the form

3.13.3   Maximize

$$\sum_{i=1}^{n} \sum_{j=1}^{n} a_{ij} x_{ij}$$

Subject to

3.13.4 $$\sum_{j=1}^{n} x_{ij} = 1 \qquad \text{for each } i$$

3.13.5 $$\sum_{i=1}^{n} x_{ij} = 1 \qquad \text{for each } j$$

3.13.6 $$x_{ij} \geq 0$$

What we have done is to replace the set of permutation matrices by the set of doubly stochastic matrices. The program (3.13.3) to

(3.13.6) is clearly feasible and bounded. It must have a maximum, which is attained at one of the extreme points of the constraint set. This extreme point will be the optimal assignment.

It may be seen that program (3.13.3) to (3.13.6) is a transportation-type problem, with availabilities and requirements all equal to 1. Unfortunately, it cannot be handled by the Koopmans algorithm because it is extremely degenerate. In fact, the constraints (3.13.4) and (3.13.5) are $2n$ equations, one of which is redundant. Thus, a b.f.p. should have $2n - 1$ non-zero variables. But we know that each b.f.p. has exactly n non-zero variables (one in each row of the matrix). This degeneracy is enough to make the Koopmans algorithm useless.

A common procedure in case of degeneracy is to solve, not the given (primal) linear program, but its dual.

Let us look, then, for the dual of (3.13.3) to (3.13.6). In Section III.11, the dual problem was defined, but only for programs with inequality constraints. This program has equation constraints. As was pointed out earlier, each equation can be replaced by two opposite inequalities; in general, we replace $\alpha = \beta$ by $\alpha \leq \beta$ and $-\alpha \leq -\beta$. Thus, two inequalities arise, in which the coefficients are negatives of each other. The dual program will have one variable for each of these inequalities; these two variables appear, in each of the constraints, with coefficients that are negatives of each other. This being so, we can deal with the difference of the two variables. The two variables are non-negative, but this places no such restriction on their difference. We have, as a rule for duality: *in duality, the variable corresponding to an equation constraint is not restricted to be non-negative.*

Using this rule, we can see that the dual of program (3.13.3) to (3.13.6) is

3.13.7 Minimize

$$\sum_{i=1}^{n} u_i + \sum_{j=1}^{n} v_j$$

Subject to

3.13.8 $$u_i + v_j \geq a_{ij}$$

The dual program attempts to put "shadow prices," u_i on the job applicants, and v_j on the several jobs. The number u_i represents, in some way, a "natural wage" for the ith applicant, while the w_j represents, perhaps, the value of the equipment used for the jth job.

Let us suppose that X^* is an optimal assignment, while $(\mathbf{u}^*, \mathbf{v}^*)$ is a solution to the dual program. We know from duality that this is equivalent to the condition

3.13.9 $$\sum \sum a_{ij} x_{ij}^* = \sum u_i^* + \sum v_j^*$$

and it is not difficult to see that a necessary condition is that,

3.13.10   if $x_{ij}^* = 1$, then $u_i^* + v_j^* = a_{ij}$

This condition may be interpreted as saying that an employee's salary, plus the value of the job's equipment, should be no greater than his utility to the firm in that job.

Condition (3.13.10) is also sufficient: if there is a permutation matrix X^* such that (3.13.10) holds, then (3.13.9) will also hold, and X^* will be the optimal assignment. Our approach to the assignment problem will be along these lines. We shall look for vectors (u,v) satisfying (3.13.8) and decrease the objective function (3.13.7), until a permutation matrix X^*, satisfying (3.13.10) is found.

We shall assume that the entries in the matrix $A = (a_{ij})$ are all integers. This is not a great loss in generality, since, whatever the numbers a_{ij} may be, they can be approximated as closely as desired by rational numbers. In turn, these rational numbers may all be multiplied by their least common denominator, thus obtaining an obviously equivalent assignment problem. The following theorem, which is intuitively reasonable, is given without proof.

III.13.4   Theorem.   Let $B = (b_{ij})$ be an $n \times n$ matrix all of whose entries are 0 or 1. A necessary and sufficient condition for the existence of a permutation matrix $X \leq B$ is that every set of k rows have non-zero entries in at least k columns.

Our method of solution of the assignment problem is as follows. A vector (u,v) satisfying (3.13.8) is found. This is most easily done by letting $v_j = 0$, and letting u_i be equal to the largest entry in the ith row. An auxiliary matrix $B = (b_{ij})$ is then formed, defined by

3.13.11
$$b_{ij} = \begin{cases} 1 & \text{if } u_i + v_j = a_{ij} \\ 0 & \text{if } u_i + v_j > a_{ij} \end{cases}$$

Note that, by (3.13.10), we can have $x_{ij}^* = 1$ only if $b_{ij} = 1$. That is, $x_{ij}^* \leq b_{ij}$. If there is a permutation matrix $X \leq B$, we know from the foregoing discussion that this is the optimal assignment. If not, then, by Theorem III.13.4, there is a set of k rows that have 1's in only $k - 1$ or fewer columns. Let us say that these k rows form the set I, and the corresponding $k - 1$ (or fewer) columns form the set J. We then define a new vector (u',v') by

3.13.12
$$u_i' = \begin{cases} u_i - 1 & \text{for } i \text{ in } I \\ u_i & \text{for other } i \end{cases}$$

3.13.13
$$v_j' = \begin{cases} v_j + 1 & \text{for } j \text{ in } J \\ v_j & \text{for other } j \end{cases}$$

Consider this vector. We notice, first of all, that $(\mathbf{u'},\mathbf{v'})$ satisfies (3.13.8). In fact, suppose $b_{ij}=0$. In this case, $a_{ij}<u_i+v_j$. Now, a_{ij},u_i and v_j are all integers, and so

$$a_{ij}\le u_i+v_j-1$$

Now, by (3.13.12), $u_i'\ge u_i-1$, and by (3.13.13), $v_j'\ge v_j$. Hence $a_{ij}\le u_i'+v_j'$.

Suppose, on the other hand, that $b_{ij}=1$. In this case, there are two possibilities. If i is in I, then j is in J, so $u_i'=u_i-1$, $v_j'=v_j+1$, and so $a_{ij}=u_i'+v_j'$. If i is not in I, then $u_i'=u_i$, $v_j'\ge v_j$, and so $a_{ij}\le u_i'+v_j'$.

In any case, we see that $(\mathbf{u'},\mathbf{v'})$ satisfies (3.13.7). Consider, next, the objective function (3.13.8). We see that

$$\sum u_i'=\sum u_i-k$$

but

$$\sum v_j'\le\sum v_j+k-1$$

since there are k rows in I, but at most $k-1$ columns in J. At each step, the objective function (3.13.7) decreases by at least 1. It follows that the minimum of the program (3.13.7) and (3.13.8) will be reached in a finite number of steps; this will give us the optimal assignment. We conclude with an example of this algorithm.

III.13.5 Example. Find the optimal assignment corresponding to the matrix

$$A=\begin{pmatrix}14&18&14&12&19&13&14\\20&17&18&15&12&14&14\\19&18&13&18&17&13&12\\12&20&12&16&13&18&18\\11&16&17&16&11&18&14\\16&14&16&18&20&15&13\\17&12&15&18&13&18&19\end{pmatrix}$$

Starting by setting $v_j=0$ and $\mathbf{u}=(19,20,19,20,18,20,19)$, we obtain the following auxiliary matrix:

$$B=\begin{pmatrix}0&0&0&0&1&0&0\\1&0&0&0&0&0&0\\1&0&0&0&0&0&0\\0&1&0&0&0&0&0\\0&0&0&0&0&1&0\\0&0&0&0&1&0&0\\0&0&0&0&0&0&1\end{pmatrix}\qquad\begin{matrix}\mathbf{u}\\ \\ \begin{pmatrix}19\\20\\19\\20\\18\\20\\19\end{pmatrix}\end{matrix}$$

$$\mathbf{v}=(0\ \ 0\ \ 0\ \ 0\ \ 0\ \ 0\ \ 0)$$

We note that the first, second, third and sixth rows of the matrix B have 1's only in the first and fifth columns. We therefore decrease u_1, u_2, u_3, u_6, and increase v_1, v_5:

$$B = \begin{pmatrix} 0 & 1 & 0 & 0 & 1 & 0 & 0 \\ 1 & 0 & 0 & 0 & 0 & 0 & 0 \\ 1 & 1 & 0 & 1 & 0 & 0 & 0 \\ 0 & 1 & 0 & 0 & 0 & 0 & 0 \\ 0 & 0 & 0 & 0 & 0 & 1 & 0 \\ 0 & 0 & 0 & 0 & 1 & 0 & 0 \\ 0 & 0 & 0 & 0 & 0 & 0 & 1 \end{pmatrix} \begin{matrix} \mathbf{u} \\ 18 \\ 19 \\ 18 \\ 20 \\ 18 \\ 19 \\ 19 \end{matrix}$$

$$\mathbf{v} = (1 \quad 0 \quad 0 \quad 0 \quad 1 \quad 0 \quad 0)$$

This time, we note that first, second, third, fourth, and sixth rows of B have 1's only in the first, second, fourth, and fifth columns. We decrease u_1, u_2, u_3, u_4, and u_6, and increase v_1, v_2, v_4, and v_5:

$$B = \begin{pmatrix} 0 & 1 & 0 & 0 & 1 & 0 & 0 \\ 1 & 0 & 1 & 0 & 0 & 0 & 0 \\ 1 & 1 & 0 & 1 & 0 & 0 & 0 \\ 0 & 1 & 0 & 0 & 0 & 0 & 0 \\ 0 & 0 & 0 & 0 & 0 & 1 & 0 \\ 0 & 0 & 0 & 0 & 1 & 0 & 0 \\ 0 & 0 & 0 & 0 & 0 & 0 & 1 \end{pmatrix} \begin{matrix} \mathbf{u} \\ 17 \\ 18 \\ 17 \\ 19 \\ 18 \\ 18 \\ 19 \end{matrix}$$

$$\mathbf{v} = (2 \quad 1 \quad 0 \quad 1 \quad 2 \quad 0 \quad 0)$$

For this scheme, we note that the first, fourth, and sixth rows of B have 1's only in the second and fifth columns. We will therefore decrease u_1, u_4, and u_6, and increase v_2 and v_5, to obtain the scheme

$$B = \begin{pmatrix} 0 & 1 & 0 & 0 & 1^* & 0 & 0 \\ 1 & 0 & 1^* & 0 & 0 & 0 & 0 \\ 1^* & 0 & 0 & 1 & 0 & 0 & 0 \\ 0 & 1^* & 0 & 0 & 0 & 1 & 1 \\ 0 & 0 & 0 & 0 & 0 & 1^* & 0 \\ 0 & 0 & 0 & 1^* & 1 & 0 & 0 \\ 0 & 0 & 0 & 0 & 0 & 0 & 1^* \end{pmatrix} \begin{matrix} \mathbf{u} \\ 16 \\ 18 \\ 17 \\ 18 \\ 18 \\ 17 \\ 19 \end{matrix}$$

$$\mathbf{v} = (2 \quad 2 \quad 0 \quad 1 \quad 3 \quad 0 \quad 0)$$

We check, and find that this matrix B does, in fact, contain a permutation matrix; this is shown by the starred elements. The optimal assignment is shown by the starred elements in the matrix A that follows:

$$
\begin{pmatrix}
14 & 18 & 14 & 12 & 19^* & 13 & 14 \\
20 & 17 & 18^* & 15 & 12 & 14 & 14 \\
19^* & 18 & 13 & 18 & 17 & 13 & 12 \\
12 & 20^* & 12 & 16 & 13 & 18 & 18 \\
11 & 16 & 17 & 16 & 11 & 18^* & 14 \\
16 & 14 & 16 & 18^* & 20 & 15 & 13 \\
17 & 12 & 15 & 18 & 13 & 18 & 19^*
\end{pmatrix}
$$

It may be checked directly that this is, indeed, the optimal assignment. In fact, we have a value of 131 for the objective functions of both programs; duality assures that these are the solutions.

We point out that this problem has 49 variables, constrained by 13 inequalities. Each simplex tableau would have 14 rows and 37 columns (counting the column of constant terms and the objective function row). If we tried, alternatively, to enumerate all the possible assignments, we would find that there are $7! = 5040$ such assignments. In either case, it is clear that this algorithm affords us a very substantial economy of time and labor.

Many types of problem can be solved by methods similar to those considered in this section. One such is the flow problem, which is that of maximizing the total flow through a network whose arcs and nodes have finite capacities. It is not, unfortunately, possible to consider all such problems here.

One that does not seem soluble by these methods is the so-called "traveling salesman" problem, which consists, essentially, of finding the shortest route through a network, touching all the nodes. Heuristically, it describes the problem faced by a salesman who wishes to visit several cities and is looking for the shortest (or cheapest) route through all of them. As such, it does not seem greatly different from the assignment problem; it cannot, however, be reduced to a linear program without the introduction of a very large number of constraints. No solution has yet been given in general (though some particular problems of this type have been solved).

PROBLEMS ON TRANSPORTATION AND ASSIGNMENT

1. Solve the following transportation problems. In each case, the matrix $C = (c_{ij})$ is the cost matrix, while $\mathbf{a} = (a_1, \ldots, a_m)^t$ and $\mathbf{b} = (b_1, \ldots, b_n)$ are the availability and requirement vectors, respectively. (In cases, in which the total availability is greater than the total requirement, introduce a *fictitious destination* to receive the remaining goods, with 0 costs from each warehouse to this destination.)

(a)
$$C = \begin{pmatrix} 3 & 1 & 4 \\ 1 & 5 & 9 \\ 2 & 6 & 5 \end{pmatrix} \quad a = \begin{pmatrix} 100 \\ 150 \\ 120 \end{pmatrix}$$
$$b = (140 \quad 90 \quad 140)$$

(b)
$$C = \begin{pmatrix} 3 & 5 & 8 & 9 \\ 7 & 9 & 3 & 2 \\ 3 & 8 & 4 & 6 \end{pmatrix} \quad a = \begin{pmatrix} 50 \\ 70 \\ 40 \end{pmatrix}$$
$$b = (30 \quad 45 \quad 30 \quad 55)$$

(c)
$$C = \begin{pmatrix} 1 & 4 & 5 & 1 & 2 \\ 2 & 6 & 1 & 3 & 5 \\ 1 & 3 & 0 & 1 & 1 \end{pmatrix} \quad a = \begin{pmatrix} 80 \\ 80 \\ 40 \end{pmatrix}$$
$$b = (50 \quad 50 \quad 20 \quad 35 \quad 45)$$

(d)
$$C = \begin{pmatrix} 1 & 4 & 1 & 5 & 2 \\ 2 & 6 & 3 & 4 & 1 \\ 5 & 8 & 5 & 6 & 4 \end{pmatrix} \quad a = \begin{pmatrix} 90 \\ 90 \\ 30 \end{pmatrix}$$
$$b = (40 \quad 60 \quad 25 \quad 35 \quad 40)$$

(e)
$$C = \begin{pmatrix} 2 & 6 & 5 & 3 & 5 \\ 2 & 7 & 1 & 8 & 2 \\ 1 & 4 & 1 & 4 & 2 \\ 1 & 7 & 3 & 2 & 3 \end{pmatrix} \quad a = \begin{pmatrix} 50 \\ 50 \\ 50 \\ 50 \end{pmatrix}$$
$$b = (40 \quad 40 \quad 40 \quad 40 \quad 40)$$

2. Formulate the duals of the foregoing problems (a) through (e). What heuristic interpretation can be given to them?

3. Solve the following assignment problems. In each problem, each row represents a candidate, and each column represents a position (If there are more candidates than positions, introduce *fictitious positions*, representing no job, with all entries 0 in the corresponding columns.) What salaries should be paid to the employees in each case?

(a)
$$\begin{pmatrix} 7 & 2 & 8 & 4 & 3 \\ 5 & 1 & 6 & 3 & 3 \\ 6 & 3 & 7 & 4 & 5 \\ 5 & 2 & 5 & 2 & 4 \\ 3 & 0 & 4 & 3 & 1 \end{pmatrix}$$

(b)
$$\begin{pmatrix} 1 & 4 & 1 & 5 & 9 & 2 & 6 \\ 5 & 3 & 5 & 8 & 9 & 7 & 9 \\ 3 & 2 & 3 & 8 & 4 & 6 & 2 \\ 6 & 4 & 3 & 3 & 8 & 3 & 2 \\ 7 & 9 & 3 & 7 & 5 & 1 & 0 \\ 6 & 0 & 3 & 9 & 9 & 2 & 7 \\ 1 & 8 & 2 & 8 & 5 & 3 & 4 \end{pmatrix}$$

(c)
$$\begin{pmatrix} 1 & 2 & 8 & 3 & 5 \\ 2 & 6 & 7 & 5 & 2 \\ 4 & 9 & 6 & 3 & 3 \\ 2 & 4 & 8 & 5 & 5 \\ 1 & 3 & 5 & 4 & 1 \\ 3 & 8 & 8 & 5 & 2 \end{pmatrix}$$

(d)
$$\begin{pmatrix} 2 & 6 & 5 & 8 & 4 & 3 \\ 1 & 3 & 5 & 7 & 9 & 2 \\ 2 & 5 & 7 & 8 & 1 & 4 \\ 3 & 6 & 9 & 9 & 2 & 5 \\ 1 & 5 & 4 & 6 & 2 & 5 \\ 4 & 7 & 8 & 3 & 1 & 6 \\ 2 & 3 & 6 & 3 & 1 & 0 \end{pmatrix}$$

CHAPTER IV

SETS,
PERMUTATIONS,
AND
COMBINATIONS

1. SETS AND FUNCTIONS

We shall start this chapter with a brief outline of the elementary theory of sets. It is quite likely that the student has already studied this subject; nevertheless, the notions that we shall cover here are so fundamental to mathematics as a whole that such an outline is worthwhile, if only to refresh the reader's memory.

A *set*, as the word is used in mathematics, is an undefined concept. The general idea is that a set contains elements—though, there is one, the empty set, that has no elements. In fact, a set is defined precisely in terms of its elements: two sets are equal if they have the same elements. (This is in contrast to the possibility that two different societies might have exactly the same members—but then a society is somewhat more than just the set of its members.) We shall use braces to denote a set (defined by its members). Thus,

$$S = \{a,b,c,d\}$$

which should be read as "S is the set whose elements are a,b,c, and d." Often, a set is defined in terms of a particular property. In such a case, the notation is, say,

$$\{x \mid x \text{ has property } P\}$$

to represent "the set of all x that have the property P." Thus,

$$\{x \mid 0 \leq x < 5\}$$

is the set of all non-negative numbers smaller than 5, while

$$\left\{ y \;\middle|\; \begin{array}{l} y \text{ is a girl} \\ y \text{ has blue eyes} \end{array} \right\}$$

is the set of all blue-eyed girls.

A special, undefined relation, exists between sets and elements. This is the relation of belonging. We use the notation

$$x \in S$$

to signify that x is an element of the set S. For instance, if R is the set of real numbers, the symbols

$$5 \in R$$

represent the (true) statement that 5 is a real number. A line drawn through the set element sign represents the negation of this relation. If Q is the set of rational numbers, then

$$\sqrt{2} \notin Q$$

represents the statement that the number 2 does not have a rational square root. Other relations are defined in terms of this:

IV.1.1 Definition. We say S is a *subset* of T, with notation

$$S \subset T$$

if every element of S is also an element of T (Figure IV.1.1).

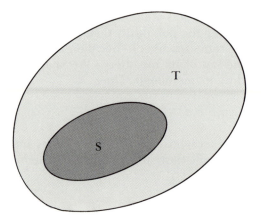

FIGURE IV.1.1 S is a subset of $T: S \subset T$.

Thus, if S is the set of girls, and T is the set of human beings, we have

$$S \subset T$$

To represent the fact that all girls are human.

IV.1.2 Definition. We say $S = T$ if $S \subset T$ and $T \subset S$.

We define two sets as equal if they have the same elements. If, for instance, S is the set of human beings, while T is the set of featherless bipeds (Plato's definition of human beings), we will (if Plato is right) find that $S = T$. We see that two sets may be actually equal even if defined by very different means.

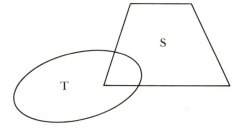

FIGURE IV.1.2 Two sets, S and T.

There are, in general, two sets that are of great importance. One of these, the *universal set*, is not a well-defined notion; it must generally be referred to the context. If we are dealing with analysis of real numbers, the universal set will be the set of all real numbers. If we are dealing with human beings, the universal set may be the set of all men. If we are dealing with plane geometry, the universal set will probably be the set of all points in the plane. In general, in the absence of any other statement, the universal set, for a given context, will be the set of all elements that are relevant to the context.

Another very important set is the *empty set*, $\varnothing$. This is a set that has no elements. As such, it is clearly unique; moreover, it is easy to see that, for any S, $\varnothing \subset S$. We define *operations* on sets as follows.

IV.1.3 Definition. The *union* of two sets, S and T is the set of all elements that belong to either S or T (or possibly both). The notation for the union of S and T is $S \cup T$ (Figure IV.1.3).

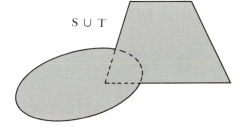

FIGURE IV.1.3 The union of S and T.

180

IV.1.4 Definition. The *intersection* of two sets, S and T, is the set of all elements of S that are also elements of T. The notation for the intersection of S and T is $S \cap T$ (Figure IV.1.4).

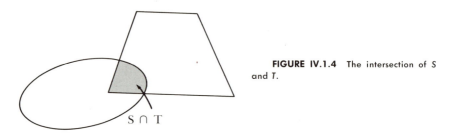

$S \cap T$

FIGURE IV.1.4 The intersection of S and T.

IV.1.5 Definition. The *complement* of a set, S, is the set of all elements of the universal set (whichever it may be in the context) that do not belong to S. The notation for the complement of S is $\overline{S}$ (Figure IV.1.5).

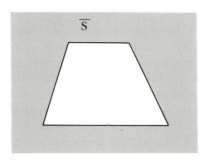

$\overline{S}$

FIGURE IV.1.5 The complement of S.

IV.1.6 Definition. The *relative complement* of S in T is the set of elements of T that are not elements of S. Its notation is $T - S$ (Figure IV.1.6).

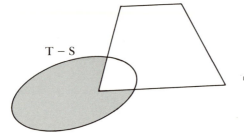

$T - S$

FIGURE IV.1.6 The relative complement of S and T.

To give an example, let us assume that S is the set of all girls, while T is the set of blue-eyed people. The universal set, in this context, is the set of all human beings. Then $S \cup T$ is the set of all human beings who either are girls or have blue eyes. $S \cap T$ is the set of all blue-eyed girls. $\overline{S}$ is the set of human beings who are not girls, i.e., of men, adult women, and boys, while $\overline{T}$ is the set of human beings whose eyes are not blue. Finally, $S - T$ is the set of girls with eyes that are not blue, while $T - S$ is the set of blue-eyed men, women, and boys.

Another very important operation on sets is the *Cartesian product*. The notion of a Cartesian product is defined in terms of an *ordered pair*. Very briefly, if a and b are any elements of any two sets (or possibly of the same set), we can put them together to form the ordered pair (a,b). This pair is *ordered* in the sense that (a,b) and (b,a) are not equal unless, of course, $a = b$.

We define equality for ordered pairs by:

$$(a,b) = (c,d) \quad \text{if and only if} \quad a = b \quad \text{and} \quad c = d$$

Two ordered pairs are equal if, and only if, corresponding terms are equal. The notion of an ordered pair should not, indeed, be new to the reader. The development of plane (two-dimensional) analytic geometry in Chapter I was based, precisely, on ordered pairs. (In effect, each point in the plane corresponds to an ordered pair of real numbers, and it is well known that the order is important. For example, the points $(0,1)$ and $(1,0)$ are distinct.)

More generally, we can talk about ordered triples (a,b,c), ordered quadruples (a,b,c,d), and, generally, ordered n-tuples. These are all things that we studied in Chapter I.

We are now able to define the Cartesian product of two sets:

IV.1.7 Definition. The *Cartesian product* of two sets A and B, denoted by $A \times B$, is the set of all ordered pairs (a,b), where $a \in A$ and $b \in B$.

The set of all vectors with two real components can be thought of as the Cartesian product of the set of real numbers, R, with itself. This is usually written R^2. More generally, the set of all vectors with n real components is usually denoted by R^n. (It is also common to use R^n to denote n-dimensional space; Chapter I explains why the same symbol can be used for these two sets.)

One of the most important concepts in mathematics is that of a *function*. We give first the more general definition of a *relation*.

IV.1.8 Definition. A *relation* from a set A to a set B is a subset of the Cartesian product $A \times B$.

If R is a relation from A to B, and $(a,b) \in R$, we usually write aRb. Some examples are:

IV.1.9 Example. Let A be the set of all men, and B, the set of all women. Let R be the relation of marriage: we will have $(a,b) \in R$ or aRb, if (and only if) a is married to b. Thus, R consists of all married couples.

IV.1.10 Example. Let A and B both be the set of real numbers. Consider the relation $<$: it consists of all pairs of numbers (a,b) such that $a < b$. This is, of course, nothing but a new way of looking at something we saw long ago.

We see from Example IV.1.9 that one element in the set A may be related to several elements in the set B, (as in the case of a man with several wives) or to none of the elements of B (as in the case of a bachelor). If, however, we restrict A to be, not the set of all men, but only the set of all married men in a monogamous country, then each element of A is related to exactly one element of B. We would say, in this case, that, if $(a,b) \in R$, then b is *the* wife of a. Each man, in this restricted set, has a unique, well-defined wife. It is this idea that gives rise to the concept of a function.

IV.1.11 Definition. A *function* from a set A into a set B is a relation f from A to B such that, for every $a \in A$, there is a *unique* $b \in B$ with $(a,b) \in f$.

A function is a very special type of relation. For any $a \in A$, we let $f(a)$ be that element of B such that $(a,f(a)) \in f$. Precisely be-because of its uniqueness, the element a determines $f(a)$, and we often think of a function as a rule that assigns an element of B to each element of A. Most often the function is defined by giving this element $f(a)$ explicitly in terms of a.

If f is a function from A into B, then A is called the *domain* of f. The set

$$f(A) = \{b \mid b = f(a) \text{ for some } a \in A\}$$

is called the *range* of f. It is also called the *image* of A under f. Clearly, $f(A) \subset B$.

For example, let the function f have as domain the set of real numbers, and let it be defined by $f(x) = x^2$. Then the range is the set of non-negative real numbers. Similarly, let the domain of g be the set of human beings, and let $g(x) =$ the father of x. Then the range of g is the set of all men who have ever fathered a child.

In general, unless a statement is made to the contrary, the functions that we consider here will have as domain the largest possible set of real numbers (subject to the demand that the range be a subset of the real numbers.)

IV.1.12 Examples. The "natural" domains and ranges of the following functions are:

(a) $f(x) = x^2 + 1$. The domain is the set of all real numbers; the range is the set of all real numbers y such that $y \geq 1$.

(b) $f(x) = \sqrt{1 - x^2}$. The domain is the set of real numbers x such that $-1 \leq x \leq 1$; the range is the set of real numbers y such that $0 \leq y \leq 1$.

(c) $f(x) = \dfrac{1}{x+1}$. The domain is the set of all real numbers x such that $x \neq -1$; the range is the set of non-zero real numbers.

(d) $f(x) = \log x$. The domain is the set of all positive real numbers; the range is the set of all real numbers.

Functions, of course, are not numbers. Nevertheless, it is natural to operate on them as if they were. If $f(x) = x^2$ and $g(x) = x - 1$ are functions, we can add them to obtain the new function $h = f + g$, defined by $h(x) = x^2 + x - 1$. This new function will have, as domain, the intersection of the domains of the two functions f and g; i.e., $h(x)$ will exist if, and only if, both $f(x)$ and $g(x)$ exist. We have, then, the following definition:

IV.1.13 Definition. Let f and g be real-valued functions with domains S and T respectively. Then the *sum* of f and g is the function h with domain $S \cap T$, defined by $h(x) = f(x) + g(x)$. We write $h = f + g$.

In a similar way, it is possible to define the function $f - g$ and fg. The quotient function f/g is defined similarly, but only for values of x such that $g(x) \neq 0$. Finally, if c is a constant, the function cf is similarly defined.

2. PERMUTATIONS AND COMBINATIONS

In mathematics it is often important to count the number of ways in which an action can be performed. This counting process can, in some cases, be quite trivial, but in others may require considerable ingenuity.

The fundamental principle used in counting can be stated in the form:

4.2.1 If there are m ways to perform act A, and if, for each of these, there are n ways to perform act B, then there are mn ways to perform the *two* acts, A and B.

We can illustrate this principle by a trivial example. There are four ways in which the suit of a card in a bridge deck may be chosen and 13 choices for the denomination of the card, giving us $52 = 4 \times 13$ ways to choose a card from a bridge deck.

Let us now consider a less trivial problem. A set S has n elements. How many subsets does it have (including S itself, and the empty set, $\varnothing$)?

With a small set, it is possible to enumerate the subsets. When n is large, however, this becomes impractical, and has the additional disadvantage that we may never be quite certain that we have left none of the subsets out of enumeration. A subtler procedure is necessary.

To see how this is done, we notice that, in choosing a subset T of S, it is sufficient to state which of the elements of S will be included in T. For each element there are two choices. Either it is included, or it is not. Thus, there are n actions to be performed (one for each element of S); each can be performed in two different ways. Repeated application of (4.2.1) tells us that the n actions can be performed in

$$2 \times 2 \times \ldots \times 2 = 2^n$$

different ways; S will have 2^n different subsets.

As an example, let $S = \{a, b, c\}$. Then $n = 3$, and we see that S must have $2^3 = 8$ subsets. These can, indeed, be enumerated:

$$\varnothing \qquad \{b\} \qquad \{c\} \qquad \{b,c\}$$
$$\{a\} \qquad \{a,b\} \qquad \{a,c\} \qquad \{a,b,c\}$$

It may be checked that there are no other subsets.

Another important counting problem deals with the different permutations of a finite set. Briefly speaking, a permutation is nothing other than an ordering of the set.

Suppose S has n elements. If we wish to order these somehow, we must first choose the first element of the ordering. This can clearly be done in n different ways. The second element must then be chosen; this can be done in $n-1$ ways (since it cannot be the same as the first). Then there are $n-2$ choices for the third element, $n-3$ for the fourth, and so on. We find, by (4.2.1) that there are

$$n(n-1)(n-2) \ldots 2 \times 1$$

different ways to order these n elements. This number is usually denoted by the symbol $n!$, and called "n factorial." In general, we define

4.2.2 $$n! = 1 \times 2 \times 3 \ldots (n-1)n.$$

whenever n is a positive integer. Thus,

$$1! = 1$$
$$2! = 1 \times 2 = 2$$
$$3! = 1 \times 2 \times 3 = 6$$
$$4! = 1 \times 2 \times 3 \times 4 = 24$$

and so on. We also define

$$0! = 1$$

While this may not seem an intuitively proper definition, it has great practical advantages, among them the fact that the recurrence relation

4.2.3 $$(n+1)! = n!(n+1)$$

is then satisfied for $n = 0$.

IV.2.1 *Example.* A class has 10 students, who are ordered according to average grades. How many possible orderings are there (discounting the possibility of ties)?

The number of orderings is clearly equal to the number of permutations of the 10 students, or 10! This is

$$10! = 10 \cdot 9 \cdot 8 \cdot 7 \cdot 6 \cdot 5 \cdot 4 \cdot 3 \cdot 2 \cdot 1 = 3,628,800$$

IV.2.2 *Example.* A baseball team consists of nine men. These must be put into a batting order. How many different batting orders are possible? How many are possible if we assume that the pitcher must bat ninth?

In the first case, the number of possible batting orders is

$$9! = 9 \cdot 8 \cdot 7 \cdot 6 \cdot 5 \cdot 4 \cdot 3 \cdot 2 \cdot 1 = 362,880$$

If we assume that the pitcher must bat ninth, then we are clearly interested only in the permutations of the remaining eight players. The number of these is

$$8! = 8 \cdot 7 \cdot 6 \cdot 5 \cdot 4 \cdot 3 \cdot 2 \cdot 1 = 40,320$$

IV.2.3 *Example.* How many different pairs of batting orders can two baseball teams (of nine men each) form, if we assume that, in both cases, the pitcher must bat ninth?

In this case, each team has 8!, or 40,320 batting orders. The number of different pairs of batting orders will, by (4.2.1), be equal to 40,320 times 40,320, or 1,625,702,400.

In some cases, it is required to know, not all the possible orderings of a set, but, rather, the number of ways in which the first k

elements of an ordering may be chosen. For example, a baseball club has 25 men on its roster. The number of different teams that its manager can field is equal only to the possible orderings of 9 men (from among the 25); it does not matter how the other 16 sit on the bench.

For a problem such as this, the first element in the ordering may be chosen in n different ways; the second element may be chosen in $n-1$ ways; the third in $n-2$ ways; and so on down to the kth, which may be chosen in $n-k+1$ ways. We obtain the number

$$P_{n,k} = n(n-1) \ldots (n-k+1)$$

or, more concisely,

4.2.4
$$P_{n,k} = \frac{n!}{(n-k)!}$$

as the number of *permutations of k elements from a set with n elements.*

As an example, let $S = \{a,b,c,d\}$ so that $n = 4$, and let us look for the permutations of two elements from S. By (4.2.4), we have

$$P_{4,2} = \frac{4!}{(4-2)!} = \frac{4!}{2!} = 12$$

possible permutations. We can, in fact, enumerate them:

(a,b)	(b,a)	(c,a)	(d,a)
(a,c)	(b,c)	(c,b)	(d,b)
(a,d)	(b,d)	(c,d)	(d,c)

Note that, for permutations, *order* is important: the pair (a,c) and the pair (c,a) are listed separately. In the concrete example of a baseball team, it is clear that order is important, since interchanging two of the players will, in effect, change the team: if the pitcher moves to center field, while the center fielder pitches, it is hardly the same team any more.

IV.2.4 *Example.* An international affairs club must choose three members to be chairmen of its committees on diplomatic relations, trade relations, and military relations. Assuming that the club has 25 members, and that no member can be chairman of more than one of the committees, in how many ways can this be done?

Here we want the number of permutations of 3 elements from a set with 25 elements. This is the number $P_{25,3}$, which is

$$P_{25,3} = 25 \cdot 24 \cdot 23 = 13,800$$

IV.2.5 Example. A baseball team has 12 outfielders. How many different outfields (of 3 men each) can it field?

Again we want the permutations of 3 elements, this time, from a set with 12 elements. This is $P_{12,3}$, or $12 \cdot 11 \cdot 10 = 1320$.

IV.2.6 Example. A psychologist is studying the relative strengths of certain personality traits in married couples. He considers seven different traits for the husband, and nine different traits for the wife. He then lists the four strongest traits in the husband and the six strongest traits in the wife (all in order of strength) and calls this the couple's profile. How many possible profiles are there?

For the husband, we must consider the number of permutations of four elements from a set of seven; this is $P_{7,4}$, or $7 \cdot 6 \cdot 5 \cdot 4 = 840$. For the wife, it is $P_{9,6}$, or $9 \cdot 8 \cdot 7 \cdot 6 \cdot 5 \cdot 4 = 60,480$. Since there are 840 possibilities for the husband's part of the profile and 60,480 for the wife's part, we conclude that there are $840 \cdot 60,480$, or $50,803,200$ possible profiles.

As we mentioned earlier, in permutations, order is important. In some cases, however, this need not be so. If the 9 men selected from 25 are chosen, not to play baseball, but to attend a course on the fine points of the game, the order in which they appear is not important. A similar consideration holds in dealing a hand for bridge: the only thing that matters is *which* 13 cards are received—not the order in which they are received. In cases such as these, we are dealing with *combinations*. Briefly speaking, a *combination of k elements from a set S is a subset of S with exactly k elements.*

Let us look, now for the number of *combinations of k elements from a set with n elements.* The number of *permutations* is given by (4.2.4), but of course the number of *combinations* must be smaller since each combination gives rise to several permutations. In fact, we see that each combination of k elements can be ordered in $k!$ different ways, and so gives rise to $k!$ permutations. This being so, the number of permutations must be $k!$ times the number of combinations. We obtain, then,

$$C_{n,k} = \frac{P_{n,k}}{k!}$$

or

4.2.5 $$C_{n,k} = \frac{n(n-1) \ldots (n-k+1)}{1 \times 2 \times 3 \ldots k}$$

as the number of combinations. This is also sometimes written in the form

4.2.6 $$\binom{n}{k} = \frac{n!}{k!(n-k)!}$$

$C_{n,k}$ and $\binom{n}{k}$ are equivalent expressions; the latter is more commonly used.

As an example, the number of possible poker hands is given by the expression

$$\binom{52}{5} = \frac{52 \times 51 \times 50 \times 49 \times 48}{1 \times 2 \times 3 \times 4 \times 5}$$

which, simplified, gives a grand total of 2,598,960 poker hands (for draw or five-card stud; it is of course different for six- or seven-card variations of the game.)

IV.2.7 Example. A congressional committee consists of 12 Democrats and 9 Republicans. A subcommittee must be chosen, containing 3 Democrats and 2 Republicans. In how many ways can this be done?

Clearly, the order in which the subcommittee members are chosen does not alter the subcommittee. We are interested here in combinations. The number of ways in which the 3 Democrats can be chosen is

$$\binom{12}{3} = \frac{12 \times 11 \times 10}{3 \times 2 \times 1} = 220$$

while the number of ways in which the 2 Republicans can be chosen is

$$\binom{9}{2} = \frac{9 \times 8}{2 \times 1} = 36$$

The total number of ways in which the subcommittee can be chosen is 220 times 36, or 7920.

IV.2.8 Example. There are 26 houses in a small town. A poll taker is to visit 7 of them. (The order in which he does it is not important.) In how many different ways can the 7 houses be chosen?

Clearly, we want combinations here. The answer is

$$\binom{26}{7} = \frac{26 \times 25 \times 24 \times 23 \times 22 \times 21 \times 20}{7 \times 6 \times 5 \times 4 \times 3 \times 2 \times 1}$$

or 657,800 different ways.

IV.2.9 Example. The town of Example IV.2.8 consists of a north side, with 16 houses, and a south side, with 10 houses. The poll taker is told to visit 4 houses in the north side, and 3 in the south side. In how many ways can this be done?

The 4 houses in the north side can be chosen in $\binom{16}{4}$, or 1820 **189** different ways. The 3 in the south side can be chosen in $\binom{10}{3}$, or 120 ways. In all, there are 1820×120, or 218,400 ways to do this.

PROBLEMS ON PERMUTATIONS AND COMBINATIONS

1. There are 7 men and 12 women in a group of people. In how many ways is it possible to choose a couple (1 man and 1 woman) from the group?

2. A classroom has 10 chairs. There are 7 children in the class. In how many different ways is it possible to assign seats to the children?

3. A Senate committee has 9 Democrats and 7 Republicans. A subcommittee of 3 Democrats and 2 Republicans must be chosen. In how many ways is it possible to do this?

4. Using the data of problem 3, in how many ways is it possible to choose the subcommittee if Senator A, a Democrat, and Senator B, a Republican, refuse to serve on the subcommittee together? In how many ways if it is Senator A and Senator C, both Democrats, who refuse to serve together? (Hint: in each case, it may be easier to compute the number of subcommittees that would have both the hostile senators and subtract this from the total number of possible subcommittees.)

5. A product must be checked to guarantee that it meets 6 different specifications. For each item produced, a card must be filled out, stating which of the 6 specifications is not met. In how many essentially different ways can a card be filled out? (Two cards are filled out in essentially different ways if there is at least one specification that is not met according to one card, but is met according to the other.)

6. A psychiatrist checks a person for 6 personality traits, and lists the 3 strongest on a card, in order of strength. In how many different ways can this card be filled out?

7. Six lines are drawn in a plane. What is the greatest possible number of points of intersection of these 6 lines?

8. A test consists of 5 multiple-choice questions, with 4 possible answers to each, and 10 true-false questions. Assuming that a student

answers all questions, in how many possible ways can he complete the answer sheet?

9. Prove the identity

$$\binom{n+1}{k} = \binom{n}{k} + \binom{n}{k-1}$$

Give a heuristic interpretation of this in terms of combinations.

10. Prove the identity

$$2^n = \binom{n}{0} + \binom{n}{1} + \cdots + \binom{n}{n}$$

by giving heuristic interpretations of the right and left sides of this equation.

3. PROPOSITIONS AND CONNECTIVES

Closely related to the theory of sets is the mathematical theory of logic. In general, it deals with the process of forming compound statements and with finding the *truth-value* of these statements (i.e., whether they are true or false), given the truth-values of the simpler statements that go into the compound.

In essence, this problem was first treated by Aristotle (384-322 B.C.). The mathematical formulation is, however, much later in date, being mainly the work of George Boole (1815-1864). This formulation, while it closely follows Aristotle's reasoning, has the advantage of mathematical strictness, which allows us to see whether the truth of one statement follows logically from another or is merely accidental.

Briefly, mathematical logic considers a *proposition* (i.e., a declarative sentence) and assigns to it a *truth-value* (true or false). It then compounds this proposition (possibly with others) and, by logical rules, assigns truth-values to the compounds.

IV.3.1 Example.   Let us consider the proposition, "I am happy." We shall denote this by the letter, p. Then, p has a truth-value T (true) if the speaker is happy, and a truth-value F (false) if the speaker is unhappy.

Consider now the proposition, "I am not happy." This is known as the *negation* of p, and is represented by the symbols $\sim p$ (the symbol $\sim$ represents negation). The proposition $\sim p$ has the value F whenever p has value T, and it has value T whenever p has value F.

Consider, next, the proposition "The sun is shining." We shall denote this by q. Again, it may have truth value T or F. We may also form the proposition $\sim q$, "The sun is not shining." The relationship of $\sim q$ to q is exactly that of $\sim p$ to p.

We may also connect the two propositions, generally by the connecting words "and" or "or," (although English usage sometimes requires a different conjunction.) In this way we may form the proposition:

"I am happy *and* the sun is shining"

known as the *conjunction* of p and q, and denoted by $p \wedge q$. Note that $p \wedge q$ has value T if, and only if, both p and q have value T. It has value F if either p, or q, or both, have value F.

On the other hand, consider the (weaker) proposition,

"I am happy *or* the sun is shining"

We shall denote this by $p \vee q$, and call it the *disjunction* of p and q. This will be T if p or q or both, are T. It will be F only if p and q are both F.

It is, of course, possible to form more complicated propositions by using more than one of these connectives. The proposition

"I am happy and the sun is not shining"

would be denoted by $p \wedge \sim q$, while

"I am happy and the sun is shining, or I am not happy"

would be denoted by $(p \wedge q) \vee \sim p$. (The parentheses are used to group $p \wedge q$ together). The reader may be interested in forming other compounds.

These, then, are the most elementary connectives. We may summarize their properties in the form of definitions.

IV.3.2 Definition. The *negation* of p, denoted $\sim p$, is true whenever p is false and false whenever p is true.

IV.3.3 Definition. The *conjunction* of p and q, denoted $p \wedge q$, is true whenever both p and q are true. It is false otherwise.

IV.3.4 Definition. The *disjunction* of p and q, denoted $p \vee q$, is true whenever either p or q, or both, are true. It is false otherwise.

PROBLEMS ON PROPOSITIONS AND CONNECTIVES

1. Find the simple parts in each of the following compound propositions:

(a) The grass is green, and the weather is hot, but it has not rained.

(b) The people are happy, although the weather is bad.

(c) Either the nation is prosperous, or there is no war.

(d) We are together, and all are good friends.

(e) We are not arming, nor will the enemy attack us.

2. Let p be the proposition "We are singing," let q be "He is playing the piano," and let r be "They are dancing." Form the following compounds.

(a) $p \lor q$.

(b) $p \land (\sim q \lor r)$.

(c) $\sim (p \land q) \lor \sim r$.

(d) $(p \lor (\sim p \lor q)) \land \sim r$.

(e) $((p \lor r) \land (p \land \sim q)) \lor (\sim p \land r)$.

4.   TRUTH-TABLES

For the most elementary compounds, such as those just described, it is very easy to determine the truth-value of the compound from the truth-values of its elements. Definitions IV.3.2, IV.3.3, and IV.3.4 give it to us immediately. For more complicated compounds, however, the result is not so obvious. The reader may, for instance, ask himself what the truth-value of the proposition

$$(p \lor (\sim p \land q)) \lor \sim q$$

will be, given that p is false and q is true. It is to deal with such problems that *truth-tables* were developed.

A *truth-table* is, briefly, a short table giving the truth-values of a compound proposition, for any possible combination of truth-values of its elementary components. The truth-table for $\sim p$ is shown in Table IV.4.1,

TABLE IV.4.1

p	$\sim p$
T	F
F	T

while those for $p \lor q$ and $p \land q$ are given in Tables IV.4.2 and IV.4.3.

TABLE IV.4.2

p	q	$p \vee q$
T	T	T
T	F	T
F	T	T
F	F	F

TABLE IV.4.3

p	q	$p \wedge q$
T	T	T
T	F	F
F	T	F
F	F	F

For more complicated compounds, a truth-table is generally constructed in a very systematic manner. First, the truth-values of the simple components are considered, then those of the slightly more complicated ones, and so on, step by step, until the truth-table of the desired compound is obtained.

IV.4.1 Example. Construct the truth-table of $(p \vee (\sim p \wedge q)) \vee \sim q$.

In our first step, we put the truth-values of each of the simple propositions, p and q, obtaining the table shown as Table IV.4.4.

TABLE IV.4.4

p	q	\multicolumn{4}{c}{$(p \vee (\sim p \wedge q)) \vee \sim q$}
T	T	T	T	T	T
T	F	T	T	F	F
F	T	F	F	T	T
F	F	F	F	F	F

Next, we compute the truth-values of the simplest compounds: $\sim p$ and $\sim q$. These are entered in the table, under the respective connectives. We have, then, Table IV.4.5.

TABLE IV.4.5

p	q	$(p$	$\vee$	$(\sim p$	$\wedge$	$q))$	$\vee$	$\sim$	q
T	T	T		F T		T		F	T
T	F	T		F T		F		T	F
F	T	F		T F		T		F	T
F	F	F		T F		F		T	F

It is now possible to compute the truth-values of the compound $(\sim p \wedge q)$. These appear in Table IV.4.6, under the connective.

TABLE IV.4.6

p	q	(p	∨	(~	p	∧	q))	∨	~	q
T	T	T		F	T	F	T		F	T
T	F	T		F	T	F	F		T	F
F	T	F		T	F	T	T		F	T
F	F	F		T	F	F	F		T	F

Next, we introduce the truth-values of the longer compound, $(p \vee (\sim p \wedge q))$. We see these in Table IV.4.7.

TABLE IV.4.7

p	q	(p	∨	(~	p	∧	q))	∨	~	q
T	T	T	T	F	T	F	T		F	T
T	F	T	T	F	T	F	F		T	F
F	T	F	T	T	F	T	T		F	T
F	F	F	F	T	F	F	F		T	F

Finally, we can compute the truth-values for the entire compound. These can be seen in Table IV.4.8:

TABLE IV.4.8

p	q	(p	∨	(~	p	∧	q))	∨	~	q
T	T	T	T	F	T	F	T	T	F	T
T	F	T	T	F	T	F	F	T	T	F
F	T	F	T	T	F	T	T	T	F	T
F	F	F	F	T	F	F	F	T	T	F

It may be seen from Table IV.4.8 that this particular compound will be true in all cases. Such a proposition is known as a *tautology*.

IV.4.2 Example. Construct the truth-table for the proposition $(p \wedge q) \vee \sim r$.

In this example, there are eight possibilities, since each of the three simple propositions p, q, and r can have two values. As in the previous example, we start by listing all these possibilities:

TABLE IV.4.9

p	q	r	(p ∧ q)	V	~r
T	T	T	T	T	T
T	T	F	T	T	F
T	F	T	T	F	T
T	F	F	T	F	F
F	T	T	F	T	T
F	T	F	F	T	F
F	F	T	F	F	T
F	F	F	F	F	F

Next, we introduce truth-values for the simpler compounds, $(p \wedge q)$, and $\sim r$, as shown in Table IV.4.10.

TABLE IV.4.10

p	q	r	(p	∧	q)	V	~	r
T	T	T	T	T	T		F	T
T	T	F	T	T	T		T	F
T	F	T	T	F	F		F	T
T	F	F	T	F	F		T	F
F	T	T	F	F	T		F	T
F	T	F	F	F	T		T	F
F	F	T	F	F	F		F	T
F	F	F	F	F	F		T	F

Finally, we compute truth-values for the entire compound. These are shown in Table IV.4.11.

TABLE IV.4.11

p	q	r	(p	∧	q)	V	~	r
T	T	T	T	T	T	T	F	T
T	T	F	T	T	T	T	T	F
T	F	T	T	F	F	F	F	T
T	F	F	T	F	F	T	T	F
F	T	T	F	F	T	F	F	T
F	T	F	F	F	T	T	T	F
F	F	T	F	F	F	F	F	T
F	F	F	F	F	F	T	T	F

PROBLEMS ON TRUTH-TABLES

1. Construct the truth-tables for the following compounds.

(a) $(p \wedge q) \vee (\sim p \wedge \sim q)$.

(b) $(\sim p \vee \sim q) \wedge (p \vee \sim q)$.

(c) $((p \lor q) \land (\sim p \lor \sim r)) \lor (p \land r)$.

(d) $(p \lor \sim q) \land (\sim q \lor r)$.

(e) $(p \land q) \land (\sim p \lor \sim q)$.

2. Show that the following pairs of compounds are equivalent:

(a) $\sim (p \lor q)$ and $\sim p \land \sim q$

(b) $\sim (p \land q)$ and $\sim p \lor \sim q$

(c) $(p \lor \sim q) \lor (\sim p \land q)$ and $p \lor \sim p$

(d) $p \land \sim p$ and $(p \land q) \land (\sim p \lor \sim q)$.

5. THE CONDITIONAL CONNECTIVES

We have, so far, studied the three elementary connectives, $\sim$, $\land$, and $\lor$. They are the simplest connectives, but there are others of great importance. These other connectives can be defined, either in terms of the elementary connectives or by their truth-tables. We choose the latter method.

The first connective we consider is the *conditional.* An example of this can be seen in the sentence: "If it rains, we will postpone the picnic." Symbolically, we would write this in the form

$$p \Rightarrow q$$

where p represents "it rains" and q is "we will postpone the picnic." p is called the *hypothesis,* and q the *consequence,* of the conditional.

Let us analyze this proposition. Under which circumstances will it be true? In the first place, it is clear that, if both p and q are true (i.e., it rains, and we postpone the picnic), then the conditional is also true. If both p and q are false (it does not rain, and we do not postpone the picnic), then the conditional will also be true.

Suppose, now, that p is false, but q is true: it does not rain, but we postpone the picnic. Does this make the conditional $p \Rightarrow q$ false? The answer is no, since, after all, $p \Rightarrow q$ states only what will happen in case of rain: we reserve the right to postpone the picnic even if it does not rain. Thus, $p \Rightarrow q$ is true in this case.

Finally, suppose that p is true, but q is false: it rains, but we do not postpone the picnic. In this case, it should be clear that the conditional is false.

We conclude, then, that the conditional $p \Rightarrow q$ is true if p is false, or if q is true, or both. It is false if p is true, but q false. We state this in the form of a definition.

IV.5.1 Definition. The truth-values of the conditional, $p \Rightarrow q$,
are given by Table IV.5.1.

TABLE IV.5.1

p	q	$p \Rightarrow q$
T	T	T
T	F	F
F	T	T
F	F	T

It should be made quite clear here that the conditional $p \Rightarrow q$ does not imply, in general, any logical connection between its *hypothesis, p,* and its *consequence, q.* A sentence such as "If dogs can talk, then the moon is made of green cheese," is true, not because of any connection between dogs' loquacity and the substance of the moon, but, quite simply, because the hypothesis ("If dogs can talk") is false. It should be noticed that this property of the conditional is sometimes used for rhetorical effects: "If that man knows what he's talking about, then my name is Napoleon." The point generally is that, since the consequence is obviously false, the hypothesis must be patently untrue. Sometimes, the conditional is used to imply the truth of the consequence by using an obvious hypothesis, even though there is no logical connection: "If he is rich, he is also honest." The man's riches have not made him honest, but the more obvious attribute (wealth) is used rhetorically to affirm the less obvious quality (honesty).

The reader should beware therefore of inferring any logical connection between consequence and hypothesis. For our purposes, we should remember that the truth of the conditional depends merely on that of its parts, as given by Definition IV.5.1.

We note also, that the truth-tables for the two compounds, $p \Rightarrow q$, and $(\sim p) \vee q$, are identical. Logically, we can see why this should be so: one will be true whenever the other is. It would have been possible to define the conditional precisely in this manner, as shown in Table IV.5.2.

TABLE IV.5.2

p	p	$(\sim$	$p)$	$\vee$	q
T	T	F	T	T	T
T	F	F	T	F	F
F	T	T	F	T	T
F	F	T	F	T	F

Another important connective is the bi-conditional. We define this also in terms of its truth-table.

IV.5.2 Definitions. The truth-values of the bi-conditional $p \Leftrightarrow q$, are given by Table IV.5.3.

TABLE IV.5.3

p	q	$p \Leftrightarrow q$
T	T	T
T	F	F
F	T	F
F	F	T

By definition, $p \Leftrightarrow q$ is true when p and q are both true or both false. It is false otherwise. We could have defined it by saying that it is identical with the compound

$$(p \wedge q) \vee (\sim p \wedge \sim q)$$

as it is easy to see that this has the same truth-table. It is more interesting, however, to point out that the compound

$$(p \Rightarrow q) \wedge (q \Rightarrow p)$$

has the same truth-table as $p \Leftrightarrow q$ (Table IV.5.4)

TABLE IV.5.4

p	q	$(p$	$\Rightarrow$	$q)$	$\wedge$	$(q$	$\Rightarrow$	$p)$
T	T	T	T	T	T	T	T	T
T	F	T	F	F	F	F	T	T
F	T	F	T	T	F	T	F	F
F	F	F	T	F	T	F	T	F

This shows the relation between the conditional and bi-conditional connectives, and indeed, the bi-conditional is generally presented in this manner: If p is "I will come," and q is "It does not rain," then the bi-conditional $p \Leftrightarrow q$ is generally expressed in the form "I will come if and only if it does not rain." It could also be expressed as "Either it will rain and I will not come, or it will not rain and I will come." The two statements are equivalent (i.e., they will be true under the same circumstances), but the first form is much more concise and effective.

Like the conditional, the bi-conditional is sometimes used even when there is no logical connection between the two parts; rhetorically, however, this is not nearly so common.

IV.5.3 *Example.* Construct the truth-table for $(p \Rightarrow q) \Rightarrow (q \Rightarrow p)$.

TABLE IV.5.5

p	q	$(p$	$\Rightarrow$	$q)$	$\Rightarrow$	$(q$	$\Rightarrow$	$p)$
T	T	T	T	T	T	T	T	T
T	F	T	F	F	T	F	T	T
F	T	F	T	T	F	T	F	F
F	F	F	T	F	T	F	T	F

IV.5.4 *Example.* Construct the truth-table for $(p \lor q) \Rightarrow (p \Rightarrow q)$.

TABLE IV.5.6

p	q	$(p$	$\lor$	$q)$	$\Rightarrow$	$(p$	$\Leftrightarrow$	$q)$
T	T	T	T	T	T	T	T	T
T	F	T	T	F	F	T	F	F
F	T	F	T	T	F	F	F	T
F	F	F	F	F	T	F	T	F

PROBLEMS ON CONDITIONAL CONNECTIVES

1. Construct the truth-tables for the following compounds.

(a) $(p \Rightarrow q) \land (q \Rightarrow p)$.

(b) $(p \land q) \Leftrightarrow (p \lor q)$.

(c) $(p \land \sim q) \Rightarrow (\sim p \lor q)$.

(d) $p \Rightarrow (q \land \sim q)$.

(e) $((p \lor q) \Rightarrow (q \lor r)) \Leftrightarrow (p \Rightarrow r)$.

(f) $(p \Rightarrow q) \Leftrightarrow (\sim p \Rightarrow \sim q)$.

(g) $(p \lor \sim q) \Rightarrow p$.

(h) $p \Rightarrow (p \lor q)$.

(i) $((p \Rightarrow q) \lor (q \Rightarrow r)) \Leftrightarrow (r \Rightarrow p)$.

(j) $(p \Leftrightarrow q) \Rightarrow (p \Rightarrow q)$.

6. TAUTOLOGIES AND LOGICAL REASONING

In Example IV.4.1, we considered the compound $(p \lor (\sim p \land q))$ $\lor \sim q$ and saw that it is true in all cases, whatever p and q may be. We use this property as a definition.

IV.6.1 Definition. A *tautology* is a statement that is logically true (i.e., whose truth-table contains only the value T).

Example IV.4.1 was a tautology. We give other examples, showing that they are so by constructing their truth-tables.

IV.6.2 Example. Show that the proposition

$$((p \Rightarrow q) \wedge (q \Rightarrow r)) \Rightarrow (p \Rightarrow r)$$

is a tautology.

We construct Table IV.6.1.

TABLE IV.6.1

p	q	r	$(p$	$\Rightarrow q)$	$\wedge$	$(q$	$\Rightarrow r))$	$\Rightarrow$	$(p$	$\Rightarrow$	$r)$
T	T	T	T	T	T	T	T	T	T	T	T
T	T	F	T	T	T	F	T	F	F	T	T	F	F
T	F	T	T	F	F	F	F	T	T	T	T	T	T
T	F	F	T	F	F	F	F	T	F	T	T	F	F
F	T	T	F	T	T	T	T	T	T	T	F	T	T
F	T	F	F	T	T	F	T	F	F	T	F	T	F
F	F	T	F	T	F	T	F	T	T	T	F	T	T
F	F	F	F	T	F	T	F	T	F	T	F	T	F

IV.6.3 Example. Show that the proposition

$$(p \wedge (p \Rightarrow q)) \Rightarrow q$$

is a tautology.

Again, we construct the truth-table (Table IV.6.2).

TABLE IV.6.2

p	q	$(p$	$\wedge$	$(p$	$\Rightarrow$	$q))$	$\Rightarrow$	q
T	T	T	T	T	T	T	T	T
T	F	T	F	T	F	F	T	F
F	T	F	F	F	T	T	T	T
F	F	F	F	F	T	F	T	F

IV.6.4   Example.   Show that $p \vee \sim p$ is a tautology.

TABLE IV.6.3

p	p	$\vee$	$\sim$	p
T	T	T	F	T
F	F	T	T	F

From a certain point of view, a tautology is a useless statement. Precisely because it is always logically true, it gives us no new information. Example IV.6.4 could be illustrated with "Either it is raining, or it is not raining" — not, indeed, a very illuminating piece of information. Yet it would be wrong to consider tautologies worthless; just because they are always true, any argument based entirely on tautologies must be valid.

Let us see how this is to be done. When we say that the proposition p *logically implies* q, we shall mean that the conditional $p \Rightarrow q$ is a tautology. Similarly, when we say that p and q are *logically equivalent*, we shall mean that the bi-conditional $p \Leftrightarrow q$ is a tautology.

Consider, now, the general idea of a valid logical argument. The following are examples of valid arguments:

IV.6.5   Examples.
(a) If it rains, the grounds will be wet.
   If the grounds are wet, the ball game will be postponed.
   Therefore, if it rains, the ball game will be postponed.

(b) If we arm, we will be attacked.
   If we do not arm, we will be attacked.
   Therefore, we will be attacked.

(c) When the sun shines, my girl-friend is happy.
   When my girl-friend is happy, she bakes cookies.
   The sun is shining.
   Therefore, my girl-friend is baking cookies.

Example IV.6.5 shows us several valid arguments. The usual form of such arguments is known as a *syllogism*. In brief, the syllogism consists of two or more propositions, known as *premises*, which have (in general) already been established or are considered obvious. From these, a *conclusion* is drawn; this is the proposition starting with "therefore" in each of these examples.

What, now, is the essence of a valid argument? If we consider the foregoing examples, we shall see that the characteristic property is that *the conjunction of the premises logically implies the conclusion.* Example IV.6.5(a) may be represented in the form:

$$((p \Rightarrow q) \wedge (q \Rightarrow r)) \Rightarrow (p \Rightarrow r)$$

which, as we saw in Example IV.6.1, is a tautology. Similarly, IV.6.5(b) will have the form:

$$((p \Rightarrow q) \wedge (\sim p \Rightarrow q)) \Rightarrow q$$

while IV.6.5(c) has the form:

$$(((p \Rightarrow q) \wedge (q \Rightarrow r)) \wedge p) \Rightarrow r$$

It may be checked directly (e.g., by constructing their truth-tables) that both these compound propositions are tautologies.

One of the most common errors in reasoning occurs because the conditional connective, unlike the other elementary connectives, is not symmetric. The propositions $p \vee q$ and $q \vee p$ are equivalent, as may be checked by constructing the truth-table of the compound

$$(p \vee q) \Leftrightarrow (q \vee p)$$

Similarly, $p \wedge q$ and $q \wedge p$ are equivalent, as are $p \Leftrightarrow q$ and $q \Leftrightarrow p$. However, the two propositions $p \Rightarrow q$ and $q \Rightarrow p$ are not equivalent. Indeed, we may construct a truth-table of their bi-conditional that is clearly not a tautology (Table IV.6.4).

TABLE IV.6.4

p	q	$(p$	$\Rightarrow$	$q)$	$\Leftrightarrow$	$(q$	$\Rightarrow$	$p)$
T	T	T	T	T	T	T	T	T
T	F	T	F	F	F	F	T	T
F	T	F	T	T	F	T	F	F
F	F	F	T	F	T	F	T	F

The statement $q \Rightarrow p$ is known as the *converse* of $p \Rightarrow q$, and the two are clearly not equivalent. Consider, then, the following argument:

When the sun shines, I am happy.
I am happy.
Therefore, the sun is shining.

Symbolically, the three propositions have the form:

$$p \Rightarrow q$$
$$q$$
Therefore, p

This would be a valid argument if the compound

$$((p \Rightarrow q) \wedge q) \Rightarrow p$$

were tautological. This is not so, however; it is

$$((q \Rightarrow p) \wedge q) \Rightarrow p$$

that is a tautology. In other words, the argument has somehow replaced $p \Rightarrow q$ by its converse, $q \Rightarrow p$. Thus, the argument is not valid.

Consider, now, the two propositions, $p \Rightarrow q$ and $\sim q \Rightarrow \sim p$. An example would be:

> If the sun shines, I am happy.
> If I am not happy, the sun is not shining.

The proposition $\sim q \Rightarrow \sim p$ is known as the *contra-positive* of $p \Rightarrow q$. It may be shown that these two propositions are equivalent. As a general rule, *it is valid to replace a statement by its contra-positive, but it is not valid to replace it by its converse.*

Some of the best-known tautologies are so commonly used that they have been given special names. Example IV.6.3, for instance, is known as the *principle of the excluded middle*: a proposition is true or false; there is nothing else left for it (so long as it makes sense). Other well-known principles appear in the following exercises.

PROBLEMS ON TAUTOLOGIES AND LOGICAL REASONING

1. Which of the following are tautologies?

(a) $(p \Rightarrow (q \wedge \sim q)) \Rightarrow \sim p$.

(b) $(p \Rightarrow \sim p) \Rightarrow \sim p$.

(c) $(p \Rightarrow q) \Rightarrow (q \Rightarrow p)$.

(d) $(p \Rightarrow \sim q) \Leftrightarrow (q \Rightarrow \sim p)$.

(e) $(p \wedge q) \Rightarrow p$.

(f) $(p \Leftrightarrow q) \Rightarrow (p \vee q)$.

(g) $(p \Leftrightarrow q) \Rightarrow ((p \vee q) \vee (\sim p \wedge \sim q))$.

(h) $(p \Leftrightarrow q) \Rightarrow ((p \wedge q) \vee (\sim p \wedge \sim q))$.

(i) $p \Leftrightarrow p$.

(j) $(p \wedge \sim p) \Rightarrow (q \vee r)$.

2. Which of the following are examples of valid logical reasoning?

(a) I am happy.
Whenever I am happy, I sing.
Therefore, I am singing.

(b) If we arm, we shall be attacked.
If we do not arm, we shall be attacked.
Therefore, we shall be attacked.

(c) When it is summer, the weather is very hot.
The weather is very hot.
Therefore, it is summer.

(d) We shall come if, and only if, you invite us.
If we come, you will see us.
You have invited us.
Therefore, you will see us.

(e) When people work hard, they are well paid.
These people do not work hard.
Therefore, these people are not well paid.

(f) If beggars were kings, the moon would be made of green cheese.
The moon is made of green cheese.
Therefore, beggars are kings.

3. Let x be a variable that is allowed to take values in a universal set $\mathcal{U}$, and let $p(x)$ and $q(x)$ be statements about the variable x. Generally speaking, these statements will have truth-value T for certain values of the variable, and truth-value F for other values of the variable. By the *truth-set* for $p(x)$, we shall mean the set

$$S = \{x \mid p(x) \text{ is true}\}$$

Similarly, the truth-set for $q(x)$ may be defined and will be denoted by T. Show that:

(a) the truth-set for $p(x) \vee q(x)$ is $S \cup T$.

(b) The truth-set for $p(x) \wedge q(x)$ is $S \cap T$.

(c) The truth-set for $\sim p(x)$ is $\overline{S}$.

4. With the same notation as in problem 3, show that:

(a) If $p(x)$ implies $q(x)$, then $S \subset T$.

(b) If $p(x)$ is a tautology, then S is the universal set $\mathcal{U}$.

CHAPTER V

THE THEORY OF PROBABILITY

1. PROBABILITIES

The theory of probability arose from the desire of men to understand the behavior of systems that cannot entirely be controlled. More exactly, it arose from the desire of a French gambler, the Chevalier de Méré, to understand why his dice had suddenly turned against him. The Chevalier's concern was a very practical one: he was losing money.

For some time, de Méré had bet (at even odds) that, in four tosses of a die, he would obtain a six at least once. This was a profitable bet, as was attested to by his systematic winnings. Unfortunately, he had been too successful and, as a result, found himself unable to obtain any takers for his bets. This led him to change the bet: he now bet that, in 24 tosses of a pair of dice, he would obtain a double-six at least once. The Chevalier reasoned that this bet would be as profitable as the previous one (since, indeed, $4/6 = 24/36$); the change in the bet, however, would enable him to find new takers. He did find new takers for the bet, but, to his surprise, he began to lose systematically.

We shall consider this example later on, and see what de Méré's mistake consisted in. For the present we will simply note that, not being a mathematician, he proceeded to consult one—Blaise Pascal (1623-1662), one of the greatest of all time. Pascal discussed the problem with Pierre de Fermat (1601-1665), and, from the analysis made by these two men, a new branch of mathematics was born.

The fundamental problem in the theory of probability is, then, as follows. A certain action, called an *experiment*, is carried out. The experiment may have any one of several possible *outcomes*; the idea is to predict, more or less, the frequency with which these outcomes (or certain sets of them, called *events*) will appear—assuming that the

205

experiment can be repeated many times. The experiment can be of many types: we give examples.

(a) A die is tossed; the *outcome* is the face that lands on the *bottom*.
(b) A dart is thrown at a bull's-eye; the *outcome* is the section of the bull's-eye in which the dart lands.
(c) A coin is tossed three times; the *outcome* is the run of heads and tails obtained.
(d) A telephone book is opened; the *outcome* is the page to which it is opened.

We give these examples to illustrate several things. Example (a) shows that the outcomes may be labeled in many ways (since it is usual to think of the *top* face of the die as the outcome of this experiment). Example (b) shows that many outcomes may be classed together as a single outcome (we could consider each point on the board as a different outcome). Example (c) shows that an experiment may consist of several other, simpler experiments. Example (d) shows that the experimenter may be able to influence, to some degree, the outcome, since he can choose, to some extent, the section of the book to which he will open (though possibly not the exact page). It is often necessary to explain the manner in which the experiment is carried out.

Now, how are probabilities to be assigned? This is an extremely complicated question—not from a mathematical, but from a practical point of view. Mathematically, we can give a system of axioms that are to be satisfied by any assignment of probabilities. The difficulty lies in our practical desire that the probability assigned to an outcome correspond to the relative frequency with which it occurs. This can be ascertained only through experimentation. But such experiments, precisely because of their nature, are always subject to random fluctuations, which tend to, but need not, cancel out over the long run. Moreover, the "long run" might be so long as to be quite impractical.

An example may be of use in explaining this difficulty. A coin is said to be *fair* if, when it is tossed, each of the two outcomes (heads and tails) has a probability 1/2 of occurring. This means (for reasons to be seen later) that, in tossing a fair coin 20 times, the probability of obtaining heads every time is $(1/2)^{20}$, or approximately 0.000001. Thus, if a coin is tossed 20 times, and turns up heads every time, we would be justified in claiming that the coin is skew (i.e., not fair).

Suppose, however, that each adult person in the United States (say, some 100,000,000 people) were to toss a (fair) coin 20 consecutive times. In this case, it is almost certain that approximately 100 of them would have the singular experience of seeing a string of 20 consecutive heads; another 100 would obtain a string of 20 consecutive tails. We might imagine each of these 200 claiming that the newly minted coins were entirely skew; in fact, however, we would claim that it was a cause for surprise if no such strings of 20 heads occurred. The idea is that, no matter how unlikely an outcome, it is likely to happen, and so we have to be careful before assigning probabilities on the basis of observed results, no matter how many times the experiment may be repeated.

A second conceptual difficulty is the fact that, in many cases, the experiment is such that it cannot or will not be repeated. The Sox meet the Cats tomorrow to determine the Baseball Championship; what is the probability that the Sox will win? A building contractor makes a bid on a projected building; what is the probability that he will be the low bidder? Mr. Jones will go hunting tomorrow for the first time in his life; what is the probability that he will shoot a fox? In each of these cases, it is clear that the "experiment," with all its implications, cannot be repeated at all (the baseball players, for example, could play a seven game championship set, but then the individual games would no longer be as important as the single championship game). Thus a probability can be assigned to these outcomes only by defining the concept of probability in a different manner.

In such cases, it is common to talk about *subjective* probability. The subjective probability of an event is, more or less, the subjective belief in its likelihood; the event has, say, a subjective probability of 1/2 if a person feels that a bet at even odds is a "fair" bet. This is not, to say the least, a very scientific idea; it is further complicated by the fact that many people may have different beliefs about the event's likelihood, and so the subjective probability is not well defined. (Practical men have, of course, found a way to solve this problem; the bookmaker, seeing that he cannot reconcile the subjective probabilities held by different people, looks for an equilibrium value that will guarantee that he wins whether, in fact, the Sox or the Cats win the championship.)

The problem of assigning probabilities to the outcomes of an experiment remains. In the absence of anything better, the rule of Pierre Simon, Marquis de Laplace (1749-1827), has been used: when there is no indication to the contrary, assume that each of the possible outcomes has the same probability. When tossing a coin, one assumes (unless there is some indication to the contrary) that each of the two sides has the same probability. If an urn has 50 balls, all identical except for color, then each of the balls has the same probability (1/50) of being drawn out by a blindfolded man. Laplace's rule does not hold when it seems reasonable that one of the outcomes is more likely than another. For instance, if a die has one face much larger than the others, it seems reasonable to assume that the die will land on that face more often than on the others.

In the rest of this chapter, we shall assume that probabilities can be assigned to the outcomes: we shall then attempt to draw logical conclusions about the outcomes of other, more complicated experiments. This is the domain of probability theory. The problem of assigning these original probabilities belongs to the field of statistical analysis and is treated elsewhere.

2. DISCRETE PROBABILITY SPACES

We shall begin our study of probability by considering only experiments with a finite number of possible outcomes.

V.2.1 Definition. A *finite probability space* is a finite set of outcomes

$$X = \{O_1, O_2, \ldots, O_n\}$$

together with function, P, which assigns a non-negative and real number to each of the outcomes O_i, in such a way that

5.2.1
$$\sum_{i=1}^{n} P(O_i) = 1$$

The number $P(O_i)$ is the probability of the outcome O_i; it is often replaced by p_i. The reason for condition (5.2.1) should be clear: since each trial of the experiment will have one (and only one) of the outcomes O_i, it follows that the sum of their relative frequencies must be equal to 1.

V.2.2 Definition. An *event* is a subset of the set X.

An event is defined as a set of outcomes. The probability function P, which is defined for outcomes, can be extended to events by saying that an event happens whenever any one of the outcomes belonging to this event occurs. Thus, the probability of the event will be equal to the sum of the probabilities of the several outcomes:

5.2.2
$$P(A) = \sum_{O_i \in A} P(O_i)$$

In particular, X is itself an event called the *certain event*, since it contains all the possible outcomes. The empty set, $\emptyset$, is also an event, called the *impossible* event. It is easy to see that

5.2.3
$$P(X) = 1$$

5.2.4
$$P(\emptyset) = 0$$

A certain event must have probability 1; an impossible event, probability 0. The converse of this is not quite true; an event may be possible and still have probability 0, or it may have probability 1 and still not be certain. For finite probability spaces, however, this will not happen.

We give now examples of finite probability spaces.

V.2.3 *Example.* Consider the experiment of tossing a die and looking at the top face. The experiment has six possible outcomes:

$$X = \{1, 2, 3, 4, 5, 6\}$$

If the die is fair, then each of these outcomes will have the same probability; since the sum of these must be 1, it follows that

$$p_i = \frac{1}{6} \text{ for } i = 1, 2, \ldots, 6$$

Consider the event "even number":

$$E = \{2,4,6\}$$

We can calculate that

$$P(E) = p_2 + p_4 + p_6 = \frac{1}{2}$$

Similarly, if $S = \{1,2\}$, we have

$$P(S) = p_1 + p_2 = \frac{1}{3}$$

V.2.4 *Example.* Let the experiment consist of three tosses of a coin; the outcome will be the run of heads and tails obtained. There are 8 possible outcomes:

H	H	H	T	H	H
H	H	T	T	H	T
H	T	H	T	T	H
H	T	T	T	T	T

If the coin is fair, each of these outcomes will have the same probability, namely, 1/8. If we consider the event "two heads," we have

$$E = \{H\,H\,T, \quad H\,T\,H, \quad T\,H\,H\}$$

and it follows that

$$P(E) = \frac{3}{8}$$

V.2.5 *Example.* Let the experiment consist of drawing a card from a deck. There are 52 possible outcomes; the event "hearts" will contain 13 of these outcomes. If we assume that each card has the same probability, 1/52, of being drawn, then

$$P\,(\text{hearts}) = \frac{13}{52} = \frac{1}{4}$$

For the same experiment, the event "ace" will contain four of the possible outcomes. Thus

$$P\,(\text{ace}) = \frac{4}{52} = \frac{1}{13}$$

It is not too difficult to see that, if A and B are disjoint events, i.e., if $A \cap B = \varnothing$, then

5.2.5 $$P(A \cup B) = P(A) + P(B)$$

In fact, $P(A \cup B)$ is the sum of the probabilities of the outcomes that belong to $A \cup B$. This sum can be split into two sums, corresponding to $P(A)$ and $P(B)$. Note that this would not be possible if $A \cap B$ were non-empty, since then any outcome that belonged to $A \cap B$ would be counted twice.

If we have several disjoint events, i.e., events A_1, A_2, ..., A_m such that $A_j \cap A_k = \varnothing$ whenever $j \neq k$, then (5.2.5) generalizes directly, to give

5.2.6 $$P(A_1 \cup A_2 \cup \ldots \cup A_m) = P(A_1) + P(A_2) + \ldots + P(A_m)$$

Again, we point out that (5.2.6) will not hold unless the events A_k are disjoint.

In case two events are not disjoint, we have the more general relation

5.2.7 $$P(A \cup B) = P(A) + P(B) - P(A \cap B)$$

To prove (5.2.7), note that the three events

$$A \cap B, \qquad A - B, \qquad B - A$$

are disjoint. Now

$$A \cup B = (A \cap B) \cup (A - B) \cup (B - A)$$

(See Figure V.2.1.) Thus, by (5.2.6),

5.2.8 $$P(A \cup B) = P(A \cap B) + P(A - B) + P(B - A)$$

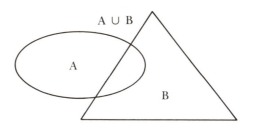

FIGURE V.2.1 Dissection of the set $A \cup B$ into the disjoint subsets $A - B$, $B - A$, and $A \cap B$.

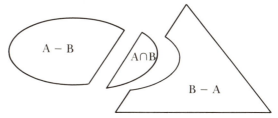

We also have

$$A = (A \cap B) \cup (A - B)$$

and

$$B = (A \cap B) \cup (B - A)$$

so that

5.2.9 $\qquad P(A) = P(A \cap B) + P(A - B)$

5.2.10 $\qquad P(B) = P(A \cap B) + P(B - A)$

Adding (5.2.9) and (5.2.10), and substituting (5.2.8), we obtain

$$P(A) + P(B) = P(A \cup B) + P(A \cap B)$$

which is equivalent to (5.2.7).

V.2.6 Example. Given the same experiment as in Example V.2.5, find the probability of obtaining *either* a heart *or* an ace.

For this problem, we can write

$$H = \text{``hearts''}$$
$$A = \text{``ace''}$$

and we are looking for $P(H \cup A)$. We already know $P(H)$ and $P(A)$; to find $P(H \cap A)$, note that $H \cap A$ consists of one single outcome (the ace of hearts) and so

$$P(H \cap A) = \frac{1}{52}$$

Applying (5.2.7), now, we will have

$$P(H \cup A) = P(H) + P(A) - P(H \cap A) = \frac{1}{4} + \frac{1}{13} - \frac{1}{52} = \frac{16}{52}$$

and so

$$P(H \cup A) = \frac{4}{13}$$

This could, of course, have been obtained directly by noticing that there are exactly 16 cards that are either hearts or aces: 13 hearts, and 3 other aces.

V.2.7 Example. Let an experiment consist of a toss of two dice. There are, all told, 36 possible outcomes (assuming that the two dice are distinguishable):

$$
\begin{array}{cccccc}
(1,1), & (1,2), & (1,3), & (1,4), & (1,5), & (1,6) \\
(2,1), & (2,2), & (2,3), & (2,4), & (2,5), & (2,6) \\
(3,1), & (3,2), & (3,3), & (3,4), & (3,5), & (3,6) \\
(4,1), & (4,2), & (4,3), & (4,4), & (4,5), & (4,6) \\
(5,1), & (5,2), & (5,3), & (5,4), & (5,5), & (5,6) \\
(6,1), & (6,2), & (6,3), & (6,4), & (6,5), & (6,6)
\end{array}
$$

Consider the event A: "the sum of the numbers on the two dice is equal to 10." We have

$$A = \{(6,4), (5,5), (4,6)\}$$

and so, assuming that each outcome has probability 1/36, we find

$$P(A) = \frac{3}{36} = \frac{1}{12}$$

In a similar way, the event B, "doubles are thrown," will contain six outcomes:

$$B = \{(1,1), (2,2), (3,3), (4,4), (5,5), (6,6)\}$$

and so

$$P(B) = \frac{6}{36} = \frac{1}{6}$$

We can see also that $A \cap B = (5,5)$; hence

$$P(A \cap B) = \frac{1}{36}$$

and so

$$P(A \cup B) = \frac{1}{12} + \frac{1}{6} - \frac{1}{36} = \frac{2}{9}$$

V.2.8 Example. There are 39 families in a small town; of these, 23 are Republican and 16 are Democrat. There are 7 families headed by a woman, and of these, 5 are Democrat.

A poll sampler chooses one family at random. Find the probability that: (a) the family is Democrat, (b) it is headed by a woman, (c) it is *either* Democrat *or* headed by a woman.

Let D be the event "the family is Democrat," and let W be the event "it is headed by a woman." Assuming that each family has a probability 1/39 of being chosen, we have:

$$P(D) = \frac{16}{39}$$

$$P(W) = \frac{7}{39}$$

while for the "joint event," "Democrat *and* headed by a woman,"

$$P(D \cap W) = \frac{5}{39}$$

We thus have

$$P(D \cup W) = \frac{16}{39} + \frac{7}{39} - \frac{5}{39} = \frac{18}{39}$$

so that the probability that the family is *either* Democrat *or* headed by a woman is 18/39, or 6/13.

PROBLEMS ON DISCRETE PROBABILITY SPACES

1. A fair die is tossed. Let E be the event $\{2,4,6\}$ and let S be the event $\{1,2\}$. Find the probabilities of the events E, S, $E \cap S$, $E - S$, $E \cup S$, and $S - E$.

2. A fair coin is tossed three times. List the eight possible outcomes. Find the probabilities of the events:

(a) The first toss is H.

(b) Either the first toss is H, or the third toss is T.

(c) Both the second and third tosses are T.

(d) No tosses are H.

(e) At least one of the tosses is T.

(f) The first toss is H, and the third toss is T.

(g) There are at least two T's.

3. A card is drawn at random from a deck (of 52 cards). What is the probability that:

(a) The card is an ace?

(b) It is the ace of hearts?

(c) It is *either* an ace *or* a heart?

4. A pair of fair dice is tossed. What is the probability that:

(a) The sum of the spots showing on the top of the dice is 8?

(b) It is 8 or more?

(c) It is 4 or less?

3. CONDITIONAL PROBABILITY. BAYES' FORMULA

As was mentioned before, we are interested in finding relationships among the probabilities of various events in order to analyze more complicated experiments. Let us consider the idea of *conditional* probability: given that an event A happens, what is the probability of a second event, B?

From a practical point of view, probability theory is applicable when the analyst lacks information about all the relevant variables. A poker player must rely on the theory of probability because he does not know what the card on top of the deck is; if he could peek, he would know whether to bet or fold. Suppose, however, that he notices the top card has a frayed corner. If he knows this deck of cards, the knowledge gives him additional information, which will make his decision easier for him. He no longer has to consider all the cards in the deck, but only those that have frayed corners.

To come back to our original question, the probability of event B, given that event A happens, is the relative frequency of B, considering only those cases in which A happens. This leads to the following definition.

V.3.1 Definition. The *conditional probability* of the event B, given the event A, is the probability $P(B \mid A)$, given by

5.3.1
$$P(B \mid A) = \frac{P(A \cap B)}{P(A)}$$

V.3.2 Example. In the experiment of Example V.2.7, find the conditional probability of obtaining doubles, given that the sum of the numbers on the two dice is 10.

In this case, we are looking for $P(B \mid A)$. We have, from V.2.7,

$$P(A \cap B) = \frac{1}{36}$$

$$P(A) = \frac{1}{12}$$

Therefore

$$P(B \mid A) = \frac{1/36}{1/12} = \frac{1}{3}$$

Note that $P(B \mid A)$ is twice $P(B)$. Whether A occurs will alter the probability of B.

V.?3 Example. Using the data of Example V.2.8, find the probability that the family sampled is Democrat, given that it is headed by a woman. Find also the probability that it is headed by a woman, given that it is Democrat. (Note that these are not the same thing.)

We are looking here for $P(D \mid W)$ and $P(W \mid D)$. Now

$$P(D \mid W) = \frac{P(D \cap W)}{P(W)} = \frac{5/39}{7/39} = \frac{5}{7}$$

while

$$P(W \mid D) = \frac{P(D \cap W)}{P(D)} = \frac{5/39}{16/39} = \frac{5}{16}$$

Two events are thought of as independent if neither has any influence on the other. Probabilistically, we can best express this by saying that the conditional probability is the same as the absolute (i.e., not conditional) probability.

V.3.4 Definition. Two events, A and B, are said to be independent if

5.3.2 $$P(A \cap B) = P(A)P(B)$$

For independent events, the *joint probability* (i.e., the probability $P(A \cap B)$ that they will *both* happen) is equal to the product of their individual absolute probabilities.

V.3.5 Example. Consider the experiment of Example V.2.4. Let A be the event "the first toss is a head," and let B be the event "the third toss is tails." We find then that

$$A = \{H\,H\,H, H\,H\,T, H\,T\,H, H\,T\,T\}$$

$$B = \{H\,H\,T, H\,T\,T, T\,H\,T, T\,T\,T\}$$

and

$$A \cap B = \{H\,H\,T, H\,T\,T\}$$

Since each outcome has probability 1/8, we find that

$$P(A) = P(B) = \frac{1}{2}$$

$$P(A \cap B) = \frac{1}{4} = \frac{1}{2} \cdot \frac{1}{2}$$

The joint probability is the product of the absolute probabilities, and we conclude that the two events are independent. This corresponds to the intuitive idea that the result of the first toss should have no influence on the third toss, or vice-versa. On the other hand, note that for Example V.2.8, the events D and W are *not* independent.

For more than two events, independence is defined in a slightly different manner:

V.3.6 Definition. Three events A, B, and C are independent if any two of them are independent and, moreover,

5.3.3 $$P(A \cap B \cap C) = P(A)P(B)P(C)$$

In general, n events A_1, A_2, . . . , A_n are independent if any $n - 1$ are independent and also

5.3.4 $\quad P(A_1 \cap A_2 \cap \ldots \cap A_n) = P(A_1)P(A_2) \ldots P(A_n)$

Definition V.3.6 is an inductive definition: it defines independence for n events in terms of independence for $n - 1$ events, plus an additional condition. An alternative definition would be to say that a collection of events is independent if, for any sub-collection, the joint probability is the product of the absolute probabilities.

Note that it is possible for three events not to be independent, although any two of them are independent. (Conversely, three events may satisfy (5.3.3) and still not be independent because two of them are not.)

V.3.7 Example. In the experiment of Example V.2.7, let A be the event "the first die shows a 6"; let B be the event "the second die shows a 3," and let C be the event "the sum of the numbers on the two dice is 7." We have, then,

$$A = \{(6,1), (6,2), (6,3), (6,4), (6,5), (6,6)\}$$
$$B = \{(1,3), (2,3), (3,3), (4,3), (5,3), (6,3)\}$$
$$C = \{(1,6), (2,5), (3,4), (4,3), (5,2), (6,1)\}$$

and so $P(A) = P(B) = P(C) = 1/6$. To check for independence, we see that

$$A \cap B = \{(6,3)\}$$
$$A \cap C = \{(6,1)\}$$
$$B \cap C = \{(4,3)\}$$

and so

$$P(A \cap B) = \frac{1}{36} = P(A)P(B)$$

$$P(A \cap C) = \frac{1}{36} = P(A)P(C)$$

$$P(B \cap C) = \frac{1}{36} = P(B)P(C)$$

We see that any two of these events are independent. On the other hand,

$$A \cap B \cap C = \varnothing$$

so that $P(A \cap B \cap C) = 0$, and it follows that the three events are not independent.

4. COMPOUND EXPERIMENTS

In the last few examples, we have assumed that we are given the joint probabilities of the events directly, i.e., that we have been given the probabilities of each of the outcomes. In general, however, this is not so: the problem usually involves computing the probabilities of the outcomes.

Let us now, for instance, consider the problem of the Chevalier de Méré. The Chevalier's original bet was that he would obtain at least one six in four tosses of a die. We can find the probability of this event by considering the complementary event, "no sixes are obtained." Assuming the die to be fair, the probability of *not* obtaining a six on a given toss is 5/6; since the four tosses are independent, we must multiply four such factors, i.e.,

$$P(\text{no sixes are obtained}) = \left(\frac{5}{6}\right)^4 = \frac{625}{1296}$$

And so the Chevalier had a probability of $1 - 625/1296 = 671/1296$, or slightly more than 1/2, of winning. He had an advantage: in the long run, he would win more often than not.

Consider now the Chevalier's second bet: the probability of *not* obtaining a double six on any given toss of the dice, assuming the dice to be fair, is 35/36. For 24 independent tosses, we have

$$P(\text{no double sixes are obtained}) = \left(\frac{35}{36}\right)^{24}$$

This probability, computed by using logarithms, is, approximately 0.512. We conclude that de Méré had only a probability 0.488 of winning. In the long run, he should lose somewhat more often than he won—a prediction verified by his experience. The two bets were by no means equivalent. Some mathematical analysis would have saved the Chevalier a great sum of money—but, as Pascal pointed out, de Méré was no mathematician.

V.4.1 Example. A test consists of five multiple-choice questions, each one giving a choice of four possible answers. A student who has not read the subject matter guesses at each answer. What is the probability that he will answer all five questions correctly?

In a sense, this problem is not entirely well defined. The difficulty is that, even though a student has not read the subject matter, the correct answer may yet be somewhat more logical than any of the others, so that an "educated guess" gives a greater probability of success than a wild guess. It may also be that the five questions are logically related, so that the student's success on one of them is not independent of success on the others. Notwithstanding these possibilities, we shall assume that the student has a 1/4 probability of success on each question (since there are four answers, of which only one is correct), and that the several questions are independent. Letting S_i be the event "success on the ith question," we have

$$P(S_i) = \frac{1}{4} \qquad \text{for } i = 1, 2, 3, 4, 5$$

We are looking for the probability of success on all the questions, i.e., for $P(S_1 \cap S_2 \cap S_3 \cap S_4 \cap S_5)$. Since the questions are assumed independent, we have, by (5.3.4),

$$P(S_1 \cap S_2 \cap S_3 \cap S_4 \cap S_5) = \left(\frac{1}{4}\right)^5 = \frac{1}{1024}$$

When events are not independent, we are quite often given their conditional probabilities. In such cases, we find joint probabilities by rewriting (5.3.1) in the form

5.4.1 $$P(A \cap B) = P(A)P(B \mid A)$$

V.4.2 *Example.* Three identical urns are labeled A, B, C. Urn A contains 5 red and 7 black balls. Urn B contains 1 red and 5 black balls. Urn C contains 12 red and 3 black balls. An urn is chosen at random, and then a ball is taken at random from the urn. What is the probability that the ball will be red?

Two assumptions are made here: the first is that each urn has the same probability, 1/3, of being chosen. The second is that, from a given urn, each of the balls has the same probability of being chosen. Thus, the conditional probability of a red ball, given that urn A is chosen, will be 5/12 (since 5 of the 12 balls in the urn are red); it is similar for the other urns. Letting R be the event "a red ball is chosen," we find that

$$P(A) = P(B) = P(C) = \frac{1}{3}$$

$$P(R \mid A) = \frac{5}{12}$$

$$P(R \mid B) = \frac{1}{6}$$

$$P(R \mid C) = \frac{4}{5}$$

The event R can be written as the union of disjoint events

$$R = (R \cap A) \cup (R \cap B) \cup (R \cap C)$$

and so, by (5.2.6),

$$P(R) = P(R \cap A) + P(R \cap B) + P(R \cap C)$$

Now, by (5.4.1),

$$P(R \cap A) = P(A)P(R \mid A) = \frac{1}{3} \cdot \frac{5}{12} = \frac{5}{36}$$

$$P(R \cap B) = P(B)P(R \mid B) = \frac{1}{3} \cdot \frac{1}{6} = \frac{1}{18}$$

$$P(R \cap C) = P(C)P(R \mid C) = \frac{1}{3} \cdot \frac{4}{5} = \frac{4}{15}$$

so that

$$P(R) = \frac{5}{36} + \frac{1}{18} + \frac{4}{15} = \frac{83}{180}$$

V.4.3 Example. A lot of 20 flashbulbs contains 17 good and 3 defective items. A sample of size 3 is taken. What is the probability that all 3 will be good?

We assume that each item in the lot has the same probability of being chosen. Let I be the event "the first item chosen is good," II be the event "the second item is good," and III be the event "the third item is good." We will apply (5.4.1) twice:

$$P(I \cap II \cap III) = P(I \cap II)P(III \mid I \cap II) = P(I)P(II \mid I)P(III \mid I \cap II)$$

Since 3 of the 20 bulbs are defective, we have, clearly,

$$P(I) = \frac{17}{20}$$

Suppose that the first item is good. In this case, there will be only 19 bulbs left, including 16 good ones, and

$$P(II \mid I) = \frac{16}{19}$$

Similarly, if the first two bulbs chosen are good, there will be 15 good bulbs among the 18 left, and so

$$P(III \mid I \cap II) = \frac{15}{18}$$

Thus,

$$P(I \cap II \cap III) = \frac{17}{20} \cdot \frac{16}{19} \cdot \frac{15}{18} = \frac{34}{57}$$

V.4.4 Example. Find the probability that, from a group of 24 people, no 2 have the same birthday.

For this problem, we shall assume that each day is as likely to be a birthday as any other. Let A_i be the event "the ith person does not have the same birthday as any of the first $i - 1$." Then we are looking for the probability of the event $A_1 \cap A_2 \ldots \cap A_n$.

Suppose, now, that all the events $A_1, A_2, \ldots, A_{i-1}$ occur. This means that no two of the first $i - 1$ persons have a common birthday; they occupy $i - 1$ different days. For event A_i to occur, the ith person must then be born in one of the remaining $365 - (i-1)$ days. Thus,

$$P(A_i \mid A_1 \cap \ldots \cap A_{i-1}) = \frac{366 - i}{365}$$

Applying rule (5.3.4), now,

$$P(A_1 \cap \ldots \cap A_n) = \frac{365}{365} \cdot \frac{364}{365} \cdot \ldots \cdot \frac{366 - n}{365}$$

For $n = 24$, we have

$$P(A_1 \cap \ldots \cap A_{24}) = \frac{365}{365} \cdot \frac{364}{365} \cdot \ldots \cdot \frac{342}{365} = 0.46$$

We see, thus, that the probability that no 2 have the same birthday is *less than 1/2!* This means that, from 24 people, 2 will have a common birthday more often than not—a strangely unintuitive result, since, indeed, 24 is much smaller than 365.

PROBLEMS ON CONDITIONAL PROBABILITIES

1. A pair of fair dice is tossed. What is the probability that one of the dice show three spots, given that the sum of the spots is 5?

2. A fair coin is tossed three times. What is the probability that the first toss will be H, given that there are at least two H's thrown?

3. A test consists of five multiple choice questions; each question has three possible answers. A student guesses at all the answers; what is the probability that he will obtain a grade of 20 per cent or better? of 100 per cent?

4. Three otherwise identical urns are labeled A, B, and C. Urn A contains 5 red and 6 black balls. Urn B contains 3 red and 1 black ball. Urn C contains 10 red and 20 black balls. An urn is chosen at random (probability 1/3 each) and then a ball is taken from the urn. What is the probability that it will be red?

5. A chewing gum company has 30 different baseball cards, and

gives 1 with each pack of gum. A boy buys 10 packs of gum. What is the probability that he will obtain 10 different cards?

6. A machine makes bolts according to specifications. When it is working well, the bolts meet the specifications with probability 0.98. However, there is a probability 0.35 that the machine will be working poorly, in which case the bolts will meet specifications with probability 0.40. What is the probability that a given bolt will meet the specifications?

7. The game of *craps* is played in the following manner: a player tosses a pair of dice. If the number obtained is 2, 3, or 12, he loses immediately; if it is 7 or 11, he wins immediately. If any other number is obtained on the first toss, then that number becomes the player's "point," and he must keep on tossing the dice until either he "makes his point" (i.e., obtains the first number again), in which case he wins; or he obtains 7, in which case he loses. Find the probability of winning.

8. A machine makes pieces to fit specifications; the probability that a given piece will be suitable is 0.95. A sample of 10 pieces is taken from the machine's production. What is the probability that all 10 pieces will be good?

9. A traveling salesman assigns 0.3 as the probability of making a sale at each call. How many calls must he make in order that the probability of making at least one sale be 0.9 or better?

10. Three urns are labeled A, B, and C. Urn A has four red and two black balls. Urn B has five red and seven black balls; urn C has two red and six black balls. An urn is chosen at random and a ball is taken from that urn. What is the probability that the ball will be black?

A very common application of conditional probabilities deals with what is known as *a priori* and *a posteriori* probabilities.

As an illustration of this type of problem, let us consider the experiment of Example V.4.2. We know that each of the three urns has the same probability, 1/3, of being chosen. Suppose that the three urns are indistinguishable; we could, nevertheless, find out which is which by the simple expedient of looking inside the urn (the number of balls of each color will determine the urn, i.e., if we find that it has 12 red and 3 black balls then we know that it must be urn C). Unfortunately, it may not be possible to look at all the balls in the urn; the question is whether seeing just one of them will give us any information. In fact, this generally does happen. An extreme example would be the case in which urns A and B had no red balls; a red ball would then, necessarily, be drawn from urn C. In our example, the situation is not so extreme, but still the contents of the three urns are sufficiently different so that some new information is gained. Since urn C has such a high proportion of red balls (12 out of 15), a red ball would most likely be drawn from urn C. In fact, we have

$$P(R \cap C) = P(C)P(R \mid C) = \frac{4}{15}$$

and

$$P(R) = \frac{83}{180}$$

Therefore

$$P(C \mid R) = \frac{4/15}{83/180} = \frac{48}{83}$$

so that the *conditional* probability $P(C \mid R)$ is considerably higher than the absolute probability $P(C)$.

In general, it is assumed that the outcomes of an experiment can be classified in two ways (as in our example, by urn and by color). This gives us two partitions of the outcome space:

$$X = A_1 \cup A_2 \cup \ldots \cup A_m$$
$$X = B_1 \cup B_2 \cup \ldots B_n$$

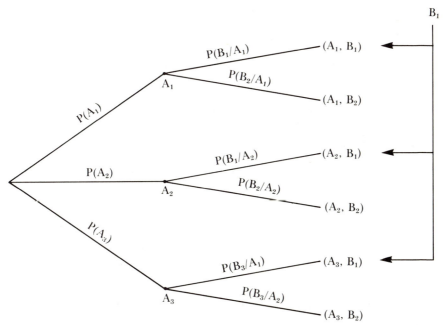

FIGURE V.4.1 Scheme of two-classification experiment.

It is assumed that, for the first classification, the *absolute* probabilities $P(A_i)$ are known. For the second partition, it is the *conditional* probabilities $P(B_j | A_i)$ that are known. Now the nature of the experiment is such that it is easy to observe the value of the B_j, whereas the value of the A_i is either impossible or extremely difficult to ascertain. The idea is then to use the observed value B_j to obtain a new (conditional) probability $P(A_i | B_j)$. This new distribution of the A_i is known as the *a posteriori* distribution, while the previous (absolute) distribution is known as the *a priori* distribution. In our example, each of the three urns had an *a priori* probability 1/3 of being chosen. After the event, i.e., assuming that a red ball has been drawn, we conclude that urn C has a considerably larger *a posteriori* probability, 48/83.

In general, we wish to compute $P(A_i | B_j)$. Now, by Definition V.3.1,

$$P(A_i | B_j) = \frac{P(A_i \cap B_j)}{P(B_j)}$$

However, we are given neither $P(A_i \cap B_j)$, nor $P(B_j)$; these must be computed. The numerator presents no trouble; it is given by

$$P(A_i \cap B_j) = P(A_i) P(B_j | A_i)$$

The denominator, on the other hand, requires some work. To simplify things somewhat, let us assume that $m = 2$, i.e., the A classification can give only two values, A_1 and A_2. We can write

$$B_j = (A_1 \cap B_j) \cup (A_2 \cap B_j)$$

Since A_1 and A_2 are disjoint, it follows that $A_1 \cap B_j$ and $A_2 \cap B_j$ are also disjoint. Then

$$P(B_j) = P(A_1 \cap B_j) + P(A_2 \cap B_j)$$

and, by (5.4.1),

$$P(B_j) = P(A_1) P(B_j | A_1) + P(A_2) P(B_j | A_2)$$

For the special case in which $m = 2$, then, we have the formula

$$P(A_i | B_j) = \frac{P(A_i) P(B_j | A_i)}{P(A_1) P(B_j | A_1) + P(A_2) P(B_j | A_2)}$$

In the more general case, the formula for $P(A_i | B_j)$ is derived analogously; it is

5.4.2
$$P(A_i \mid B_j) = \frac{P(A_i) P(B_j \mid A_i)}{\sum\limits_{k=1}^{m} P(A_k) P(B_j \mid A_k)}$$

Equation (5.4.2) is known as *Bayes' formula.*

V.4.5 Example. Of all applicants for a job, it is felt that 75 per cent are able to do the job, and 25 per cent are not. To aid in the selection process, an aptitude test is designed such that a capable applicant has probability 0.8 of passing the test, while an incapable applicant has probability 0.4 of passing the test. An applicant passes the test. What is the probability that he will be able to do the job?

This is a straightforward application of Bayes' formula. Let A_1 be the event "the applicant is able to do the job," and A_2 the complementary event. Let B_1 be the event "the applicant passes the test," and B_2 the complement. Then

$$P(A_1) = 0.75 \qquad P(A_2) = 0.25$$
$$P(B_1 \mid A_1) = 0.8 \qquad P(B_1 \mid A_2) = 0.4$$

By Bayes' formula,

$$P(A_1 \mid B_1) = \frac{(0.75)(0.8)}{(0.75)(0.8) + (0.25)(0.4)} = \frac{0.6}{0.6 + 0.1} = \frac{6}{7}$$

The test is a reasonably effective screening device in this case; only 1/7 of the persons passing the test will be incapable.

We may also check for the probability $P(A_1 \mid B_2)$, i.e., the probability that an applicant who fails the test would still be able to do the job. We have

$$P(B_2 \mid A_1) = 0.2 \qquad P(B_2 \mid A_2) = 0.6$$

and so

$$P(A_1 \mid B_2) = \frac{(0.75)(0.2)}{(0.75)(0.2) + (0.25)(0.6)} = \frac{0.15}{0.15 + 0.15} = \frac{1}{2}$$

Thus, half the rejected applicants are able to do the work. Note that the test is not infallible: incapable applicants may pass; capable applicants may fail.

V.4.6 Example. A machine produces bolts to meet certain specifications. When the machine is in good working order, 90 per cent of the bolts meet the specifications. Occasionally, however, the machine goes out of order, and only 50 per cent of the bolts produced then meet the specifications. It is calculated that the machine is in good order 80 per cent of the time. A sample of two bolts is taken from the machine's current production, and both are found to meet the specifications. What is the probability that the machine is out of order?

Once again, we must use Bayes' formula. Let A_1 be the event "the machine is in good order," and A_2 the complementary event. Let B_1 be the event "both bolts are good." We know that

$$P(A_1) = 0.8 \qquad P(A_2) = 0.2$$

Next, we must compute $P(B_1 \mid A_i)$. If the machine is working well, the probability that a bolt will meet the specifications is 0.9. The probability that both bolts meet the specifications is, then, $(0.9)^2 = 0.81$:

$$P(B_1 \mid A_1) = 0.81$$

and similarly, $(0.5)^2 = 0.25$, so

$$P(B_1 \mid A_2) = 0.25$$

Thus,

$$P(A_1 \mid B_1) = \frac{(0.8)(0.81)}{(0.8)(0.81) + (0.2)(0.25)} = \frac{0.648}{0.648 + 0.05} = 0.93$$

V.4.7 Example. Under the same conditions as in Example V.4.6, a sample of two bolts is taken and both are found defective. What is the probability that the machine is out of order?

Let B_2 represent the event "both bolts are defective." As in the previous example, we compute

$$P(B_2 \mid A_1) = (0.1)^2 = 0.01$$
$$P(B_2 \mid A_2) = (0.5)^2 = 0.25$$

Applying Bayes' law, we have

$$P(A_2 \mid B_2) = \frac{(0.2)(0.25)}{(0.8)(0.01) + (0.2)(0.25)} = \frac{0.05}{0.008 + 0.05} = 0.86$$

V.4.8 Example. A person is either suffering from tuberculosis (with probability 0.01) or not. To test for this disease, chest x-rays are taken. If the person is ill, the test will be positive with probability 0.999; if he is well, the test will be positive with probability 0.2. A person takes the x-ray test, which gives a positive result. What is the probability that the person is well?

Again, let A_1 be the event "tuberculosis," A_2 the complementary event. Let B_1 be the event "positive." We have

$$P(A_1) = 0.01 \qquad P(A_2) = 0.99$$
$$P(B_1 \mid A_1) = 0.999 \qquad P(B_1 \mid A_2) = 0.2$$

then

$$P(A_2 \mid B_1) = \frac{(0.2)(0.99)}{(0.01)(0.999) + (0.2)(0.99)} = \frac{0.198}{0.00999 + 0.198} = 0.95$$

Note that even with a positive test result, the person tested is still more likely to be well than ill. The reason for this is simply that it is considered better for a healthy patient to undergo further examinations than for a sick man to go undetected. The x-ray test is made purposely difficult to pass, i.e., the smallest dark spot is considered a positive result.

PROBLEMS ON COMPOUND EXPERIMENTS

1. A company has a position open; the probability that an applicant be capable of doing the required work is 0.65. The firm's personnel department devises an aptitude test that a capable applicant will pass with probability 0.8, whereas an incapable applicant will pass with probability 0.4. An applicant passes the test; what is the probability that he will be able to do the work?

2. Three identical urns are labeled A, B, and C. Urn A contains 5 red and 3 black balls. Urn B contains 6 red and 10 black balls; urn C contains 12 red and 4 black balls. An urn is chosen at random, and then a ball is taken from the urn, also at random. If the ball is red, what is the probability that the urn was urn B?

3. A person is either well (probability 0.9) or ill (probability 0.1). If he is well, his temperature will be normal with probability 0.95, whereas if he is ill, his temperature will be normal with probability 0.35.

  (a) If a person's temperature is normal, what is the probability that he is ill?

  (b) If the temperature is not normal, what is the probability that he is well?

4. A machine makes bolts to fit specifications. When the machine is working well (probability 0.8) the bolts fit specifications with probability 0.95; when it is not working well, the bolts fit specifications with probability 0.5. Two bolts taken from the machine's production are both defective. What is the probability that the machine is working well?

5. A contractor must build a road through a piece of ground. The soil is either clay (probability 0.68) or rock. If it is rock, a geological test will give a positive result with probability 0.75; if it is clay, the same test will give a positive result with probability 0.25. Given that the test shows a positive result, what is the probability that the soil is rock?

5. REPETITION OF SIMPLE EXPERIMENTS: THE BINOMIAL DISTRIBUTION

Let us consider a very elementary type of experiment: one with only two outcomes. These outcomes are generally called *success* (S) and *failure* (F). Such an experiment is called a *simple experiment*. It may be described entirely in terms of the single number p, where

5.5.1 $$p = P(S)$$

Since the probability of success is p, it follows that the probability of failure is $1 - p$; we write

5.5.2 $$q = 1 - p = P(F)$$

For a single trial of this experiment, there is very little else to say. Suppose, however, that the experiment is repeated many times. We will obtain a *run* of successes and failures, which may look like this:

SFFSSFFSSFSFFSSS

We are interested, generally, not so much in the exact run of successes and failures, but rather, in the number of successes obtained.

Consider the run just given. The probability of each success is p; that of each failure is q. If we assume (as we shall) that the trials are independent, then the probability of this run is

pqqppqqppqpqqppp

obtained by multiplying the probabilities for each trial. This product may, of course, be rewritten as $p^9 q^7$. Note that any other run of nine successes and seven failures would have the same probability. In general, we find that, in n independent trials of a simple experiment, any run that contains k successes and $n - k$ failures has probability $p^k q^{n-k}$. To find the probability of obtaining exactly k successes in n trials, we must find the number of runs having k successes. This number is obtained by noticing that the k successes can appear in any k of the n trials; thus, the number of such runs must be equal to the number of combinations of k elements that can be taken from a set of n elements, i.e., $\binom{n}{k}$. Multiplying this number by the probability of each of the runs, we obtain the *binomial coefficient*

5.5.3
$$B(n,k;p) = \binom{n}{k} p^k q^{n-k}$$

This expression is also known as a *Bernoulli* coefficient, after Jacob Bernoulli (1654-1705), who analyzed the problem. A sequence of independent trials is also known as a sequence of Bernoulli trials.

V.5.1 *Example.* A fair coin is tossed 10 times. What is the probability of obtaining exactly seven heads?

The coin is fair, so $p = 1/2$. According to formula (5.5.3), the probability of this event is

$$B\left(10,7;\frac{1}{2}\right) = \frac{10!}{3!\,7!} \left(\frac{1}{2}\right)^7 \left(\frac{1}{2}\right)^3 = \frac{120}{1024}$$

V.5.2 *Example.* A fair die is tossed five times. What is the probability of obtaining three aces?

In this case, we have $p = 1/6$. Hence

$$B\left(5,3;\frac{1}{6}\right) = \frac{5!}{3!\,2!} \left(\frac{1}{6}\right)^3 \left(\frac{5}{6}\right)^2 = \frac{250}{7776}$$

V.5.3 *Example.* A machine produces bolts to meet certain specifications; 90 per cent of the bolts produced meet those specifications. A sample of five bolts is taken from the machine's production. What is the probability that two or more of these fail to meet the specifications?

In this problem we are looking for the event "at least two fail to meet specifications," or more concisely, "at least 2 failures." The formula that we have developed, however, will only give the probability of exactly k successes (or failures). Thus we must divide the event into the disjoint events "2 failures," "3 failures," "4 failures," and "5 failures." We have $n = 5$, and $p = 0.9$, and so the relevant probabilities are

$$P(2\text{ failures}) = B(5,3;0.9) = \frac{5!}{3!\,2!} (0.9)^3 (0.1)^2 = 0.0729$$

$$P(3\text{ failures}) = B(5,2;0.9) = \frac{5!}{2!\,3!} (0.9)^2 (0.1)^3 = 0.0081$$

$$P(4 \text{ failures}) = B(5,1;0.9) = \frac{5!}{1!\,4!}(0.9)(0.1)^4 = 0.00045$$

$$P(5 \text{ failures}) = B(5,0;0.9) = \frac{5!}{0!\,5!}(0.1)^5 = 0.00001$$

and therefore

$$P(\text{at least 2 failures}) = 0.0729 + 0.0081 + 0.00045 + 0.00001 = 0.08146$$

This same problem could be solved by considering the complementary event, "at most 1 failure." This can be written as the union of the two events "1 failure" and "no failures." We have

$$P(1 \text{ failure}) = B(5,4;0.9) = \frac{5!}{4!\,1!}(0.9)^4(0.1) = 0.32805$$

$$P(\text{no failures}) = B(5,5;0.9) = \frac{5!}{5!\,0!}(0.9)^5 = 0.59049$$

so that

$$P(\text{at most 1 failure}) = 0.32805 + 0.59049 = 0.91854$$

Since the events "at most 1 failure" and "at least 2 failures" are complementary, we have

$$P(\text{at least 2 failures}) = 1 - 0.91854 = 0.08146$$

Note that, as expected, we obtain the same answer both ways.

V.5.4 *Example.* In a large city, 60 per cent of the heads of household are Democrats. A poll taker visits eight houses, and asks the party affiliation of the head of household. What is the probability that at least six are Democrats?

Because the city is large, it may reasonably be assumed that the probabilities remain equal at each trial, i.e., the trials are independent. The probability of *at least six Democrats* is then

$$B(8,6;0.6) + B(8,7;0.6) + B(8,8;0.6)$$

Now

$$B(8,6;0.6) = \frac{8!}{6!\,2!}(0.6)^6(0.4)^2 = 0.209$$

$$B(8,7;0.6) = \frac{8!}{7!\,1!}(0.6)^7(0.4) = 0.089$$

$$B(8,8;0.6) = \frac{8!}{8!\,0!}(0.6)^8 = 0.017$$

so that

$$P(\text{at least 6 Democrats}) = 0.315$$

6. DRAWINGS WITH AND WITHOUT REPLACEMENT

Let us now consider the following problem:

An urn contains 10 red and 4 black balls. Three balls are drawn from the urn. What is the probability that exactly two of the balls drawn will be red?

In a problem such as this, in which elements are taken from a finite set (drawings from a finite population), there are two possible and very distinct methods of carrying out the experiment. The experimenter may, if he wishes, take a ball from the urn, record its color, and *replace it* before drawing the next ball. This method is known as *drawing with replacement.* Alternatively, the experimenter may draw the three balls, simultaneously or one at a time, without replacing them. This method is known as *drawing without replacement.* It is clear that the two methods will give different results; for instance, if the urn contains only two balls, it will be impossible even to make three drawings without replacement, whereas drawings with replacement can always be made.

Suppose that the drawings are made *with* replacement. In this case, it is not too difficult to see that, at each of the drawings, the urn contains the original number of red and black balls. The probability of obtaining a given colored ball will be the same at each drawing, regardless of what might have happened in previous drawings, i.e., the drawings are independent. It follows that such drawings are, in effect, Bernoulli trials. Thus, for drawings *with replacement,* the solution to the problem is as follows. Since 10 of the 14 balls are red, we have $p = 5/7$. Then the probability of 2 red balls is

$$B\left(3,2;\frac{5}{7}\right) = \frac{3!}{2!\,1!}\left(\frac{5}{7}\right)^2\left(\frac{2}{7}\right) = \frac{150}{343}$$

Let us consider, now, the case of drawings *without* replacement. For this method, it is clear that the drawings are no longer independent, and so they cannot be treated as Bernoulli trials. A new analysis of the problem is therefore necessary.

We analyze the problem of *drawings without replacement* by assuming that all the possible samples of three balls that can be drawn from the urn have the same probability. It is then merely a matter of seeing how many such samples there are, and how many of those belong to the desired event (i.e., satisfy the requirement of containing two red and one black ball). The desired probability is then the quotient of these two numbers.

In fact, the number of possible samples is equal to the number of combinations of 3 elements from a set of 14 elements, i.e., the binomial coefficient $\binom{14}{3}$. In turn, the number of samples containing two red balls and one black ball is obtained by multiplying the number of ways in which two red balls can be taken by the number of ways in which one black ball can be taken. These are, respectively, the number of combinations of 2 elements from a set of 10 elements, and the number of combinations of 1 element from a set of 4 elements, i.e., $\binom{10}{2}$ and $\binom{4}{1}$. There are, then, $\binom{14}{3}$ possible samples, of which $\binom{10}{2}\binom{4}{1}$ belong to our event. Thus

$$P(\text{exactly 2 red balls}) = \frac{\binom{10}{2}\binom{4}{1}}{\binom{14}{3}} = \frac{\frac{10!\,4!}{2!\,8!\,1!\,3!}}{\frac{14!}{3!\,11!}}$$

$$= \frac{144}{364} = \frac{36}{91}$$

Note that the probability in this case is slightly less than in the case of drawings with replacement.

In general, we may postulate an urn containing m balls, of which m_1 are red and $m_2 = m - m_1$ are black. A sample of n balls is taken (without replacement) from the urn; we wish to know the probability that exactly k of them are red.

There are, in all, $\binom{m}{n}$ combinations of n elements from a set of m elements. This is the number of possible samples. Of these, we consider those that have exactly k red and $n - k$ black balls. There are $\binom{m_1}{k}$ combinations of k red balls, and $\binom{m_2}{n-k}$ combinations of $n - k$ black balls. The product of these two numbers will be the number of samples in the event. Thus,

5.6.1 $$P(k \text{ red balls}) = \frac{\binom{m_1}{k}\binom{m_2}{n-k}}{\binom{m}{n}}$$

V.6.1 Example.   A poker hand consists of 5 cards, taken at random from a deck of 52 (which is divided into 4 suits of 13 cards each). What is the probability that all the cards in a hand will be of the same suit (a hand known as a flush)?

We can best solve this problem by considering the probability that all five cards in the hand are, say, spades. The probability of a flush will be four times this, since the flush can be in any one of the four suits.

It is easy to see that this is a problem of sampling without replacement. We will have $m = 52$, $m_1 = 13$ (the number of spades), and $m_2 = 39$. Also, $n = k = 5$.

$$P(\text{flush in spades}) = \frac{\binom{13}{5}\binom{39}{0}}{\binom{52}{5}} = \frac{\frac{13!\,39!}{5!\,8!\,0!\,39!}}{\frac{52!}{5!\,47!}} = \frac{33}{66640}$$

and, therefore

$$P(\text{flush}) = \frac{4 \cdot 33}{66640} = \frac{33}{16660}$$

The probability of a flush is slightly less than 1 in 500. (This probability is not exactly the same as that given in poker primers because we include straight flushes among the flushes.)

V.6.2 Example. A company tests a lot of 50 flash-bulbs by taking a sample of 10. If the sample contains 2 or more duds, the lot is rejected. Given that a lot contains 13 duds, what is the probability that it will be rejected?

As in Example V.6.1, we shall do this by considering the complementary event, "at most one dud." This is, in turn, divided into the two events, "no duds," and "exactly one dud." Now

$$P(\text{no duds}) = \frac{\binom{37}{10}\binom{13}{0}}{\binom{50}{10}} = 0.033$$

$$P(\text{1 dud}) = \frac{\binom{37}{9}\binom{13}{1}}{\binom{50}{10}} = 0.153$$

and so the probability that the lot will be rejected is

$$1 - 0.033 - 0.153 = 0.814$$

V.6.3 Example. A small town has 20 houses. In 12 of these, the head of household is a Democrat. A poll taker visits 8 houses. What is

THE THEORY OF PROBABILITY

FINITE MATHEMATICS

234 the probability that in at least 6, the head of household be a Democrat?
(Compare Example V.5.4.)

This is clearly an example of sampling without replacement.

We have

$$P(6 \text{ Democrats}) = \frac{\binom{12}{6}\binom{8}{2}}{\binom{20}{8}} = 0.205$$

$$P(7 \text{ Democrats}) = \frac{\binom{12}{7}\binom{8}{1}}{\binom{20}{8}} = 0.050$$

$$P(8 \text{ Democrats}) = \frac{\binom{12}{8}\binom{8}{0}}{\binom{20}{8}} = 0.004$$

so that

$$P(\text{at least 6 Democrats}) = 0.259$$

PROBLEMS ON DRAWINGS WITH AND WITHOUT REPLACEMENT

1. A fair coin is tossed 12 times. What is the probability of obtaining:

(a) Exactly six heads?

(b) At least eight heads?

(c) Between five and seven heads?

2. A fair die is tossed six times. What is the probability of obtaining:

(a) Exactly two aces?

(b) At least three aces?

3. A multiple-choice test consists of eight questions, each one having three possible answers. What should the passing mark be to ensure that someone who guesses at the answers will have a probability smaller than 0.05 of passing?

4. A machine produces pieces to meet specifications; the probability that a given piece be suitable is 0.95. A sample of 10 pieces is taken. What is the probability that 2 or more be defective?

5. Two baseball teams meet in the world series; team A, the stronger team, has probability 0.6 of winning in each game. What is the probability that team A will win the series (i.e., win at least four of seven games)?

6. A package of flash bulbs contains 16 good and 4 defective bulbs. A sample of 5 bulbs is taken. What is the probability that at least three of the bulbs will be good?

7. An urn contains 15 red and 10 black balls. Three balls are taken from the urn; what is the probability that 2 of these be red?

8. A lot of 16 tires contains 13 good and 3 defective tires. A sample of 6 is taken. What is the probability that it contains at least 4 good tires?

9. Lots of 20 elements are tested in the following manner: a sample of size 4 is taken, and the number of defective elements in this sample is called x. If $x = 0$, the lot is accepted, if $x \geq 3$, it is rejected. If $x = 1$ or 2, a further sample of size 3 is taken and the number of defectives in this lot is called y. Then the lot is accepted if $x + y \leq 3$, and rejected if $x + y \geq 4$. What is the probability that a lot will be accepted if:

(a) It contains 15 good and 5 defective elements?

(b) It contains 12 good and 8 defective?

(c) It contains 17 good and 3 defective?

10. A simple experiment, with probability of success p, is repeated until m successes are obtained. Show that the probability of obtaining the mth success on the kth trial is given by the *negative binomial distribution*

$$P_m(k) = \binom{k-1}{m-1} p^m (1-p)^{k-m}$$

236   7.   RANDOM VARIABLES

Very often there is a numerical quantity associated with each outcome of an experiment. It may represent money, as in the case of a gamble, or it may represent some other measurable quantity—the strength of a rope, the number of games won by a baseball team, or the life of a light bulb. Such a numerical quantity is known as a *random variable*. We give a precise definition.

V.7.1   Definition.   A *random variable* is a function whose domain is the set of outcomes of an experiment and whose range is a subset of the real numbers.

We shall generally denote a random variable by a capital letter: X, Y, and so on. The values of the random variable will be denoted by lower-case letters: x, y, and so on.

Since, by its very nature, a random variable takes its values according to a randomization scheme, it becomes natural to look for the probability of each of these values. These probabilities make up what is known as the *distribution* of the random variable.

Associated with the random variable X, there are two functions that describe its distribution. The first is the *probability function*, $f(x)$, defined by

5.7.1 $$f(x) = P(X = x)$$

The second is the *cumulative distribution function*, $F(x)$, defined by

5.7.2 $$F(x) = P(X \leq x)$$

Thus f gives the probabilities of each of the several possible values of X, while F gives the probability that X will be smaller than or equal to a given number. We shall, almost invariably, deal with the probability function, f, although the cumulative distribution function will be of use on some occasions.

V.7.2   *Example.*   Let X be the number of spots appearing on one toss of a fair die. Then X has the probability function

$$f(x) = \begin{cases} \dfrac{1}{6} & \text{for } x = 1,2,\ldots,6 \\ 0 & \text{otherwise} \end{cases}$$

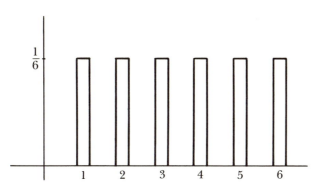

FIGURE V.7.1   Probabilities on the toss of one die.

and cumulative distribution function

$$F(x) = \begin{cases} 0 \text{ for } x < 1 \\[2mm] \dfrac{1}{6} \text{ for } 1 \le x < 2 \\[2mm] \dfrac{1}{3} \text{ for } 2 \le x < 3 \\[2mm] \dfrac{1}{2} \text{ for } 3 \le x < 4 \\[2mm] \dfrac{2}{3} \text{ for } 4 \le x < 5 \\[2mm] \dfrac{5}{6} \text{ for } 5 \le x < 6 \\[2mm] 1 \text{ for } \qquad x \ge 6 \end{cases}$$

See Figures V.7.1 and V.7.2 for these two functions.

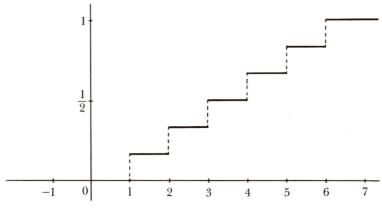

FIGURE V.7.2   Cumulative distribution for the toss of one die.

FINITE MATHEMATICS

238

V.7.3 Example. Let Y be the number of spots appearing on two independent tosses of a fair die. The distribution of Y may be found by studying the table in Example V.2.7; it is

$$f(2) = \frac{1}{36}$$

$$f(3) = \frac{2}{36} = \frac{1}{18}$$

$$f(4) = \frac{3}{36} = \frac{1}{12}$$

$$f(5) = \frac{4}{36} = \frac{1}{9}$$

$$f(6) = \frac{5}{36}$$

$$f(7) = \frac{6}{36} = \frac{1}{6}$$

$$f(8) = \frac{5}{36}$$

$$f(9) = \frac{4}{36} = \frac{1}{9}$$

$$f(10) = \frac{3}{36} = \frac{1}{12}$$

$$f(11) = \frac{2}{36} = \frac{1}{18}$$

$$f(12) = \frac{1}{36}$$

and $f(x) = 0$ for all other x (see Figure V.7.3).

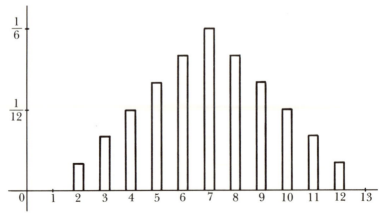

FIGURE V.7.3 Probabilities for toss of two dice.

V.7.4 Example. Let X be the number of heads obtained in 10 independent tosses of a fair coin. Its distribution is obtained by using formula (6.4.3) with $n = 10$, $p = 1/2$.

$$f(0) = \binom{10}{0}\left(\frac{1}{2}\right)^{10} = \frac{1}{1024}$$

$$f(1) = \binom{10}{1}\left(\frac{1}{2}\right)\left(\frac{1}{2}\right)^{9} = \frac{10}{1024}$$

$$f(2) = \binom{10}{2}\left(\frac{1}{2}\right)^{2}\left(\frac{1}{2}\right)^{8} = \frac{45}{1024}$$

$$f(3) = \binom{10}{3}\left(\frac{1}{2}\right)^{3}\left(\frac{1}{2}\right)^{7} = \frac{120}{1024}$$

$$f(4) = \binom{10}{4}\left(\frac{1}{2}\right)^{4}\left(\frac{1}{2}\right)^{6} = \frac{210}{1024}$$

$$f(5) = \binom{10}{5}\left(\frac{1}{2}\right)^{5}\left(\frac{1}{2}\right)^{5} = \frac{252}{1024}$$

$$f(6) = \binom{10}{6}\left(\frac{1}{2}\right)^{6}\left(\frac{1}{2}\right)^{4} = \frac{210}{1024}$$

$$f(7) = \binom{10}{7}\left(\frac{1}{2}\right)^{7}\left(\frac{1}{2}\right)^{3} = \frac{120}{1024}$$

$$f(8) = \binom{10}{8}\left(\frac{1}{2}\right)^{8}\left(\frac{1}{2}\right)^{2} = \frac{45}{1024}$$

$$f(9) = \binom{10}{9}\left(\frac{1}{2}\right)^{9}\left(\frac{1}{2}\right) = \frac{10}{1024}$$

$$f(10) = \binom{10}{10}\left(\frac{1}{2}\right)^{10} = \frac{1}{1024}$$

In Figure V.7.4, note that while 5 is the most probable value of the random variable, its probability is still less than 1/4. On the other hand, there is approximately a 0.66 probability that the variable will lie between 4 and 6.

If X and Y are random variables that depend on the outcome of the same experiment, we consider their *joint distribution*, given by the function f:

5.7.3 $$f(x,y) = P(X = x, Y = y)$$

Thus in the experiment of tossing a fair die twice, we may let X and Y be the number of spots showing on the first and second tosses of the die, respectively. Then the joint probability of X and Y is given by

$$f(x,y) = \begin{cases} \dfrac{1}{36} & \text{for } x,y = 1, 2, \ldots, 6 \\ 0 & \text{otherwise} \end{cases}$$

FINITE MATHEMATICS

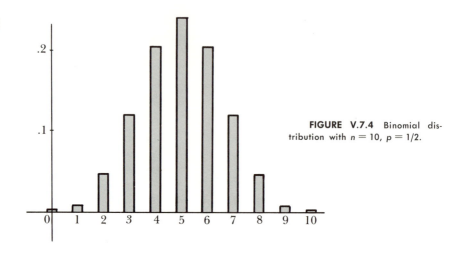

FIGURE V.7.4 Binomial distribution with $n = 10$, $p = 1/2$.

The *absolute* distribution of X may be obtained from the joint distribution of X and Y; it will be given by

5.7.4
$$g(x) = \sum_j f(x, y_j)$$

where the summation is taken over all the possible values, y_j, of the variable Y. Similarly, the distribution of Y is given by

5.7.5
$$h(y) = \sum_i f(x_i, y)$$

where the summation is taken over all the possible values, x_i, of X.

Closely connected with these are the *conditional* distributions of X and Y. For a possible value, y_j, of Y, we define the conditional distribution of X by

5.7.6
$$g(x \mid y_j) = \frac{f(x, y_j)}{h(y_j)}$$

Similarly, for one of the possible values, x_i, of X, the conditional distribution of Y is given by

5.7.7
$$h(y \mid x_i) = \frac{f(x_i, y)}{g(x_i)}$$

Note the similarity of these definitions to those of Section V.3.

V.7.5 Example. Let X be the number of heads obtained in three tosses of a fair coin, and let Y be either 1 or 0, depending on whether the first toss is heads or tails. Then the joint distribution of X and Y is given by Table V.7.1.

TABLE V.7.1

Y	X				$h(y)$
	0	1	2	3	
0	1/8	1/4	1/8	0	1/2
1	0	1/8	1/4	1/8	1/2
$g(x)$	1/8	3/8	3/8	1/8	

The marginal row and marginal column give the absolute distributions of X and Y, respectively. The conditional distributions are easily obtained; we might give them in the form of two tables. The first, Table V.7.2, gives the conditional distribution, $g(x \mid y_j)$; the second, Table V.7.3, gives $h(y \mid x_i)$:

TABLE V.7.2

Y	X				
	0	1	2	3	
0	1/4	1/2	1/4	0	$g(x \mid y_j)$
1	0	1/4	1/2	1/4	

TABLE V.7.3

Y	X				
	0	1	2	3	
0	1	2/3	1/3	0	$h(y \mid x_i)$
1	0	1/3	2/3	1	

As may be seen in Example V.7.5, the conditional distributions of a random variable may be quite different from its absolute distribution. In some cases, however, these are all equal: the random variables are then independent. We give a more formal definition as follows.

V.7.6 Definition. Two random variables, X and Y, are *independent* if, for every pair (x,y) of values of the two variables,

$$f(x,y) = g(x) \, h(y)$$

An equivalent definition of independence would be to say that two variables are independent if their joint probability function can be factored into two functions, each of which depends on only one of the variables.

For instance, the random variables X,Y, of Example V.7.4, are independent. On the other hand, the variables of Example V.7.5 are not independent.

V.7.7 Example. Consider the data of Example V.2.7. We can let X be 0 or 1 according to whether the family is Republican or Democrat, and Y be 0 if the family is not headed by a woman, 1 if it is. Then the joint distribution of (X,Y) is shown in Table V.7.4.

TABLE V.7.4

Y	X		$h(y)$
	0	1	
0	21/39	11/39	32/39
1	2/39	5/39	7/39
$g(x)$	23/39	16/39	

The two marginal distributions, $g(x)$ and $h(y)$, are given at the bottom and to the right, respectively, of the joint distribution table. The conditional distributions are shown in Tables V.7.5 and V.7.6.

TABLE V.7.5

Y	X		
	0	1	
0	21/32	11/32	$g(x \mid y_j)$
1	2/7	5/7	

TABLE V.7.6

Y	X		
	0	1	
0	21/23	11/16	$h(y \mid x_i)$
1	2/23	5/16	

Note that X and Y are *not* independent.

8. EXPECTED VALUES. MEANS AND VARIANCES

In general, if a random variable has any practical interpretation (if, for instance, it represents money or some measurable service or commodity) it becomes natural to look, in some fashion, for an *average* value of the variable. An example will perhaps explain this best.

Consider the game of roulette, as played in European casinos. The roulette wheel has 37 numbers ranging from 0 to 36. There are many types of bet possible, but the simplest is a bet on a single number. If the ball lands on the chosen number, the player gets back his bet, plus 35 times his bet. If it lands on any other number, he loses his bet.

Let us assume that the player bets $1 on a single number. Let X be the amount won by the player; then X can have the values -1 (if the player loses) and $+35$ (if he wins.) The probabilities of these two values are obtained by assuming that the roulette wheel is perfectly symmetric, so that the ball has the same probability of landing on each of the 37 numbers. Thus,

$$f(+35) = \frac{1}{37}$$

$$f(-1) = \frac{36}{37}$$

Let us suppose, now, that the player makes this bet, not once, but very many times. If the total number of bets made is n, then, by the frequency definition of probability, he should win approximately $n/37$ times, and lose approximately $36n/37$ times. Since each win is $35, while each loss is $1

$$\text{Total winnings} = \frac{n}{37} \cdot 35 = \frac{35n}{37}$$

$$\text{Total losses} = \frac{36n}{37} \cdot 1 = \frac{36n}{37}$$

$$\text{Net winnings} = -\frac{n}{37}$$

It follows that, in a large number, n, of bets, the player will probably lose approximately $n/37$ dollars. (It may be, of course, that he will have a lucky streak and win, or a bad streak and lose much more, but, for large n, these will be rare occurrences.) On the average, he will lose approximately 1/37 dollars per bet. Since a loss is interpreted as a negative win, we obtain a value of $-1/37$ as the average value of the random variable X. We call this the *expected value* of X.

244

In the general case, let us assume that the random variable X has possible values $x_1, x_2, \ldots, x_m$, with probabilities $f(x_1)$, $f(x_2), \ldots, f(x_m)$ respectively. If the experiment is repeated a large number, n, of times, then X should take on the value x_i approximately $nf(x_i)$ times. If X represents money, or has some such practical interpretation (so that there is a point to adding the values of X obtained), we find that the sum of the values of X, for the (approximately) $nf(x_i)$ times that it has the value x_i, is $nx_if(x_i)$. We now take the sum of these expressions for all possible values of x_i, and obtain (approximately),

$$\sum_{i=1}^{m} nx_if(x_i) = n \sum_{i=1}^{m} x_if(x_i)$$

Since we are generally interested, rather, in the average value (per trial of the experiment) we divide this expression by n, obtaining the somewhat simpler form

5.8.1
$$E[X] = \sum_{i=1}^{m} x_if(x_i)$$

This expression is called the *expected value*, or *expectation*, of the random variable X.

V.8.1 Example. Let X be the number of spots showing after a toss of a fair die. Its expectation can be calculated from Table V.8.1.

TABLE V.8.1

x_i	$F(x_i)$	$x_if(x_i)$
1	1/6	1/6
2	1/6	2/6
3	1/6	3/6
4	1/6	4/6
5	1/6	5/6
6	1/6	6/6

Adding the entries in the last column of this table, we obtain the expectation

$$E[X] = \frac{7}{2}$$

V.8.2 Example. Let X be the number of heads obtained in 10 tosses of a fair coin. We form Table V.8.2.

TABLE V.8.2

x_i	$f(x_i)$	$x_i f(x_i)$
0	1/1024	0
1	10/1024	10/1024
2	45/1024	90/1024
3	120/1024	360/1024
4	210/1024	840/1024
5	252/1024	1260/1024
6	210/1024	1260/1024
7	120/1024	840/1024
8	45/1024	360/1024
9	10/1024	90/1024
10	1/1024	10/1024

Once again, we add the entries in the last column, to obtain

$$E[X] = 5$$

If X is a random variable, and U is a function of X, say

$$U = \varphi(X)$$

then U is also a random variable. We may obtain the expected value of U by computing its distribution (which can be done if we know the distribution of X). This is not generally necessary, however, as it is not difficult to see that

5.8.2 $$E[\varphi(X)] = \sum_{i=1}^{m} \varphi(x_i) f(x_i)$$

V.8.3 Example. Let X be as in Example V.8.1, and let $U = X^2$. We compute its expectation from Table V.8.3.

TABLE V.8.3

x_i	x_i^2	$f(x_i)$	$x_i^2 f(x_i)$
1	1	1/6	1/6
2	4	1/6	4/6
3	9	1/6	9/6
4	16	1/6	16/6
5	25	1/6	25/6
6	36	1/6	36/6

Adding the entries in the last column, we obtain

$$E[U] = \frac{91}{6}$$

FINITE MATHEMATICS

Two of the most important descriptive properties of a random variable are the *mean* and the *variance*. These are defined in terms of expectations.

V.8.4 Definition. Let X be a random variable. Then the *mean*, μ_X, and the *variance*, σ_X^2, of X are defined by

5.8.3
$$\mu_X = E[X]$$

5.8.4
$$\sigma_X^2 = E[(X - \mu_X)^2]$$

The mean is the same as the expected value of X. The *variance*, which is always non-negative, is a measure of the *dispersion* of X: if σ_X^2 is small, then X will usually be close to μ; whereas if σ_X^2 is large, then X will often differ considerably from its mean. The square root of the variance, σ_X, is known as the *standard deviation*.

9. RULES FOR COMPUTING THE MEAN AND VARIANCE

Computation of the mean and variance of a random variable is often simplified by using the following theorem:

V.9.1 Theorem. Let X and Y be random variables, and α a constant. Then

5.9.1
$$E[\alpha] = \alpha$$

5.9.2
$$E[\alpha X] = \alpha E[X]$$

5.9.3
$$E[X + Y] = E[X] + E[Y]$$

Moreover, if X and Y are independent, then

5.9.4
$$E[XY] = E[X]\,E[Y]$$

Proof. If α is constant (i.e., not a variable), then it will have the value α with probability 1, and no other values. This proves (5.9.1). To prove (5.9.2), note that

$$E[\alpha X] = \sum \alpha x_i f(x_i) = \alpha \sum x_i f(x_i) = \alpha E[X]$$

To prove (5.9.3), let X and Y have the joint distribution f; let g and h be the absolute distributions of X and Y respectively. Then

$$E[X+Y] = \sum_{i,j} (x_i + y_j)f(x_i,y_j)$$

$$= \sum_{i,j} x_i f(x_i,y_j) + \sum_{i,j} y_j f(x_i,y_j)$$

$$= \sum_{i} \left\{ x_i \sum_{j} f(x_i,y_j) \right\} + \sum_{j} \left\{ y_j \sum_{i} f(x_i,y_j) \right\}$$

$$= \sum_{i} x_i g(x_i) + \sum_{j} y_j h(y_j)$$

$$= E[X] + E[Y]$$

Finally, suppose X and Y are independent. This means $f(x,y) = g(x)h(y)$. Then

$$E[XY] = \sum_{i,j} x_i y_j f(x_i,y_j)$$

$$= \sum_{i,j} (x_i g(x_i))(y_j h(y_j))$$

$$= \sum_{i} \left\{ x_i g(x_i) \sum_{j} y_j h(y_j) \right\}$$

$$= \left\{ \sum_{i} x_i g(x_i) \right\} \left\{ \sum_{j} y_j h(y_j) \right\}$$

$$= E[X] E[Y]$$

V.9.2   Corollary.   If X has mean μ and variance σ^2, then

5.9.5
$$\sigma^2 = E[X^2] - \mu^2$$

Proof.   We have

$$\sigma^2 = E[(X-\mu)^2] = E[X^2 - 2\mu X + \mu^2]$$
$$= E[X^2] - 2\mu E[X] + \mu^2$$
$$= E[X^2] - 2\mu^2 + \mu^2$$
$$= E[X^2] - \mu^2$$

V.9.3   Corollary.   If X and Y are independent random variables with variances σ_X^2 and σ_Y^2 respectively, then the variance of $X+Y$ is

5.9.6
$$\sigma_{X+Y}^2 = \sigma_X^2 + \sigma_Y^2$$

Proof.   We have

$$E[(X+Y)^2] = E[X^2 + 2XY + Y^2]$$
$$= E[X^2] + 2E[XY] + E[Y^2]$$

248

and

$$\mu^2_{X+Y} = (\mu_X + \mu_Y)^2 = \mu^2_X + 2\mu_X \mu_Y + \mu^2_Y$$

Now, by independence, $E[XY] = \mu_X \mu_Y$, and so

$$\sigma^2_{X+Y} = E[(X+Y)^2] - \mu^2_{X+Y}$$
$$= E[X^2] - \mu^2_X + E[Y^2] - \mu^2_Y$$

$$= \sigma^2_X + \sigma^2_Y$$

V.9.4 Corollary. Let X be a random variable with mean μ_X and variance μ^2_X, and let $Y = \alpha X + \beta$. Then Y has mean

5.9.7 $$\mu_Y = \alpha\mu_X + \beta$$

and variance

5.9.8 $$\sigma^2_Y = \alpha^2\sigma^2_X$$

We will omit the proof of Corollary V.9.4.

10. TWO IMPORTANT THEOREMS

The importance of the mean and standard deviation of a random variable is given by the following two theorems.

V.10.1 Theorem (Chebyshev's Inequality). Let X have mean μ and standard deviation σ. Then, for any $t > 0$,

5.10.1 $$P(|X-\mu| \geq t\sigma) \leq \frac{1}{t^2}$$

Proof. Assume that the event

$$|X - \mu| \geq t\sigma$$

or, equivalently,

5.10.2 $$(X - \mu)^2 \geq t^2\sigma^2$$

has probability p. We may partition the possible values of X into two sets; those for which (5.10.2) holds and those for which it does not.

We may write

$$\sigma^2 = \sum_i (x_i - \mu)^2 f(x_i)$$

$$= \sum_i{}' (x_i - \mu)^2 f(x_i) + \sum_i{}'' (x_i - \mu)^2 f(x_i)$$

where the first sum is over those x_i for which (5.10.2) holds, and the second over the other values of X. Now, the second sum is non-negative, and so

$$\sigma^2 \geq \sum_i{}' (x_i - \mu)^2 f(x_i)$$

and, as (5.10.2) holds for this sum,

$$\sigma^2 \geq \sum_i{}' t^2 \sigma^2 f(x_i)$$

$$= t^2 \sigma^2 \sum_i{}' f(x_i)$$

The sum here is precisely the probability of (5.10.2); thus

$$\sigma^2 \geq t^2 \sigma^2 p$$

and so

$$p \leq \frac{1}{t^2}$$

Chebyshev's inequality states that the probability of a large deviation from the mean can never be too probable; this is given in terms of the standard deviation.

V.10.2 Theorem (The Law of Large Numbers). Let X be a random variable with mean μ. Let $X_1, X_2, X_3, \ldots$ be independent repetitions of the variable X, and let

$$\overline{X}_n = \frac{1}{n} \sum_{i=1}^{n} X_i$$

Then, for any $\epsilon > 0$, and any $\delta > 0$, there exists n such that

5.10.3 $$P(\,|\overline{X}_n - \mu| \geq \epsilon) < \delta$$

Proof. Assume X has variance σ^2. Then

$$Y_n = \sum_{i=1}^{n} X_i$$

will have variance $n\sigma^2$ and so $\overline{X}_n = \frac{1}{n} Y_n$ will have variance σ^2/n, and standard deviation $\sigma/\sqrt{n}$. Applying Chebyshev's inequality, we see that

$$P(\,|\overline{X}_n - \mu| > \epsilon) \leq \frac{\sigma^2}{n\epsilon^2}$$

and it is easy to see that, as n increases, the right side of this inequality approaches 0.

The Law of Large Numbers states that the idea of an expected value is really a valid idea: if the experiment is repeated often enough, it is almost certain that $\overline{X}_n$, the *average observed value* of X, will be quite close to μ. Note that this does *not* say that any particular value of $\overline{X}_n$ will be very probable, only that those values that are close to μ will be much more probable than those that are not. For example, in 100 tosses of a fair coin, the most probable number of heads is 50 (giving a value $\overline{X}_n = 0.5$). Yet the probability of having *exactly* 50 heads is only 0.08. This is not large, but, on the other hand, the probability of exactly 40 heads is only about 0.01, while that of exactly 35 heads is only 0.001. The probability of having between 40 and 60 heads (i.e., of $0.4 \leq \overline{X}_n \leq 0.6$), however, will be about 0.96.

V.10.3 Example. Find the mean and standard deviation of the number of successes, X, in n trials of a simple experiment with probability of success p.

Let us assume first that $n = 1$. In this case, X can have only the values 0 and 1, with probabilities q and p, respectively.

$$E[X] = 0 \cdot q + 1 \cdot p = p$$

and

$$E[X^2] = 0^2 \cdot q + 1^2 \cdot p = p$$

It follows that X has mean $\mu = p$, and variance

$$\sigma^2 = E[X^2] - \mu^2 = p - p^2 = pq$$

Thus, for a single trial, $\mu = p$ and $\sigma = \sqrt{pq}$.

For larger values of n, we define the random variable Y_i, where

$$Y_i = \begin{cases} 0 & \text{if } i\text{th trial is a failure} \\ 1 & \text{if } i\text{th trial is a success} \end{cases}$$

It is not difficult to see that

$$X = Y_1 + Y_2 + \ldots + Y_n$$

Now, the Y_i are independent random variables, each having mean p and variance pq. By Theorem V.9.7 and its corollaries, X must have mean np and variance npq. Thus

5.10.4 $$\mu = np; \ \sigma = \sqrt{npq}$$

Note that the standard deviation increases only as the *square root* of the number of trials.

V.10.4 Example. Let X be the number of spots showing on n independent tosses of a die. Find the mean and variance of X.

Once again, we start by letting $n = 1$. In this case, we know (from previous problems) that $E[X] = 7/2$ and $E[X^2] = 91/6$. Thus

$$\mu = E[X] = \frac{7}{2}$$

$$\sigma^2 = E[X^2] - \mu^2 = \frac{91}{6} - \frac{49}{4} = \frac{35}{12}$$

For other values of n, we use the fact that the tosses are independent, obtaining

$$\mu = \frac{7n}{2}$$

$$\sigma^2 = \frac{35n}{12}$$

V.10.5 Example. A gambler bets $1 on a simple experiment that has probability p of success. If the outcome is a success, he wins $1 and stops playing. If the outcome is a failure, he loses $1 and bets $2 on the next trial of the experiment. He continues in this manner, doubling his bet after each failure, until he obtains a success. He stops betting if he either obtains a success or loses n times in a row (an event that he considers extremely improbable). What is the expected value of his winnings?

The gambler's bet on the kth trial (assuming he is still playing) will be 2^{k-1} dollars. If he loses on each of the first k trials, he will have lost

$$1 + 2 + 2^2 + \ldots + 2^{k-1} = 2^k - 1$$

dollars before betting on the $(k+1)$th trial. The bet then will be 2^k dollars so that, if he wins, he will win back all his losses plus an additional dollar. If, however, he loses n consecutive times, he will wind up losing $2^n - 1$ dollars. Letting X be his winnings, we see that X can have the two values 1 and $1 - 2^n$. The probability of losing n consecutive times is q^n, and so

$$\begin{aligned} E[X] &= 1(1 - q^n) + (1 - 2^n) q^n \\ &= 1 - 2^n q^n \\ &= 1 - (2q)^n \end{aligned}$$

Now, if $q > 1/2$, then $2q > 1$, and so $(2q)^n$ increases (without limit) as n increases. For large values of n, therefore, the variable X will have an expected value that is negative and very large. This is true even though large values of n mean that the probability q^n of losing will be very small. We conclude that this system of betting (known as a *martingale*) is not a very good system.

V.10.6 Example. In a certain large city, 55 per cent of the population are Democrats. A pollster takes a sample of n people, asking their party affiliation. How large should n be, in order that there be a probability of at least 0.95 that between 54 per cent and 56 per cent of those sampled be Democrats?

Since the city is large, the trials may be considered independent. In a sample of size n, the number of Democrats, X, will be a binomial random variable, with mean

$$\mu = 0.55 \, n$$

and variance

$$\sigma^2 = (0.55)(0.45) \, n = 0.2475 \, n$$

The event "between 54 per cent and 56 per cent are Democrats" can be expressed as

$$\left| \frac{X - \mu}{n} \right| \le 0.01$$

or, equivalently,

$$|X - \mu| \le 0.01 \, n$$

We apply Chebyshev's inequality, with $t\sigma = 0.01 \, n$, or

$$t^2 = \frac{0.0001 \, n^2}{0.2475 \, n} = \frac{n}{2475}$$

The probability of the desired event is at least $1 - 1/t^2$; we therefore want $1/t^2 \leq 0.05$. It is sufficient to take

$$\frac{2475}{n} \leq 0.05$$

or, equivalently, $n \geq 49,500$. A sample of 49,500 will be adequate for our purpose.

In reality, this value of n is unnecessarily large; it may be shown that $n = 9604$ would be sufficient. This is because Chebyshev's inequality uses only the mean and variance of the distribution; it does not take special features of the distribution (i.e., that it is a binomial distribution) into account.

Theorem V.10.2 (The Law of Large Numbers) nearly guarantees that, in the long run, the average value of a random variable will approach its expectation. It becomes natural, then, *if an experiment can be repeated many times,* to take the course of action that will maximize the decision-maker's *expected* profit. This is of importance when the profit (or cost) depends on both the decision-maker's action and the outcome of the experiment.

V.10.7 *Example.* An automobile company produces a DeLuxe Model that costs \$4000 to produce and sells for \$8500. The demand for this model is uncertain; the company estimates (from past experience) that as many as four cars might be sold; the probability that k cars will be sold is given by Table V.10.1.

TABLE V.10.1

k	$f(k)$
0	0.1
1	0.4
2	0.3
3	0.15
4	0.05

How many cars should the company produce to maximize its profits?

Let us assume that the company produces n cars, and the demand is for k cars. Assuming a profit of \$4500 for each car sold, and a loss of \$4000 for each car produced but not sold, we obtain the table of profits (Table V.10.2).

TABLE V.10.2

k	n					f(k)
	0	1	2	3	4	
0	0	−4000	−8000	−12000	−16000	0.1
1	0	4500	500	−3500	−7500	0.4
2	0	4500	9000	5000	1000	0.3
3	0	4500	9000	13500	9500	0.15
4	0	4500	9000	13500	18000	0.05
$E[X]$	0	3650	3900	1600	−2075	

The right-hand column gives, once again, the probabilities of each value. The bottom row gives the expected profits for each choice of n. We see that this expectation gives a maximum for $n = 2$. We conclude that the company should produce two cars every year.

In dealing with expectations, it is assumed, at least tacitly, that the experiment can be repeated many times. Otherwise the Law of Large Numbers has no chance to work. It follows that, when the experiment cannot be repeated, the criterion of maximizing the expected profit cannot be applied indiscriminately. Consider, for example, a person who is given the possibility of investing $5000 in a new company. The company has, say, a 0.5 probability of succeeding, in which case the investor will receive $20,000 in profits. On the other hand, the company might also fail with probability 0.5, in which case the investor will lose his $5000. It is easy to see that the expected profit from this investment is

$$\frac{1}{2}(20{,}000) + \frac{1}{2}(-5000) = 7500$$

so that, to maximize the expected profit, the investor should choose this gamble rather than some other course of action that might give him a certain profit of, say, $6000. On the other hand, if the investor is not too rich, he might be unwilling to risk $5000; he could, quite reasonably, feel that it is more money than he can afford to lose and that, moreover, he will have no chance to recoup his losses if the company fails (he will then be broke). For a very rich investor, however, the gamble might be worth-while: he can afford a $5000 loss. It is clear that it is not sufficient to look at money only; economists have suggested that, instead of maximizing the expected monetary profit, the decision maker should maximize the expected *utility* of the outcome. Why this should be, and what is meant by utility, is studied in a subsequent chapter of this book.

PROBLEMS ON RANDOM VARIABLES

1. A fair die is tossed 100 times; let X be the number of aces obtained. Find the mean and variance of X.

2. A player tosses a fair coin 10 times. If the number X of heads is even, then the player wins X dollars; if X is odd, then the player loses X dollars. What are the player's expected winnings?

3. A merchant stocks units of a perishable item. Each unit costs him \$2 and sells for \$5. The merchant estimates that the demand k for this item has distribution $f(k)$ given by

$$
\begin{aligned}
f(0) &= 0.1 \\
f(1) &= 0.15 \\
f(2) &= 0.25 \\
f(3) &= 0.3 \\
f(4) &= 0.15 \\
f(5) &= 0.05
\end{aligned}
$$

How many units should the merchant stock to maximize his expected profits?

4. A fair coin is tossed 10,000 times. Using Chebyshev's inequality, find an upper bound for the probability that the number of heads obtained be smaller than 4500 or larger than 5500.

5. If X and Y are random variables with mean μ_X and μ_Y respectively, and variance σ_X^2 and σ_Y^2 respectively we define the *covariance*

$$
\sigma_{XY} = E[(X - \mu_X)(Y - \mu_Y)]
$$

Prove that

$$
\sigma_{XY} = E[XY] - E[X]E[Y]
$$

and that, if $U = aX + b$, and $V = cY + d$ (where a, b, c, and d are constants) then

$$
\sigma_{UV} = ab\,\sigma_{XY}
$$

6. If X and Y are random variables with variance σ_X^2 and σ_Y^2 respectively, and covariance σ_{XY}, we define the *correlation coefficient*

$$
\rho_{XY} = \frac{\sigma_{XY}}{\sigma_X \sigma_Y}
$$

256

Prove that, if $U = aX + b$, and $V = cY + d$, then

$$\rho_{UV} = \rho_{XY}$$

7. Let X, Y, and Z be mutually independent random variables with variance σ_X^2, σ_Y^2, and σ_Z^2, respectively. Let

$$U = X + Z$$
$$V = Y + Z$$

Find σ_U^2, σ_V^2, σ_{UV} and ρ_{UV} in terms of σ_X^2, σ_Y^2, and σ_Z^2.

8. A coin is tossed 10 times. Let X be the number of heads obtained, and let $Y = X^2$. Find the correlation coefficient ρ_{XY}.

9. An urn contains 100 balls each painted with two colors. One of the colors is either red or black; the other color is either green or yellow. The distribution of colors is given by the following table:

	Green	Yellow
Red	38	15
Black	15	32

A ball is drawn from the urn; let $X = 0$ if the first color is red, and 1 if it is black. Let $Y = 0$ if the second color is green, and 1 if it is yellow. Find the correlation coefficient ρ_{XY}.

10. A machine is either working well (probability 0.8) or it is not. If it is working well, its product will be good with probability 0.75; if not working well, its product will be good with probability 0.4. Find the correlation coefficient ρ_{XY}, where $X = 0$ or 1 depending on whether the machine is working well or not, and $Y = 0$ or 1 depending on whether its product is good or defective.

11. MARKOV CHAINS

A Markov chain is a probabilistic process by which a system changes from one of several possible states to another at intervals of time. Examples of Markov processes are the manner in which the length of a waiting line changes, the way in which a gambler's fortune varies, or the economic growth of a country. The characteristic property of such processes is that past history is unimportant, i.e., the future behavior of the system depends (up to a point) on its present state, but is independent of its state at any previous moment of time.

Thus, knowledge of the present state gives as much information (for the purposes of prediction) as knowledge of the system's entire history.

We shall assume here that the system has only a finite number of possible states and that the *transitions* (changes from one state to another) can happen only at discrete intervals of time (say, every minute or every hour).

Let us suppose that a Markov process has n states. It is clear from our discussion that the process is entirely described by giving the n^2 probabilities of transition from one state to another: these n^2 numbers can be arranged to form an nth order matrix

$$A = (a_{ij}) \quad i,j = 1,2, \ldots, n$$

where a_{ij} is the conditional probability that the system will be in state j at time $t+1$, given that it is in state i at time t. The matrix A satisfies the conditions

5.11.1
$$a_{ij} \geq 0 \quad \text{for all } i,j$$

5.11.2
$$\sum_{j=1}^{n} a_{ij} = 1 \quad \text{for each } i = 1, \ldots, n$$

The reasons for conditions (5.11.1) and (5.11.2) should be clear. Any matrix that satisfies these conditions is known as a *stochastic matrix*. (Compare this with the definition of a *doubly* stochastic matrix in Chapter III.)

V.11.1 Example. Two gamblers, G_1 and G_2, have a total of n units in their possession. They carry out a sequence of independent trials of a simple experiment with probability p of success. If the experiment is a success, G_2 pays 1 unit to G_1; if a failure, G_1 pays 1 unit to G_2. This process terminates if either of the two gamblers is ever wiped out (i.e., if he has no units left).

This is a Markov process with $n+1$ states, the ith state (for $i = 0, 1, \ldots, n$) coming when G_1 has i units. It is easy to see that, for $1 \leq i \leq n-1$, the system has a probability p of moving to state $i+1$, and q of moving to state $i-1$. For $i = 0$ or n, however, there are no further changes: the process is said to have an *absorbing barrier* at states 0 and n. We have, thus:

$$a_{i,i+1} = p \quad \text{for } i = 1, \ldots, n-1$$
$$a_{i,i-1} = q \quad \text{for } i = 1, \ldots, n-1$$
$$a_{00} \quad = 1$$
$$a_{nn} \quad = 1$$
$$a_{ij} \quad = 0 \quad \text{for all other } (i,j)$$

For $n = 5$, the transition matrix has the form

$$A = \begin{pmatrix} 1 & 0 & 0 & 0 & 0 & 0 \\ q & 0 & p & 0 & 0 & 0 \\ 0 & q & 0 & p & 0 & 0 \\ 0 & 0 & q & 0 & p & 0 \\ 0 & 0 & 0 & q & 0 & p \\ 0 & 0 & 0 & 0 & 0 & 1 \end{pmatrix}$$

V.11.2 *Example.* Consider the following process: an urn contains n balls, k of which are red, and $n - k$, black. At each step, there is a probability k/n that one of the red balls will become black, and an independent probability $(n - k)/n$ that one of the black balls will become red.

Let the kth state hold when the urn has k red balls. We see that we have $n + 1$ possible states. To compute the transition probabilities, note that we pass from state k to $k + 1$ if one of the black balls "mutates" (i.e., becomes red) but none of the red balls mutates. Therefore

$$a_{k,k+1} = \frac{n - k}{n} \left(1 - \frac{k}{n} \right) = \left(\frac{n - k}{n} \right)^2$$

Similarly, we find

$$a_{k,k-1} = \left(\frac{k}{n} \right)^2$$

and, finally,

$$a_{kk} = 1 - \left(\frac{n - k}{n} \right)^2 - \left(\frac{k}{n} \right)^2$$

We thus have:

$$a_{k,k+1} = \left(\frac{n - k}{n} \right)^2$$

$$a_{k,k-1} = \left(\frac{k}{n} \right)^2$$

$$a_{kk} = \frac{2nk - 2k^2}{n^2}$$

$$a_{ij} = 0 \quad \text{otherwise}$$

For $n = 5$, the transition matrix would have the form

$$A = \begin{pmatrix} 0 & 1 & 0 & 0 & 0 & 0 \\ 1/25 & 8/25 & 16/25 & 0 & 0 & 0 \\ 0 & 4/25 & 12/25 & 9/25 & 0 & 0 \\ 0 & 0 & 9/25 & 12/25 & 4/25 & 0 \\ 0 & 0 & 0 & 16/25 & 8/25 & 1/25 \\ 0 & 0 & 0 & 0 & 1 & 0 \end{pmatrix}$$

In dealing with Markov processes it is often of interest to predict the state of the system, not in the time period that follows immediately, but at some later period. It becomes natural to look for "s-period transition probabilities, $a_{ij}^{(s)}$: we define $a_{ij}^{(s)}$ as the probability that the system will be in state j at time $t + s$ given that it is in state i at time t. Then

$$A^{(s)} = (a_{ij}^{(s)})$$

is the *s-period transition matrix.*

It is clear that the matrices $A^{(s)}$ depend on the matrix A. Let us see how they are to be computed. What, for instance, is $a_{ij}^{(2)}$? We compute this by considering all the possible ways in which the system can reach state j at time $t + 2$, given that it is in state i at time t.

Generally speaking, the system can pass from state i to state k at the intermediate time $t + 1$ and then from k to j at time $t + 2$. The probability of this is $a_{ik}a_{kj}$. Since k is arbitrary we must add all these terms, obtaining

5.11.3
$$a_{ij}^{(2)} = \sum_{k=1}^{n} a_{ik}a_{kj}$$

But this is simply the rule (studied in Chapter II) for the multiplication of matrices (in this case, for multiplying A with itself.) We conclude that

$$A^{(2)} = A \ A = A^2$$

and, in general,

5.11.4
$$A^{(s)} = A^s$$

so that the *s-period transition matrix is simply the sth power of the single-period transition matrix.*

V.11.3 Example. The Markov chain with transition matrix

$$A = \begin{pmatrix} q & 1 - q \\ 0 & 1 \end{pmatrix}$$

has the 2-period transition matrix

$$A^2 = \begin{pmatrix} q^2 & 1-q^2 \\ 0 & 1 \end{pmatrix}$$

and, in general, it will have the s-period matrix

$$A^s = \begin{pmatrix} q^s & 1-q^s \\ 0 & 1 \end{pmatrix}$$

V.11.4 Example. The Markov chain with transition matrix

$$A = \begin{pmatrix} 0.3 & 0.6 & 0.1 \\ 0.1 & 0.5 & 0.4 \\ 0.4 & 0.1 & 0.5 \end{pmatrix}$$

has the 2-period matrix

$$A^2 = \begin{pmatrix} 0.19 & 0.49 & 0.32 \\ 0.24 & 0.35 & 0.41 \\ 0.33 & 0.34 & 0.33 \end{pmatrix}$$

and the 3-period matrix

$$A^3 = \begin{pmatrix} 0.234 & 0.391 & 0.375 \\ 0.271 & 0.360 & 0.369 \\ 0.265 & 0.401 & 0.334 \end{pmatrix}$$

12. REGULAR AND ABSORBING MARKOV CHAINS

For most Markov chains, the state in the near future will depend considerably on the present state. On the other hand, it is not inconceivable that (for many such chains at least) the importance of the initial state should vanish over a long period of time. This should be especially so in a Markov chain such as that of Example V.11.2, in which the system's characteristics seem to push it away from the extremes of many red or many black balls. Example V.11.4 also seems to behave in this manner: note how the three rows, which are quite dissimilar in A, are nearly equal in A^3. The pattern is even more obvious in the 4-period matrix:

$$A^4 = \begin{pmatrix} 0.259 & 0.374 & 0.367 \\ 0.265 & 0.380 & 0.355 \\ 0.253 & 0.393 & 0.354 \end{pmatrix}$$

Further multiplications would show that for large s, the rows of $A^{(s)}$ are almost equal; in fact it would be seen that they approach the matrix

$$V = \begin{pmatrix} 0.259 & 0.382 & 0.359 \\ 0.259 & 0.382 & 0.359 \\ 0.259 & 0.382 & 0.359 \end{pmatrix}$$

Thus, for large s, the state at time $t + s$ depends only slightly on the state at time t. The vector $v = (0.259, 0.382, 0.359)$ is a solution of the equation

$$vA = v$$

and shows the "steady state" behavior of the system.

We shall show, now, that this convergence of the matrices A^s occurs quite often. Let us assume, first, that all the entries a_{ij} are positive. Let a be the smallest of all these:

$$a = \min_{i\,j} a_{ij}$$

and let $r = 1 - 2a < 1$. Now let $d(s)$ be the largest difference between entries in the same column of the matrix A^s:

$$d(s) = \max_{i,j,k} \{a_{ij}^{(s)} - a_{kj}^{(s)}\}$$

Intuitively, it is easy to see that, if $d(s)$ is small then the rows of A^s are all nearly equal.

We use now the fact that $A^{s+1} = AA^s$ and so

$$a_{ij}^{(s+1)} = \sum_{l=1}^{n} a_{il} a_{lj}^{(s)}$$

We know that A is a stochastic matrix, so

$$\sum_{l=1}^{n} a_{il} = 1$$

Let $a_{pj}^{(s)}$ be the largest entry, and $a_{qj}^{(s)}$ the smallest entry in the jth row of A^s. We have then

$$a_{ij}^{(s+1)} = a_{ip} a_{pj}^{(s)} + \sum_{l \neq p} a_{il} a_{lj}^{(s)}$$

Now $a_{ip} \geq a$, and, for all l, $a_{lj}^{(s)} \geq a_{qj}^{(s)}$. It will follow that

5.12.1 $$a_{ij}^{(s+1)} > a \cdot a_{pj}^{(s)} + (1 - a) a_{qj}^{(s)}$$

In a similar way, we will see that:

5.12.2
$$a_{kj}^{(s+1)} \leq (1-a)\,a_{pj}^{(s)} + a \cdot a_{gj}^{(s)}$$

and so

$$a_{kj}^{(s+1)} - a_{ij}^{(s+1)} \leq r(a_{pj}^{(s)} - a_{qj}^{(s)})$$

or

$$a_{kj}^{(s+1)} - a_{ij}^{(s+1)} \leq r\,d(s)$$

Since this holds for all i, j, and k;

5.12.3
$$d(s+1) = \max_{i,j,k}\{a_{kj}^{(s+1)} - a_{ij}^{(s+1)}\} \leq r\,d(s)$$

We conclude (from (5.12.1) and (5.12.2)), that each entry in the jth column of A^{s+1} is at least as large as the smallest entry in the jth row of A^s, and at least as small as the largest entry in the jth row of A^s. We conclude from (5.12.3) that $d(s) \leq r^s$, and so, as s increases, $d(s)$ tends toward zero, so that, for large s, the rows of $A^{(s)}$ are nearly equal to each other and also nearly equal to the rows of $A^{(s+1)}$ and all higher powers of A. It will follow that the matrices A^s approach a limiting matrix, V, all of whose rows are equal: the rows of V represent the "steady state" condition of the Markov chain.

In general, it is not necessary that all the entries in A be positive; it is sufficient that all the entries in one of the matrices $A^{(s)}$ be positive: convergence can be proved similarly for such matrices.

V.12.1 Definition. A stochastic matrix A is said to be *regular* if one of its powers, A^s, has all positive entries.

We give the following theorem without further proof:

V.12.2 Theorem. Let A be a regular matrix. Then

(a) The matrices A^s approach a matrix, V, as s grows.
(b) The rows of V are all equal to the vector v.
(c) $v\,A = v$.

Generally speaking, the easiest way to find the limit matrix V is to solve the equation $vA = v$. If A is regular, this equation will have only one solution that satisfies also the stochastic condition

$$\sum v_i = 1$$

V.12.3 *Example.* The cigarette company that manufactures brand C starts an aggressive advertising campaign. The results of this campaign are such that, of people smoking brand C in a given week,

80 per cent continue to smoke it the following week; of those smoking any other brands in a given week, 40 per cent are won over to brand C the following week. In the long run, what fraction of the smokers will be smoking brand C?

We have here a Markov chain with transition matrix

$$A = \begin{pmatrix} 0.8 & 0.2 \\ 0.4 & 0.6 \end{pmatrix}$$

Consider the equation

$$\mathbf{v}A = \mathbf{v}$$

or

$$0.8v_1 + 0.4v_2 = v_1$$
$$0.2v_1 + 0.6v_2 = v_2$$

This has the solution

$$v_2 = 2v_1$$

Since we want $v_1 + v_2 = 1$, we obtain

$$\mathbf{v} = \left(\frac{2}{3}, \frac{1}{3} \right)$$

It follows that

$$V = \begin{pmatrix} \dfrac{2}{3} & \dfrac{1}{3} \\ \dfrac{2}{3} & \dfrac{1}{3} \end{pmatrix}$$

In the long run the company will have 2/3 of the market. Note that the initial state does not matter; even starting from scratch, the company will, eventually, "nearly" capture its share of the market (though, of course, starting from scratch, it will take longer to do so).

If the transition matrix is not regular, there is still a possibility that the process may behave in this manner (see Example V.11.3), but generally this will not happen. A special case arises when some of the states in the process form *absorbing barriers*: once the system reaches such a state, there is no further change. Example V.11.1 has this property, inasmuch as the game terminates as soon as either of the two gamblers is wiped out. It is not too difficult to see, from this example, that, if s is large enough, then the game will "very probably" have terminated before s trials of the experiment; one of the two will have been wiped out. It may be shown (by a process similar to the proof of Theorem V.12.2) that, as s increases, the matrices A^s approach, as limit, a matrix W, which has non-zero entries only in the first and last columns. The entry w_{i0} then represents the probability that G_1 will eventually be wiped out, given that he starts with i units. We shall, in a subsequent chapter, study this problem.

PROBLEMS ON MARKOV CHAINS

1. When a company president finds an item on his desk, he may:

(a) send it to the vice-president, to act on it the following day;

(b) leave it on his own desk for the following day, or

(c) act on it immediately; the probabilities of these three options are 0.3, 0.6, and 0.1 respectively. Similarly, the vice-president has the options of keeping the paper on his own desk, sending it to the president, or acting on it immediately; these choices have probability 0.5, 0.2, and 0.3 respectively.

Represent this as a Markov chain, and form a transition matrix. If the president has the paper, what is the probability that some action will be taken within three days?

2. Three soap brands, I, II, and III, dominate their field. There is a continuous switching among customers, which can be represented probabilistically. If a customer is now using brand I, there is 0.6 probability that he will continue to use it the following week, 0.3 probability that he will switch to brand II, and 0.1 that he will switch to III. If he is using II, there is 0.5 probability that he will continue to use II, 0.4 that he will switch to I, and 0.1 that he will switch to III. If he is using brand III, there is 0.7 probability that he will continue to use it, and 0.3 that he will switch to brand II.

Given that a customer is now using brand II, what is the probability that he will be using brand I, three weeks from now?

3. Using the data of Problem 2, assume that brands I, II, and III are presently holding 50 per cent, 30 per cent and 20 per cent, respectively, of the market. What shares will they hold two weeks from now?

4. Again using the data of Problem 2, what share of the market will the three brands hold in the long run?

5. Customers arrive at random at a service counter that can hold four people. In an interval of time, there is a probability 0.3 that a new customer will arrive (though he will be turned away if there are already four present) and a 0.2 probability that the customer at the head of the line (if there is such a customer) will finish service and leave. The probabilities that more than one person will arrive or that more than one will finish service in such an interval are considered negligible.

(a) Express this situation as a Markov chain, in which the states are the number of persons at the counter.

(b) What is the probability that the counter be idle (i.e., no customers present)?

(c) What is the probability that a customer be turned away?

6. Consider a Markov chain with $m + n$ states $\{S_1, S_2, \ldots, S_{m+n}\}$. Suppose that, for $n + 1 \leq i \leq m + n$, $a_{ii} = 1$, i.e., if the chain reaches one of the last m states, it remains thereafter in that state. Suppose, moreover, that, for $1 \leq i \leq n$,

$$t_i = \sum_{=n+1}^{m+n} a_{ij} > 0$$

i.e., if the chain is in one of the remaining n states, there is a positive probability that it will pass to one of the last m states. Then the states $\{S_{n+1}, \ldots, S_{m+n}\}$ are called *absorbing states*, while $\{S_1, \ldots, S_n\}$ are called *transient states*. Show that, as k approaches infinity, the powers A^k of the transition matrix approach a matrix, W, which is such that $w_{ij} = 0$ for $1 \leq j \leq n$. (Essentially, this means that, with probability 1, the Markov process will eventually reach one of the absorbing states, and remain there. Prove this by showing that, if t is the minimum of the t_j, then the entries in the first n columns of A^k are all smaller than $(1 - t)^k$. Since t is positive, these powers all approach 0.)

7. Show that the conclusions of Problem 6 will hold if it is possible to reach one of the absorbing states from each of the transient states in not more than p moves in which p is an integer. (Do this by considering the p-period transition matrix A^p.)

8. With the assumptions of Problem 6, let $B = (b_{ij})$ be the $n \times n$ matrix formed by the first n rows and columns of A, i.e.,

$$b_{ij} = a_{ij}$$

for $i, j = 1, \ldots, n$. Then B is called the *transient part* of A. The matrix

$$C = (I - B)^{-1}$$

(which may be shown to exist) is known as the *fundamental* matrix of the chain. Show that, if k is large, C is approximately equal to

$$I + B + B^2 + B^3 + \ldots + B^k$$

and that the entries c_{ij} represent the expected number of periods which the chain will spend in the jth (transient) state, given that it is in the ith (transient) state.

9. With the assumptions of Problem 6, let d_i be the expected

number of periods before the system reaches an absorbing state, given that it is now in state i. Show then that, for $n+1 \leq i \leq m+n$,

$$d_i = 0$$

whereas, for $1 \leq i \leq n$,

$$d_i = 1 + \sum_{=1}^{m+n} p_{ij} d_j$$

10. Two gamblers, with a total of 5 units between them, play a fair game (i.e., one in which each has probability 0.5 of winning) for stakes of 1 unit each time; the game will terminate whenever either one of the gamblers holds all 5 units. Represent this as a Markov chain.

(a) Given that one gambler holds 2 units now, what is the expected number of times that he will hold 3 units, before the game terminates?

(b) What is the expected duration of the game?

(c) What is the probability that he will eventually be wiped out?

CHAPTER VI

THE THEORY
OF GAMES

1. GAMES: EXTENSIVE AND
NORMAL FORMS

The theory of games is a comparatively new branch of mathematics, designed to analyze the behavior of people in situations of conflict of interest. The exact nature of these conflicts is best evaluated in the case of parlor games (in which such things as winning and losing are clearly defined) and it is this fact that has given the theory its name. Notwithstanding, the theory is applicable to many serious situations: a game, as the term shall be used here, may be an extremely serious affair, including such things as cutthroat economic competition and all-out nuclear warfare.

Research into the theory of games was started independently during the 1920's by J. von Neumann and E. Borel. Some small progress was made in the 1930's, but it was not until the Second World War that it reached the status of an independent branch of mathematics (closely allied to economics); von Neumann and O. Morgenstern collaborated on its strict formulation. Since then, the close relationship between game theory and linear programming has contributed significantly to its development; progress has been rapid.

The basic elements of a game are best seen in parlor games such as bridge, chess, or poker. These include:

(a) An alternation of moves, some of which may be random moves (such as shuffling a pack of cards), while others are personal moves.
(b) Well-defined probability distributions for the random moves

(note that attempts to change these distributions, as by stacking a deck or loading dice, are considered "cheating").
(c) Some knowledge (which may be perfect, as in chess, or very poor, as in poker) of the outcomes of prior moves.
(d) A function that assigns a payoff (in such terms as money, fame, or satisfaction) to each terminal position, or *play*, of the game.

A strict mathematical definition of this structure could be given, but would serve no purpose here. For this, the interested reader is referred to the literature.

It is generally possible to represent a game by a graph (a set of arcs and nodes). The nodes (vertices) of the graph represent the many possible positions in the game; the arcs represent the moves. There will be an arc from node A to node B if the game can be changed from position A to position B in a single move. An elementary, but instructive, example is the game of "matching pennies."

VI.1.1 Example (Matching Pennies). A player, P_1, puts a coin either "heads up" (H) or "tails up" (T). A second player, P_2, in ignorance of P_1's move, calls either "heads" or "tails." If P_2 guesses correctly (i.e., if their choices match), then P_1 pays P_2 1 unit; otherwise, P_2 pays P_1 1 unit.

Figure VI.1.1 shows the game tree for matching pennies. Each of the vertices is labeled to which the player moves at the corresponding position. Vertex A "belongs" to P_1, while B and C "belong" to P_2. The other vertices represent terminal positions and therefore belong to no player; instead, a "payoff vector" is assigned to each of them. Vertex A is "distinguished"; it represents the initial position. Note how vertices B and C are connected: we use this to represent the fact that P_2 will not (unless he peeks) know whether he is at B or at C; they are said to belong to the same *information set*.

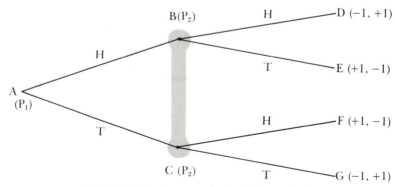

FIGURE VI.1.1   Game tree for "matching pennies."

A second easy example is a rudimentary form of poker:

VI.1.2 Example. A deck contains $m + n$ cards, m of which are marked "high" (H), and n "low" (L). One of these cards is given at random to player P_1; he can then "fold," in which case he pays 1 unit to player P_2, or bet a units. If P_1 bets, then P_2 can fold, in which case he pays P_1 1 unit, or "call" (i.e., match the bet), in which case P_1 pays P_2 $1 + a$ units if the card is L, and P_2 pays P_1 $1 + a$ units if the card is H.

Figure VI.1.2 shows the tree (graph) for this game. Note that a probability distribution ($m/(m + n)$ for H and $n/(m + n)$ for L) is given at vertex A. Note also how vertices E and F are connected: they belong to the same *information set*, since P_2 does not know whether P_1 really has a high card (vertex E) or is bluffing (vertex F).

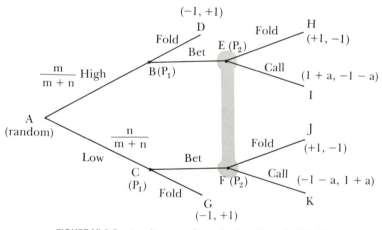

FIGURE VI.1.2 A rudimentary form of poker (Example VI.1.2).

In our attempt to analyze these games, we shall assume, of course, that each player attempts to maximize his return, given in units, say, of money or possibly of utility. This is complicated by the fact that certain games involve randomizations, so that, even if we know what the players do, we still do not know, except in probability, what the outcome of the game will be. We shall overcome this difficulty by assuming (somewhat unjustifiedly, perhaps) that the players will attempt to maximize their *expected* returns. It will then be sufficient to deal with expected values only.

Let us suppose that some persons wish to play a game, but, unfortunately, can find no time that is convenient for all of them. If the game is not too difficult, it is possible for them to play it by proxy: each one can simply give detailed instructions to his secretary; if the instructions are comprehensive enough, the secretaries can then play the game without ever needing to make a single decision because these will all have been made by the original players. Such a comprehensive set of instructions—one that tells a player what to do in each conceivable position of the game—is known as a *strategy*. For instance, we see that in Example VI.1.1 each player has two strategies, which we will label H and T, respectively. In Example VI.1.2, P_1 has four strategies, since he can distinguish two positions (vertices B and C) and has two choices at each of them. These strategies are: (a) to bet in any case; (b) to fold in any case; (c) to bet with H, and fold with L; and (d) to fold with H, and bet with L. The second player, P_2, has only two strategies: he can either call a bet or fold. Note that he would like to call bluffs only, but this is not possible, since he cannot distinguish between a bluff and an honest bet; if he could, he would have four strategies.

The game trees (graphs) given earlier are known as the *extensive form* of the game. We will not deal greatly with this, but instead, with a more concise form.

VI.1.3 Definition. The *normal form* of a game is a list of all strategies for its players, together with the expected payoffs for each possible choice of strategies (one strategy for each player).

In the two examples just given, the amounts won by one player are lost by the other; i.e., the two players represent a closed system. Such games are called *zero-sum games,* since the sum of the amounts won by the two players (treating a loss as a negative win) is always zero. A zero-sum game for two players is also known as a *strictly competitive game,* since the situation is such as to preclude cooperation by the players. If, in a strictly competitive game, each of the two players has only a finite number of strategies, then the normal form of the game can be given as a matrix $A = (a_{ij})$: each row represents a strategy for P_1; each column, a strategy for P_2. The entry a_{ij} will then be the expected amount won by P_1 (and hence lost by P_2), assuming that they choose their ith and jth strategies respectively. Because of this, such games are known as *matrix games.* As examples, we give the game matrices for the two games just described.

VI.1.4 Example. For the game described in Example VI.1.1, each of the two players has two strategies, H and T. The game matrix is

$$
\begin{array}{c}
\quad H \quad T \\
\begin{array}{c} H \\ T \end{array}
\begin{pmatrix} -1 & 1 \\ 1 & -1 \end{pmatrix}
\end{array}
$$

(We repeat that positive entries mean gains for P_1, negative entries are losses for P_1.)

VI.1.5 Example. For the rudimentary poker game of Example VI.1.2, the first player has four strategies, which we will call BB (bet on both cards); BF (bet on H, fold on L); FB (fold on H, bet on L), and FF (fold in both cases). The second player has the strategies C (call) and F (fold). If P_1 chooses BB, and P_2 chooses C, then the payoff (to P_1) will be $1 + a$ if he has H (probability $m/m + n$) and $-1 - a$ if he has L (probability $n/m + n$). The expected payoff is then

$$\frac{m}{m+n}(1+a) - \frac{n}{m+n}(1+a) = \frac{(m-n)(1+a)}{m+n}$$

If P_1 chooses BB, and P_2 chooses F, then P_1 will always win 1 unit. In a similar way, we can compute the remaining entries in the matrix to obtain

$$\begin{pmatrix} & C & F \\ BB & \dfrac{(m-n)(1+a)}{m+n} & 1 \\ BF & \dfrac{m(1+a)-n}{m+n} & \dfrac{m-n}{m+n} \\ FB & \dfrac{-m-n(1+a)}{m+n} & \dfrac{n-m}{m+n} \\ FF & -1 & -1 \end{pmatrix}$$

If we write $p = m/m + n$, $q = n/m + n$, this can be simplified to

$$\begin{matrix} & C & F \\ BB & (p-q)(1+a) & 1 \\ BF & p(1+a)-q & p-q \\ FB & -1-qa & q-p \\ FF & -1 & -1 \end{matrix}$$

2. SADDLE POINTS

Let us consider the following matrix game (in which, as before, the entries represent payoffs from the column player, P_2, to the row player, P_1):

$$\begin{pmatrix} 2 & 2 & 1 & 4 \\ 0 & 2 & 5 & 3 \\ 4 & 2 & 3 & 2 \end{pmatrix}$$

We shall try to analyze the game by duplicating the two players' thinking process. We assume, as before, that P_1 wishes to maximize the payoff, while P_2 wishes to minimize it.

Let us consider first P_2's point of view. He could, of course, choose the second column, which will guarantee that he loses no more than 2 units. On the other hand, by choosing some other column, he could hope to lose nothing at all, or only 1 unit. What should he do? At first, this is not entirely clear. It does seem to be quite clear that he should *not* choose the fourth column, since the second column is distinctly better. In fact, we see that, whatever P_1 may do, P_2 will always do at least as well and possibly better if he chooses the second column rather than the fourth. (This is because each entry in the second column is smaller than or equal to the corresponding entry in the fourth column). P_2 will therefore discard the fourth column and consider only the other three. This gives us a reduced game

$$\begin{pmatrix} 2 & 2 & 1 & 4 \\ 0 & 2 & 5 & 3 \\ 4 & 2 & 3 & 2 \end{pmatrix}$$

Let us now consider P_1's point of view. He knows that, assuming his opponent to be rational, the fourth column can be discarded. Looking only at the remaining columns, he can see that the first row can be discarded in favor of the third. In fact, each entry in the third row is larger than or equal to the corresponding entry in the first row. This leaves us a smaller matrix:

$$\begin{pmatrix} 2 & 2 & 1 & 4 \\ 0 & 2 & 5 & 3 \\ 4 & 2 & 3 & 2 \end{pmatrix}$$

In this 2×3 matrix, P_2 sees that he can discard his third column, which is uniformly worse than the second. This leaves

$$\begin{pmatrix} 2 & 2 & 1 & 4 \\ 0 & 2 & 5 & 3 \\ 4 & 2 & 3 & 2 \end{pmatrix}$$

Now it is P_1's turn; in this matrix, the last row is at least as good as the other, and sometimes better. We will be left with

$$\begin{pmatrix} 2 & 2 & 1 & 4 \\ 0 & 2 & 5 & 3 \\ 4 & 2 & 3 & 2 \end{pmatrix}$$

Thus, if P_1 is rational, and if he assumes that his opponent is rational, he will always choose the bottom row of the matrix. This means that P_2 should choose the second column.

We see then that a very rational analysis of the situation will lead P_1 to choose the third row, while P_2 will choose the second column. This is, essentially, a very pessimistic point of view: each player assumes that the other is capable of carrying out the analysis and acting rationally on this basis. If P_1 were very optimistic, he might decide to choose the second row (which gives him a chance to win 5 units if P_2 slips up); similarly, an optimistic P_2 might choose the first column, hoping that P_1 would make a poor choice. Game Theory is pessimistic, at least to this extent: opponents are assumed to be rational and intelligent.

The method by which we have deleted rows and columns is known as *row* and *column domination*.

VI.2.1 Definition. Let $A = (a_{ij})$ be a matrix game. We say that the *i*th row *dominates* the *k*th row if $a_{ij} \geq a_{kj}$ for each *j* and, moreover, $a_{ij} > a_{kj}$ for at least one value of *j*. Similarly, the *j*th column *dominates* the *l*th column if $a_{ij} \leq a_{il}$ for each *i*, and, moreover, $a_{ij} < a_{il}$ for at least one value of *i*.

In essence, one strategy dominates another if it is never worse and is sometimes better. The importance of this notion is that dominated strategies can always be discarded; optimal behavior will never require them.

VI.2.2 Example. In the game with matrix

$$\begin{pmatrix} 3 & 1 & 5 \\ 2 & 0 & 6 \end{pmatrix}$$

the second column dominates both the other columns. Once the first and third columns are deleted, the first row dominates the second row (note that this does not happen if the third column is not deleted). We conclude that P_1 should choose the first row, while P_2 chooses the second column.

Let us consider the matrix game

$$A = \begin{pmatrix} 6 & 2 & 1 \\ 4 & 3 & 5 \\ 0 & 1 & 6 \end{pmatrix}$$

As before, we look for row or column dominations, but there are none. We cannot reduce the matrix in this fashion. Let us, nevertheless, analyze the game from a pessimistic point of view.

Let us suppose P_1 chooses his first strategy. Then, depending on what P_2 does, P_1 will win 6, 2, or 1 units. A pessimistic P_1 will consider the worst that can happen to him: he can count on winning 1 unit. This is the *security level* that his first strategy will give him. In a similar way, his second strategy gives him a security level of 3 units, while the third strategy gives him a security level of 0 units. We see that P_1 can maximize his security level by choosing his second strategy.

Let us now analyze the game from P_2's point of view. If he chooses his first strategy, he might lose 6, 4, or 0 units. His security level (the worst that can happen to him) is 6 units lost. Similarly, his second and third strategies yield security levels of 3 and 6 units lost, respectively. We conclude that his best choice, as far as security level is concerned, is the second column.

Let us see what this means. P_1 has a "gain-floor" of 3 units, obtained by using the second row. P_2 has a "loss-ceiling" of 3 units, obtained by using the second column. In other words, P_1 can assure himself a gain of 3 units, while P_2 can insure himself against any greater loss. This being so, it seems reasonable to demand that, for this game, rational play by intelligent players should bring about this payoff of 3 units. P_1 will choose the second row, while P_2 chooses the second column. Nor is this all: if P_1 suspects that P_2 will choose the second column, this will only make it more imperative that he choose the second row, and, conversely, if P_2 suspects P_1 of choosing the second row, he will have all the more reason to choose the second column. The reasons for this outcome reinforce each other. The outcome of the game is, in a sense, determined; we shall say that the game is *strictly determined* and has a *value* of 3 units.

The fundamental property of this game is that the two strategies (second row, second column) are *in equilibrium*: neither player can gain by a unilateral change from these strategies. This is clear since the entry corresponding to these two strategies is both the largest in its column (so that P_1 cannot gain by changing) and the smallest in its row (so that P_2 cannot gain by changing). Such an entry is known as a *saddle point*:

VI.2.2 Definition. Let A be a matrix game. The entry a_{ij} is said to be a *saddle point* if, for any value of k,

6.2.1 $$a_{ij} \geq a_{kj}$$

and, for every value of l,

6.2.2 $$a_{ij} \leq a_{il}$$

The name, saddle point, is derived by analogy from the surface of a saddle, which curves up in one direction and down in another (Figure VI.2.1).

Our analysis suggests that, if a matrix game has one, both players should choose the strategies that produce the saddle point. A question automatically arises about the game with two saddle points. It is not inconceivable that, for such a game, P_1 might try for one saddle point while P_2 tried for another, and (conceivably) something very different from a saddle point might arise. There are, in fact, games that show such pathologies, but these are not matrix games.

FINITE MATHEMATICS

FIGURE VI.2.1 A saddle-shaped surface.

VI.2.3 Theorem. Let a_{ij} and a_{kl} be saddle points for the matrix game A. Then a_{kj} and a_{il} are also saddle points, and moreover,

$$a_{ij} = a_{kl}$$

Proof. Since a_{ij} is a saddle point, we must have

$$a_{ij} \leq a_{il}$$
$$a_{ij} \geq a_{kj}$$

Since a_{kl} is also a saddle point,

$$a_{kl} \geq a_{il}$$
$$a_{kl} \leq a_{kj}$$

Thus we have

$$a_{ij} \leq a_{il} \leq a_{kl}$$

and

$$a_{ij} \geq a_{kj} \geq a_{kl}$$

But this means that

$$a_{ij} = a_{il} = a_{kj} = a_{kl}$$

Now, a_{ij} is the smallest entry in the ith row, while a_{kl} is the largest entry in the lth column. The entry a_{il} is in the ith row and lth column, and is equal to both a_{ij} and a_{kl}. But this means a_{il} is also the smallest entry in the row (tied with a_{ij}) and the largest in its column. Therefore, a_{il} is a saddle point. Similarly a_{kj} is a saddle point.

Theorem VI.2.3 shows that saddle points for matrix games are always equivalent (i.e., the payoff is the same) and interchangeable; if P_1 aims for a saddle point a_{ij}, while P_2 aims for a different saddle point a_{kl}, the saddle point a_{il} will be obtained. The existence of several saddle points does not present a problem: the outcome will always be the same.

276 PROBLEMS ON SADDLE POINTS

1. Solve the following matrix games:

(a) $\begin{pmatrix} 3 & 5 \\ 4 & 6 \end{pmatrix}$

(b) $\begin{pmatrix} 3 & 6 & 4 \\ 1 & 0 & 8 \end{pmatrix}$

(c) $\begin{pmatrix} 2 & 0 & 6 \\ 5 & 1 & 3 \\ 7 & -2 & 1 \end{pmatrix}$

(d) $\begin{pmatrix} 3 & 5 & 1 & 2 \\ 4 & 8 & 3 & 0 \\ 5 & 4 & 3 & 6 \end{pmatrix}$

(e) $\begin{pmatrix} 7 & 6 & 1 & 2 \\ 5 & 5 & 4 & 6 \\ 4 & 8 & 3 & 0 \end{pmatrix}$

2. Two competing companies must build their next branch at one of three cities, A, B, and C, whose location and distances are shown in Figure VI.2.2. If both companies build in the same city, they will split the business evenly. If, however, they build in different cities, the company that is closer to a given city will get all that city's business. If all three cities have the same amount of business, where should the companies build?

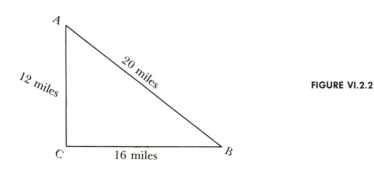

FIGURE VI.2.2

3. Show that, if all the entries in a 2×2 matrix are positive, and the matrix is not invertible, then it must have a saddle point.

4. Show that, in a 2×2 matrix, if both entries in one row are equal, then the matrix must have a saddle point.

5. Show that, if a 2×2 or 2×3 matrix has a saddle point, then either one row dominates the other, or one column dominates another. Show by a counterexample that this statement does not generalize to 3×3 matrices.

3. MIXED STRATEGIES

In the previous section we studied the properties of saddle points; arguments were given for the use of saddle point, or equilibrium, strategies, by both players. Unfortunately, only comparatively few games have saddle points. A game as simple as matching pennies (Examples VI.1.1 and VI.1.4), with matrix

$$\begin{pmatrix} -1 & 1 \\ 1 & -1 \end{pmatrix}$$

fails to have a saddle point. For such a game, it is difficult to see what either player should do. P_1 can obtain a gain-floor of -1 unit, while P_2 can obtain a loss-ceiling of 1 unit, but these both represent definite losses. Nor does it help to try to outguess the opponent: P_1 can be hurt just as much by crediting P_2 with too much intelligence as by not crediting him with enough. In fact, we can imagine P_1 reasoning along these lines: "People generally choose H, so P_2 will expect me to choose H, and he himself will choose H. Therefore I should choose T. But perhaps he has reasoned that way and expects me to choose T. Therefore I had better choose H. But perhaps that is P_2's reasoning, too, and so I should choose T. But then ...." It is clear that this sort of thing will get the players nowhere. The difficulty, in brief, is this: for a strategy to be "good" (whatever meaning may be attached to this word), it must be rationally chosen. But, if a player chooses rationally, his opponent (who is also rational) can reproduce his thinking process and guess what he plans to do. Yet the essence of a game such as this is, precisely, a need for secrecy: a player loses if his opponent discovers what he plans to do. This is in sharp contrast to the situation for games with saddle points, in which knowledge of each other's plans does not cause the players to change them. (The chief trouble is that there is no equilibrium here.)

The apparent conflict between rationality and unpredictability was solved, almost simultaneously, by von Neumann and Borel. In essence their suggestion was quite simple: it had been in use, more or less unconsciously, by most gambling establishments. The suggestion, in brief, was to choose the strategy by means of some random mechanism. There was, however, a new element introduced: the random mechanism would behave according to a rationally chosen scheme. In other words, the probabilities of each of the possible strategies should be chosen rationally to be "good." But, while the opponent can figure out the exact form of the randomization scheme (by somehow reproducing our manner of thinking), he cannot decide which strategy is to be used at each play of the game (since the randomization mechanism acts irrationally).

This gives us, then, the idea of a mixed strategy: it is a probability distribution on the set of a player's strategies (henceforth called *pure strategies* to distinguish them from the mixed strategies).

VI.3.1 Definition. Let A be an $m \times n$ matrix. Then, a *mixed strategy* for P_1 (in the matrix game A) is a vector $x = (x_1, x_2, \ldots, x_m)$ satisfying

6.3.1
$$\sum_{i=1}^{m} x_i = 1$$

6.3.2
$$x_i \geq 0 \qquad i = 1, \ldots, m$$

Similarly, a mixed strategy for P_2 is a vector $y = (y_1, \ldots, y_n)$ satisfying

6.3.3
$$\sum_{j=1}^{n} y_j = 1$$

6.3.4
$$y_j \geq 0 \qquad j = 1, \ldots, n$$

For the game of "matching pennies," described earlier the obvious suggestion is to decide on H or T by tossing a coin. Assuming the coin to be fair, this gives the mixed strategy $(1/2, 1/2)$ for P_1. It is easy to see that, whatever P_2 does, the game will have an expected payoff 0. In other words, use of a mixed strategy gives P_1 a gain-floor of 0 compared to the gain-floor of -1 that he can obtain by using pure strategies. In a similar way, P_2 can use the mixed strategy $(1/2, 1/2)$, which gives him a loss-ceiling of 0 compared to the loss-ceiling of 1 obtained by using pure strategies. We see that gain-floor and loss-ceiling are equal; it follows that the two mixed strategies that guarantee the value of 0 are in equilibrium, and we say that the game has a *saddle point in mixed strategies*. Thus, if we admit mixed strategies and agree to the use of expectations (as we said earlier that we would), it follows that this game is almost as well determined as the games studied in Section 2 of this chapter: both players should use the mixed strategies $(1/2, 1/2)$; moreover, knowledge of the opponent's mixed strategy should not cause either player to change his own.

The question naturally arises whether every game will have a saddle point in mixed strategies. The answer is affirmative. We shall leave its proof until later; first, we give an analysis of the situation.

Let us assume that, for the matrix game A, player P_1 chooses the mixed strategy $\mathbf{x}$, while P_2 chooses $\mathbf{y}$. Since the two randomizations are clearly independent, the probability that P_1 chooses his ith (pure) strategy, while P_2 chooses his jth (pure) strategy, will be $x_i y_j$. The payoff in this case is a_{ij}, and so the expected payoff will be

6.3.5
$$A(\mathbf{x},\mathbf{y}) = \sum_{i=1}^{m} \sum_{j=1}^{m} a_{ij} x_i y_j$$

or, in matrix notation,

6.3.6
$$A(\mathbf{x},\mathbf{y}) = \mathbf{x}A\mathbf{y}^t$$

Let us suppose, now, that P_1 is considering using the mixed strategy $\mathbf{x}$. As before, we assume that he is cautious; therefore he will at least look at the worst that can happen to him with such a strategy. His security level, then, is

6.3.7
$$u(\mathbf{x}) = \min_{\mathbf{y} \in Y} \mathbf{x} A \mathbf{y}^t$$

where the minimum is taken over all possible mixed strategies, $\mathbf{y}$, for P_2.

Similarly, if P_2 uses the strategy $\mathbf{y}$, his security level is

6.3.8
$$w(\mathbf{y}) = \max_{\mathbf{x} \in X} \mathbf{x} A \mathbf{y}^t$$

where the maximum is taken over all possible strategies, $\mathbf{x}$, for P_1.

It is not difficult to see that, if $\mathbf{x}$ and $\mathbf{y}$ are mixed strategies for P_1 and P_2 respectively, then

6.3.9
$$u(\mathbf{x}) \le w(\mathbf{y})$$

In fact, P_1 can assure himself of the amount $u(\mathbf{x})$ (in expectation), while P_2 can insure himself against losing more than $w(\mathbf{y})$. Thus (6.3.9) must hold. Suppose, however, that there exist strategies $\mathbf{x}^*$ and $\mathbf{y}^*$ such that

6.3.10
$$u(\mathbf{x}^*) = w(\mathbf{y}^*) = v$$

In that case, it is clear that $\mathbf{x}^*$ and $\mathbf{y}^*$ are *optimal* in the sense that they maximize and minimize, respectively, the two players' security levels. (It may be noted that the situation here resembles the one we observed when dealing with dual linear programs; that this is indeed the case will be established.) The number v, which is the common value of the security levels, will be known as the *value* of the game, and it is clear that, in expectation, both players can guarantee the value of the game by using the mixed strategies $\mathbf{x}^*$ and $\mathbf{y}^*$ respectively. Neither can hope to gain from switching unilaterally; these two strategies are in equilibrium.

Let us consider the expression (6.3.7). This defines $u(\mathbf{x})$ as the minimum of the terms $\mathbf{x}A\mathbf{y}^t$. This can best be evaluated if we write

$$\sum_{i=1}^{m}\sum_{j=1}^{n} x_i a_{ij} y_j = \sum_{j=1}^{n} \left\{ y_j \sum_{i=1}^{m} x_i a_{ij} \right\}$$

It may be seen that $\Sigma x_i a_{ij}$ is the jth component of the vector $\mathbf{x}A$. We shall write this $(\mathbf{x}A)_j$, and so

$$\mathbf{x}A\mathbf{y}^t = \sum_{j=1}^{n} y_j (\mathbf{x}A)_j$$

so that $\mathbf{x}A\mathbf{y}^t$ is an average (i.e., expected value) of the components of the vector $\mathbf{x}A$. It will follow that (for a fixed $\mathbf{x}$) this expectation $\mathbf{x}A\mathbf{y}^t$ can be made as small as the smallest component of $\mathbf{x}A$, and no smaller. In other words, the security level $u(\mathbf{x})$ is the smallest component of the vector $\mathbf{x}A$. (This suggests that P_2 can get along using pure strategies only; as, in fact, he could if he knew P_1's choice of strategy. But P_2 must also use mixed strategies to prevent P_1 from finding out.) Thus,

6.3.11 $$u(\mathbf{x}) = \min_{j} \sum_{i=1}^{m} x_i a_{ij}$$

Similarly,

6.3.12 $$w(\mathbf{y}) = \max_{i} \sum_{j=1}^{n} a_{ij} y_j$$

We obtain, then, the following definitions:

VI.3.2 Definition. Let A be a matrix game. We say that v is the *value* of A if there exist strategies $\mathbf{x}^*$ and $\mathbf{y}^*$ (for P_1 and P_2, respectively) such that

6.3.13 $$\sum_{i=1}^{m} x_i^* a_{ij} \geq v \qquad \text{for } j = 1, \ldots, n$$

and

6.3.14 $$\sum_{j=1}^{n} a_{ij} y_j^* \leq v \qquad \text{for } i = 1, \ldots, m$$

In this case, the strategies $\mathbf{x}^*$ and $\mathbf{y}^*$ are known as *optimal strategies*. A *solution* of the game is a triple $(v, \mathbf{x}^*, \mathbf{y}^*)$ consisting of the value and a pair of optimal strategies.

In terms of matrix operations, the conditions (6.3.13) and (6.3.14) can be restated in the form

6.3.15 $$xA \geq vJ$$

6.3.16 $$Ay^t \leq vJ^t$$

where J is the row vector $(1,1,\ldots,1)$ of desired size (m components in (6.3.15), and n components in (6.3.16)). More briefly, we might say that each component of xA is at least v, and each component of Ay^t is at most v.

It should be clear that the value, if it exists, must be unique (as it represents, simultaneously, a gain-floor for P_1 and a loss-ceiling for P_2). We shall prove that it always does exist.

Applying this definition to the game of matching pennies we find that the value is 0, with optimal strategies $(1/2, 1/2)$ for both players. In fact,

$$\left(\frac{1}{2}, \frac{1}{2}\right)\begin{pmatrix} -1 & 1 \\ 1 & -1 \end{pmatrix} = (0,0)$$

and

$$\begin{pmatrix} -1 & 1 \\ 1 & -1 \end{pmatrix}\begin{pmatrix} \frac{1}{2} \\ \frac{1}{2} \end{pmatrix} = \begin{pmatrix} 0 \\ 0 \end{pmatrix}$$

so that all the components of xA and Ay^t are 0. It is not too difficult to see that the optimal strategies are unique. In fact, suppose x is any strategy for P_1. We have

$$(x_1, x_2)\begin{pmatrix} -1 & 1 \\ 1 & -1 \end{pmatrix} = (x_1 - x_2, x_2 - x_1)$$

and it is clear that one of the two components of xA will be negative unless $x_1 = x_2$. Thus, only $(1/2, 1/2)$ has the security level 0 for P_1. A similar argument shows the uniqueness of P_2's optimal strategy.

VI.3.3 Example (Scissors, Stone, and Paper). In a well-known Oriental game, the two players simultaneously expose a hand, which, depending on its position, represents scissors (two open fingers), stone (a clenched fist), or paper (an open hand). There is a payoff of 1 unit according to the rule: scissors cut paper, paper covers stone, stone breaks scissors. If both show the same, the game is a standoff.

The matrix for this game is

$$
\begin{array}{c c c}
 & St & Sc & P \\
\begin{matrix} St \\ Sc \\ P \end{matrix} &
\begin{pmatrix}
0 & 1 & -1 \\
-1 & 0 & 1 \\
1 & -1 & 0
\end{pmatrix}
\end{array}
$$

It is not difficult to guess (from the symmetry of the game) that the value must be 0, with optimal strategies (1/3, 1/3, 1/3) for both players. In fact,

$$
\left(\frac{1}{3},\frac{1}{3},\frac{1}{3}\right)
\begin{pmatrix}
0 & 1 & -1 \\
-1 & 0 & 1 \\
1 & -1 & 0
\end{pmatrix} = (0,0,0)
$$

and

$$
\begin{pmatrix}
0 & 1 & -1 \\
-1 & 0 & 1 \\
1 & -1 & 0
\end{pmatrix}
\begin{pmatrix}
\frac{1}{3} \\ \frac{1}{3} \\ \frac{1}{3}
\end{pmatrix} =
\begin{pmatrix}
0 \\ 0 \\ 0
\end{pmatrix}
$$

4. SOLUTION OF 2 × 2 GAMES

Suppose we are given a 2 × 2 game matrix. To look for optimal strategies, we look first of all for a saddle point (in pure strategies); this is easy, as it is merely a question of checking whether P_1's gain-floor (the maximum of the row minima) and P_2's loss-ceiling (the minimum of the column maxima) in pure strategies are equal. If such a saddle point exists, there is no further problem; this gives us both the optimal strategies (which are then pure strategies) and the value. Let us assume, then, that no saddle point exists. This means that the optimal strategies (if any such exist) must be mixed strategies in which each player uses both his pure strategies with positive probability.

Conditions (6.3.13) can be restated

$$
a_{11}x_1 + a_{21}x_2 \geq v
$$
$$
a_{12}x_1 + a_{22}x_2 \geq v
$$

Suppose, however, that strict inequality held in either of these two conditions. Say, for instance,

$$
a_{11}x_1 + a_{21}x_2 > v
$$

It is not difficult to see that, in this case, P_2 should never use his first pure strategy. For, if $y_1 > 0$, we will have

$$\mathbf{x}A\mathbf{y}^t = (a_{11}x_1 + a_{21}x_2)y_1 + (a_{12}x_1 + a_{22}x_2)(1-y_1) > vy_1 + v(1-y_1)$$

or

$$\mathbf{x}A\mathbf{y}^t > v$$

which contradicts the optimality of **y**. (Heuristically, we are saying here that it cannot be optimal for P_2 to consider using a strategy that gives him an expectation smaller than v against P_1's optimal strategy.) We conclude that, if P_2 is to use both his strategies with positive probability we must have

6.4.1 $\qquad\qquad a_{11}x_1 + a_{21}x_2 = v$

6.4.2 $\qquad\qquad a_{12}x_1 + a_{22}x_2 = v$

Similarly, if P_1 uses both his strategies with positive probability, we will have

6.4.3 $\qquad\qquad a_{11}y_1 + a_{12}y_2 = v$

6.4.4 $\qquad\qquad a_{21}y_1 + a_{22}y_2 = v$

We have, additionally, the "strategy" conditions

6.4.5 $\qquad\qquad x_1 + x_2 = 1$

6.4.6 $\qquad\qquad y_1 + y_2 = 1$

This gives us a system of six equations in only five variables (x_1, x_2, y_1, y_2, and v). It so happens, however, that one of the equations is always redundant, and so a solution will always exist. Moreover, this solution will be unique and will additionally satisfy the non-negativity constraints $\mathbf{x} \geq 0$, $\mathbf{y} \geq 0$ unless the game has a saddle point in pure strategies. In fact, it is not difficult to see that the solution of this system is

6.4.7 $\qquad\qquad x_1 = \dfrac{a_{22} - a_{21}}{a_{11} - a_{12} - a_{21} + a_{22}}$

6.4.8 $\qquad\qquad x_2 = \dfrac{a_{11} - a_{12}}{a_{11} - a_{12} - a_{21} + a_{22}}$

6.4.9 $\qquad\qquad y_1 = \dfrac{a_{22} - a_{12}}{a_{11} - a_{12} - a_{21} + a_{22}}$

6.4.10

$$y_2 = \frac{a_{11} - a_{21}}{a_{11} - a_{12} - a_{21} + a_{22}}$$

6.4.11

$$v = \frac{a_{11} a_{22} - a_{21} a_{12}}{a_{11} - a_{12} - a_{21} + a_{22}}$$

Unless the game has a saddle point (in pure strategies), the numerators in equations (6.4.7) to (6.4.10) will all have the same sign, and so the non-negativity constraints are satisfied.

A simpler way of expressing these equations would be as follows: if the game has no saddle point, take the game matrix

$$A = \begin{pmatrix} a_{11} & a_{12} \\ a_{21} & a_{22} \end{pmatrix}$$

Interchange the two elements on the main diagonal and change the signs of the other two elements to obtain the *adjoint* matrix:

$$A^* = \begin{pmatrix} a_{22} & -a_{12} \\ -a_{21} & a_{11} \end{pmatrix}$$

Form the vectors JA^* and A^*J^t, where $J = (1,1)$. (This corresponds to adding the rows and adding the columns of the matrix, A^*, respectively):

$$JA^* = (1,1) \begin{pmatrix} a_{22} & -a_{12} \\ -a_{21} & a_{11} \end{pmatrix} = (a_{22} - a_{21}, a_{11} - a_{12})$$

$$A^*J^t = \begin{pmatrix} a_{22} & -a_{12} \\ -a_{21} & a_{11} \end{pmatrix} \begin{pmatrix} 1 \\ 1 \end{pmatrix} = \begin{pmatrix} a_{22} - a_{12} \\ a_{11} - a_{21} \end{pmatrix}$$

The optimal strategies will then be *proportional* to JA^* and A^*J^t, respectively, i.e., JA^* and A^*J^t must be multiplied by some constant to guarantee that the sum of their components is equal to 1.

We give the following examples of the use of these equations to solve games.

VI.4.1 *Example (Modified Matching Pennies).* Let us consider the game of matching pennies, with the following modification: if P_2 fails to guess P_1's call correctly, then he must pay P_1 2 units. If both P_1 and P_2 call H, then P_1 must pay P_2 3 units. If both call T, then P_1 pays P_2 1 unit.

The game matrix is

$$A = \begin{pmatrix} -3 & 2 \\ 2 & -1 \end{pmatrix}$$

At first glance, this game seems to be fair (i.e., the value seems to be zero). Application of equations (6.4.7) to (6.4.10), however, gives the optimal strategies and value

$$x_1 = \frac{-1-2}{-3-2-2-1} = \frac{3}{8}$$

$$x_2 = \frac{-3-2}{-3-2-2-1} = \frac{5}{8}$$

$$y_1 = \frac{-1-2}{-3-2-2-1} = \frac{3}{8}$$

$$y_2 = \frac{-3-2}{-3-2-2-1} = \frac{5}{8}$$

$$v = \frac{(-1)(-3) - (2)(2)}{-3-2-2-1} = \frac{1}{8}$$

revealing that the game is favorable to P_1 as it has positive value $v = 1/8$. It is easily checked that the strategy $(3/8, 5/8)$ is optimal for both players, guaranteeing the value.

VI.4.2 Example. A secret service agent must carry a message from his headquarters to a friend's office. He can take either of two routes, one of them deserted, the other leading through the center of a city. His opponent, a corrupt law-enforcement officer, wishes to prevent him from delivering the message and must guess which route to guard. If the corrupt officer catches the secret agent on the deserted road, he will kill him; if he catches him on the city street, he will merely be able to arrest him and delay him for some time. We shall assign to the agent a utility of $+5$ if he delivers the message, of -1 if he is arrested, and of -100 if he is killed.

The matrix for this game is

$$\begin{pmatrix} -1 & 5 \\ 5 & -100 \end{pmatrix}$$

Application of equations (6.4.7) to (6.4.10) to this game gives us the optimal strategies and value

$$x = \left(\frac{35}{37}, \frac{2}{37}\right)$$

$$y = \left(\frac{35}{37}, \frac{2}{37}\right)$$

$$v = -\frac{25}{37}$$

Note that, according to the solution, P_1 takes the safe route (first strategy) with very high probability; he must, however, take the riskier route with some small probability, if only to make his opponent guess. (A randomization scheme such as this is best effected by putting in a hat 37 pieces of paper, 35 of them black and 2 white. Then a paper is drawn from the hat; if black, the safe route is chosen.)

5. $2 \times n$ AND $m \times 2$ GAMES

In case one player has two strategies, but the other has more than two, it is nevertheless possible to solve the game by applying the foregoing rules with some modification. The fact is that, in any $m \times 2$ or $2 \times n$ matrix game, there is always a 2×2 sub-matrix that will serve to give us the solution of the game.

As an example, consider the 3×2 game

$$\begin{pmatrix} 0 & 3 & 2 \\ 4 & 2 & 3 \end{pmatrix}$$

which has no row or column dominations and no saddle points. Let us suppose that P_2 restricts himself to two of his three strategies. Depending on which strategy he discards, we will obtain one of the three matrix games

$$A_1 = \begin{pmatrix} 0 & 3 \\ 4 & 2 \end{pmatrix} \quad A_2 = \begin{pmatrix} 0 & 2 \\ 4 & 3 \end{pmatrix} \quad A_3 = \begin{pmatrix} 3 & 2 \\ 2 & 3 \end{pmatrix}$$

with values

$$v_1 = \frac{12}{5} \qquad v_2 = \frac{8}{3} \qquad v_3 = \frac{5}{2}$$

respectively.

Of these values, v_1 is the smallest. Hence, if P_2 is going to restrict himself to two strategies, he should choose those that give this matrix: the first and second columns. Under these restrictions, the optimal strategies, obtained by applying rules (6.4.7) to (6.4.10) to the 2×2 matrix A_1 are

$$x = \left(\frac{2}{5}, \frac{3}{5} \right)$$

$$y = \left(\frac{1}{5}, \frac{4}{5}, 0 \right)$$

It is now a simple matter to check that these are optimal strategies for the larger game A. In fact,

$$x A = \left(\frac{2}{5}, \frac{3}{5}\right) \begin{pmatrix} 0 & 3 & 2 \\ 4 & 2 & 3 \end{pmatrix} = \left(\frac{12}{5}, \frac{12}{5}, \frac{13}{5}\right)$$

so that x gives P_1 the security level 12/5. Similarly,

$$A y^t = \begin{pmatrix} 0 & 3 & 2 \\ 4 & 2 & 3 \end{pmatrix} \begin{pmatrix} \frac{1}{5} \\ \frac{4}{5} \\ 0 \end{pmatrix} = \begin{pmatrix} \frac{12}{5} \\ \frac{12}{5} \end{pmatrix}$$

and the value of the game is, indeed, 12/5.

It is sometimes even possible, by using the symmetry of a game, to reduce larger games to the form of a $2 \times n$ or $m \times 2$ matrix.

VI.5.1 Example. The well-known military strategist, Colonel Blotto, has three divisions, which he must use to capture as many as possible of three mountain passes, against an opponent with two divisions. We shall assume, first, that divisions cannot be split, and, second, that a superiority of one division (or more) at a pass will guarantee capture of that pass. Blotto scores one point for each pass he captures, and one point for each enemy division overpowered. He loses one point for each contrary instance.

We will designate each strategy as a triple denoting the number of divisions sent to each pass. Thus, 201 represents Blotto's strategy of sending two divisions to the first pass, and one to the third pass. If, for example, Blotto uses 201, while his opponent uses 002, we find that Blotto captures the first pass, scoring one point, but loses the third pass, and additionally has a unit overpowered on that pass, giving him a net of −1 point. We obtain, thus, the 10×6 matrix

	200	020	002	110	101	011
300	3	0	0	1	1	−1
030	0	3	0	1	−1	1
003	0	0	3	−1	1	1
210	1	−1	1	2	2	0
201	1	1	−1	2	2	0
120	−1	1	1	2	0	2
021	1	1	−1	2	0	2
102	−1	1	1	0	2	2
012	1	−1	1	0	2	2
111	0	0	0	1	1	1

288

This game is much too large to solve by the methods we have studied so far. It is not too difficult, however, to see that Blotto's strategies can be classified into three types, which can be called, respectively, 3, 21, and 111 according to the way his forces are divided (but regardless of their destination). By symmetry, it is also possible to see that strategies of the same type should be used with the same probability. Similarly, his opponent's strategies can be classified into two types, called 2 and 11 respectively. Again, symmetry suggests that strategies of the same type should be used with equal probabilities.

We can think of this as a two-stage decision process: first, each player decides the *class* of strategy he will use; later, the particular strategy. If Blotto decides to use type 21, and his opponent decides to use type 11, we find that the expected payoff to Blotto is 4/3 points (2 points in 2/3 of the cases, and 0 in 1/3 of the cases). In this manner, we obtain the *reduced* matrix

$$\begin{array}{cc} & \begin{array}{cc} 2 & 1,1 \end{array} \\ \begin{array}{c} 3 \\ 2,1 \\ 1,1,1 \end{array} & \begin{pmatrix} 1 & \frac{1}{3} \\ \frac{1}{3} & \frac{4}{3} \\ 0 & 1 \end{pmatrix} \end{array}$$

In this matrix, we see that the second row dominates the third. We have then the 2×2 game

$$\begin{pmatrix} 1 & \frac{1}{3} \\ \frac{1}{3} & \frac{4}{3} \end{pmatrix}$$

with solution $x = (3/5, 2/5)$; $y = (3/5, 2/5)$; $v = 11/15$. The 3×2 game will have optimal strategies

$$x = \left(\frac{3}{5}, \frac{2}{5}, 0 \right)$$

$$y = \left(\frac{3}{5}, \frac{2}{5} \right)$$

and remembering that strategies of the same type must have the same probability, we obtain the optimal strategies

$$\bar{x} = \left(\frac{1}{5}, \frac{1}{5}, \frac{1}{5}, \frac{1}{15}, \frac{1}{15}, \frac{1}{15}, \frac{1}{15}, \frac{1}{15}, \frac{1}{15}, 0 \right)$$

$$\bar{y} = \left(\frac{1}{5}, \frac{1}{5}, \frac{1}{5}, \frac{2}{15}, \frac{2}{15}, \frac{2}{15} \right)$$

FINITE MATHEMATICS

and value

$$v = \frac{11}{15}$$

for the original 10×6 game.

Note that the reduction here was possible only because of the complete symmetry existing among strategies of the same type: in a sense, the three mountain passes are equivalent.

PROBLEMS ON $2 \times n$ AND $m \times 2$ GAMES

1. Solve the following matrix games.

(a) $\begin{pmatrix} 3 & 5 \\ 8 & 1 \end{pmatrix}$

(b) $\begin{pmatrix} 4 & 6 & 3 \\ 1 & 2 & 5 \end{pmatrix}$

(c) $\begin{pmatrix} 2 & 4 & 3 \\ 1 & 0 & 2 \end{pmatrix}$

(d) $\begin{pmatrix} 1 & 5 & 7 \\ 6 & 2 & 8 \end{pmatrix}$

(e) $\begin{pmatrix} 6 & 2 & 1 \\ 3 & 4 & 5 \end{pmatrix}$

2. A game is played as follows. Player P_1 is given the aces of spades and diamonds, and the deuce of hearts. P_2 is given the aces of hearts and clubs, and the deuce of spades. Each then chooses a card, and these are shown simultaneously, with payoffs as follows. If both cards are of the same suit, then the ace beats the deuce. If the cards are of different suits and different denominations, then the black card beats the red. If they are of the same denomination, then the red card beats the black. If they are both black aces or both red aces, there is a standoff. In any case, the player winning will win as many dollars as there are spots on his opponent's card. Find optimal strategies and the value of this game.

3. Colonel Blotto, commanding two divisions, must capture an outpost defended by one division. To do this, he must have a numerical superiority of one division when he attacks the outpost. It may be, however, that while Blotto attacks, his opponent simultaneously attacks Blotto's camp. If so, a superiority of one division is also sufficient to win. Blotto gains 1 unit if he captures the opponents's camp without losing his own; he loses 1 unit if the opponent captures his camp. Otherwise the game is a standoff. What is Blotto's optimal strategy?

FINITE MATHEMATICS

6. RELATION TO LINEAR PROGRAMMING

It was mentioned earlier that matrix games are related to linear programs. The connection is quite apparent, since the two *hemigames* (the problems faced by each of the two players in a game) are a maximizing and a minimizing problem. We shall now formulate a linear program for the computation of optimal strategies.

As we know, an optimal strategy can be defined as one that gives a security level equal to the value of the game. Since no higher security level is available for P_1, it follows that P_1's search for optimal strategies reduces to the maximization of his security level. The strategy x gives a security level of λ (or better) if every component of the vector xA is at least equal to λ, i.e., if

6.6.1 $$\sum_{i=1}^{m} a_{ij}x_i \geq \lambda \qquad j=1,\ldots,n$$

P_1 must maximize λ, subject to (6.6.1) and the additional constraint that x must be a strategy. We therefore obtain the linear program

6.6.2 Maximize $\qquad\qquad \lambda$
Subject to

6.6.3 $$-\sum_{i=1}^{m} a_{ij}x_i + \lambda \leq 0 \qquad j=1,\ldots,n$$

6.6.4 $$\sum_{i=1}^{m} x_i = 1$$

6.6.5 $$x_i \geq 0 \qquad i=1,\ldots,m$$

(Note that λ is unrestricted, i.e., it may take negative values.)

In a similar way, P_2 is faced with the problem of minimizing his security level (loss-ceiling) μ. We have, then, the program

6.6.6 Minimize $\qquad\qquad \mu$
Subject to

6.6.7 $$-\sum_{j=1}^{n} a_{ij}y_j + \mu \geq 0 \qquad i=1,\ldots,m$$

6.6.8 $$\sum_{j=1}^{n} y_j = 1$$

6.6.9 $$y_j \geq 0 \qquad j=1,\ldots,n$$

(Again, note that μ is an unrestricted variable.)

As with most linear programs, a question arises about the duals. What, for instance, is the dual of (6.6.2) to (6.6.5)? If we go back to our study of duality in Chapter III (including the fact that an unrestricted variable corresponds to an equation, as was said with respect to transportation problems) we will see that the dual of (6.6.2) to (6.6.5) is precisely (6.6.6) to (6.6.9). In other words, the two hemi-games are mutually dual problems. We saw in Chapter III that, in general, the dual of a linear program can be interpreted only in terms of "shadow prices," but here we have a case in which a linear program and its dual both have practical (and, in fact, very similar) interpretations.

We know, of course, that both the programs (6.6.2) to (6.6.5) and (6.6.6) to (6.6.9) are feasible. We can therefore apply Theorem III.11.5 (the principal theorem of linear programming) to obtain a very important theorem.

VI.6.1 Theorem. Every matrix game has a value.

Theorem VI.6.1. is generally known as the *Minimax theorem,* since it was originally given in the form

6.6.10 $$\min_{y \in Y} \max_{x \in X} xAy^t = \max_{x \in X} \min_{y \in Y} xAy^t$$

The fact that the two hemi-games are mutually dual programs has one additional advantage: not only are the two security levels equal, but computation of P_1's optimal strategy (by using the simplex algorithm) will automatically give us P_2's optimal strategy as well.

VI.6.2 Example. Compute the optimal strategies for the matrix game

$$\begin{pmatrix} 3 & 6 & 1 & 4 \\ 5 & 2 & 4 & 2 \\ 1 & 4 & 3 & 5 \end{pmatrix}$$

We write the maximizing problem in the form

Maximize λ
Subject to

$$3x_1 + 5x_2 + x_3 \geq \lambda$$
$$6x_1 + 2x_2 + 4x_3 \geq \lambda$$
$$x_1 + 4x_2 + 3x_3 \geq \lambda$$
$$4x_1 + 2x_2 + 5x_3 \geq \lambda$$
$$x_1 + x_2 + x_3 = 1$$
$$x_1, x_2, x_3 \geq 0$$

This gives us the tableau

	x_1	x_2	x_3	λ	
	-3	-5	-1	1	$= -u_1$
	-6	-2	-4	1	$= -u_2$
	-1	-4	-3	1	$= -u_3$
	-4	-2	-5	1^*	$= -u_4$
	-1	-1	-1	0	$= -1$

This tableau is not in standard form for linear programs: λ is the objective function and therefore should be put among the basic variables (actually, on the bottom row of the tableau). On the other hand, the 1 should be taken from among the basic variables and put at the top of the right-hand column. Unfortunately, the entry at the bottom right-hand corner is 0 and cannot be used as a pivot, so that we will need two pivot steps to accomplish this. We pivot on the starred entries to obtain the following sequence of tableaux:

	x_1	x_2	x_3	u_4	
	1	-3	4	-1	$= -u_1$
	-2	0	1	-1	$= -u_2$
	3	-2	2	-1	$= -u_3$
	-4	-2	-5	1	$= -\lambda$
	-1	-1	-1^*	0	$= -1$

	x_1	x_2	1	u_4	
	-3	-7	4	-1	$= -u_1$
	-3	-1	1	-1	$= -u_2$
	1	-4	2	-1	$= -u_3$
	1	3	-5	1	$= -\lambda$
	1	1	-1	0	$= -x_3$

We now interchange the third and fourth columns, and also the fourth and fifth rows, to place the 1 and λ in the desired positions. We also change the signs in the row corresponding to λ:

	x_1	x_2	u_4	1	
	-3	-7	-1	4	$= -u_1$
	-3	-1	-1	1	$= -u_2$
	1	-4	-1^*	2	$= -u_3$
	1	1	0	-1	$= -x_3$
	-1	-3	-1	5	$= \lambda$

We proceed now according to the usual rules until the solution of this program is obtained:

x_1	x_2	u_3	1	
-4	-3	-1^*	2	$=-u_1$
-4	3	-1	-1	$=-u_2$
-1	4	-1	-2	$=-u_4$
1	1	0	-1	$=-x_3$
-2	1	-1	3	$=\lambda$

x_1	x_2	u_1	1	
4	3	-1	-2	$=-u_3$
0	6^*	-1	-3	$=-u_2$
3	7	-1	-4	$=-u_4$
1	1	0	-1	$=-x_3$
2	4	-1	1	$=\lambda$

This gives us a basic feasible point. We now proceed as always in Stage II:

x_1	u_2	u_1	1	
4^*	$-1/2$	$-1/2$	$-1/2$	$=-u_3$
0	$1/6$	$-1/6$	$-1/2$	$=-x_2$
3	$-7/6$	$1/6$	$-1/2$	$=-u_4$
1	$-1/6$	$1/6$	$-1/2$	$=-x_3$
2	$-2/3$	$-1/3$	3	$=\lambda$

	u_3	u_2	u_1	1	
v_1	$1/4$	$-1/8$	$-1/8$	$-1/8$	$=-x_1$
v_2	0	$1/6$	$-1/6$	$-1/2$	$=-x_2$
y_4	$-3/4$	$-19/24$	$13/24$	$-1/8$	$=-u_4$
v_3	$-1/4$	$-1/24$	$7/24$	$-3/8$	$=-x_3$
-1	$-1/2$	$-5/12$	$-1/12$	$13/4$	$=\lambda$
	$=y_3$	$=y_2$	$=y_1$	$=\mu$	

This last tableau gives us the solution of the game: P_1's optimal strategy is

$$x = \left(\frac{1}{8}, \frac{1}{2}, \frac{3}{8}\right)$$

The dual variables, which were omitted in the intermediate steps, tell us that P_2's optimal strategy is

$$y = \left(\frac{1}{12}, \frac{5}{12}, \frac{1}{2}, 0\right)$$

The value of the game is 13/4. (This may be checked directly by forming the products xA and Ay^t.)

While it is normal to reduce a matrix game to a linear program (for purposes of solution) it is, conversely, possible to reduce a linear program to a matrix game. To show how this is done, we give first a definition.

VI.6.3 Definition. A square matrix, A, is *skew-symmetric* if $A = -A^t$.

Examples of skew-symmetric matrices are

$$\begin{pmatrix} 0 & 1 \\ -1 & 0 \end{pmatrix} \qquad \begin{pmatrix} 0 & 1 & -3 \\ -1 & 0 & 2 \\ 3 & -2 & 0 \end{pmatrix} \qquad \begin{pmatrix} 0 & 4 & 1 \\ -4 & 0 & 2 \\ -1 & -2 & 0 \end{pmatrix}$$

A game with skew-symmetric matrix is said to be *symmetric*, and it is not difficult to prove that the value of such a game is 0.

Take, now, the pair of dual linear programs

6.6.11   Maximize xc^t
          Subject to

6.6.12 $xA \leq b$

6.6.13 $x \geq 0$

and

6.6.14   Minimize yb^t
          Subject to

6.6.15 $yA^t \geq c$

6.6.16 $y \geq 0$

and consider the $(m+n+1) \times (m+n+1)$ game matrix

$$M = \begin{pmatrix} \theta & A^t & -b^t \\ -A & \theta & c^t \\ b & -c & \theta \end{pmatrix}$$

where the θ's are zero-matrices of correct size and shape, and A, b, c, are as just given. It is easy to see that the matrix M is skew-symmetric, and so must, as a game, have value 0.

Let

$$p = (p_1, \ldots, p_n; p_{n+1}, \ldots, p_{n+m}; p_{n+m+1})$$

be an optimal strategy (for P_1) for this game, such that

$$p_{n+m+1} > 0$$

and define x and y by

6.6.17
$$x_i = \frac{p_{n+i}}{p_{n+m+1}} \qquad i = 1, \ldots, m$$

6.6.18
$$y_j = \frac{p_j}{p_{n+m+1}} \qquad j = 1, \ldots, n$$

Then, it is not difficult to show that x and y solve the dual problems (6.6.11) to (6.6.13) and (6.6.14) to (6.6.16) respectively.

If $p_{n+m+1} = 0$, it is of course impossible to do this. However, it may be shown that, if both programs are feasible, there will always be an optimal strategy, p, for the game, M, such that $p_{n+m+1} > 0$.

PROBLEMS ON LINEAR PROGRAMMING SOLUTION OF GAMES

1. Solve the following matrix games.

(a) $\begin{pmatrix} 3 & 5 & 1 & 2 \\ 4 & 6 & 0 & 1 \\ 2 & 1 & 5 & 3 \end{pmatrix}$

(d) $\begin{pmatrix} 3 & 0 & 1 \\ 1 & 5 & 2 \\ 2 & 3 & 0 \end{pmatrix}$

(b) $\begin{pmatrix} 3 & 5 & 2 \\ 4 & 2 & 6 \\ 1 & 6 & 3 \end{pmatrix}$

(e) $\begin{pmatrix} 0 & 1 & -2 \\ -1 & 0 & 3 \\ 2 & -3 & 0 \end{pmatrix}$

(c) $\begin{pmatrix} 1 & 3 & 1 & 5 \\ 2 & 0 & 4 & 3 \\ 5 & 2 & 3 & 0 \end{pmatrix}$

2. An $n \times n$ matrix $A = (a_{ij})$ is said to be *skew-symmetric* if, for each pair (i,j)

$$a_{ij} = -a_{ji}$$

Show that, if a game has a skew-symmetric matrix, the value of the game must be zero. (Hint: compute the products xAx^t, where x is any strategy.)

3. Two duelists, each holding a pistol with one shot, advance towards each other. After taking n steps (if neither has been shot before this) they will be next to each other.

Assume that, if one of the duelists shoots after taking k steps, there is a probability k/n that he will hit his opponent. Assume also that the duel is "silent," i.e., a player does not know whether his opponent has shot, unless he is hit. Then each player has n strategies, corresponding to the number of steps taken before shooting. Let the payoff be $+1$ if P_1 hits his opponent without being himself hit, and -1 in the contrary instance. The game is a standoff if both fire and hit each other simultaneously or if neither is hit.

(a) Compute the game matrix.

(b) Solve this game for $n = 2, 3, 4$, and 5.

4. Two players, P_1 and P_2, are given respectively the ace, deuce and trey of spades, and the ace, deuce, and trey of hearts. The ace, deuce, and trey of diamonds are placed, face up, on the board. Each player then places one of his cards (face down) next to one of the diamonds. The cards are then turned, and the player with the higher card captures the corresponding diamond. The payoff is the difference in points scored; the points are equal to the number of spots on the diamond cards captured. Find optimal strategies for this game.

5. Modify Problem 4 to the case in which the cards are the deuce, trey, and four of each suit.

6. Modify Problem 3, page 289 to the case in which Blotto has three divisions, and his opponent, two divisions. Find optimal strategies and the value.

7. SOLUTION OF GAMES BY FICTITIOUS PLAY

Let us consider the following hypothetical situation: a pair of players, perhaps ignorant of game theory, play a game a large number of times. While they may know nothing about game theory, they are, nevertheless statistically inclined, and so each keeps a complete record of strategies used by his opponent on the previous trial of the game. Each then proceeds to play according to the (rather naïve) rule of choosing that pure strategy that is optimal against his opponent's observed mixture of pure strategies.

To express this mathematically, let us suppose that, during the first k trials of the game, P_2 has used his jth strategy Y_j^k times. Then, on the $(k+1)$th trial, P_1 should choose that row which maximizes the sum

6.7.1
$$\sum_{j=1}^{n} a_{ij} Y_j^k$$

Similarly, if P_1 uses his ith strategy X_i^k times during the first k trials, then P_2 should, at the $(k+1)$th trial, choose that column which minimizes

6.7.2
$$\sum_{i=1}^{m} a_{ij} X_i^k$$

There are two things to be said about this procedure. The first is that it is very quickly carried out, since it is merely a question of keeping track of the running totals (6.7.1) and (6.7.2). The second is that it is wonderfully naïve, as it seems to suggest that each of the two players believes he is playing against an irrational opponent (say, nature). And yet, the fact is that, naïve as the procedure seems, it nevertheless solves the game.

Mathematically, we give the following theorem:

VI.7.1 Theorem. Let X^k and Y^k be generated as just explained, and define

6.7.3
$$\mathbf{x}^k = \frac{1}{k} (X_1^k, \ldots, X_m^k)$$

6.7.4
$$\mathbf{y}^k = \frac{1}{k} (Y_1^k, \ldots, Y_n^k)$$

Then $\mathbf{x}^k$ and $\mathbf{y}^k$ are mixed strategies, and moreover, $u(\mathbf{x}^k)$ and $w(\mathbf{y}^k)$ both tend to v as k increases, where v is the value of the game.

Theorem VI.7.1, which is too difficult to prove here, was originally conjectured by G. W. Brown and later proved by J. Robinson. It states that, as k increases, the mixed strategies generated by this method will be nearly optimal. Thus, sufficiently many iterations will yield an approximate solution of the game.

To give an example of this procedure, consider the game with matrix

$$A = \begin{pmatrix} 3 & 1 & 6 & 2 \\ 5 & 8 & 4 & 2 \\ 2 & 3 & 1 & 6 \\ 1 & 5 & 3 & 4 \end{pmatrix}$$

The initial choice may be made arbitrarily; the best way is perhaps to choose the pure strategies with best security level, so perhaps P_1 will choose row 2 while P_2 chooses column 1. Against P_1's choice of row 2, column 4 is best, and against P_2's choice of column 1, row 2 is best. On the second trial, P_1 chooses row 2, while P_2 chooses column 4.

FINITE MATHEMATICS

298 Proceeding in this manner, we find that, on the first 20 trials, P_1's choices are:

$$2,2,3,3,3,1,1,1,1,1,2,2,2,2,2,2,2,2,3,3,3$$

while P_2's choices are

$$1,4,4,3,3,3,3,1,1,1,2,4,4,4,4,4,4,4,4,4$$

which gives us the "empirical" strategies

$$\mathbf{x}^{20} = (0.25,0.45,0.30,0)$$
$$\mathbf{y}^{20} = (0.20,0.05,0.20,0.55)$$

It is easy to calculate that

$$\mathbf{x}A = (3.6,4.75,3.6,3.2)$$

and

$$A\mathbf{y}^t = (2.95,3.3,4.05,3.25)^t$$

so that $u(\mathbf{x}) = 3.2$ and $w(\mathbf{y}) = 4.05$; the "spread" $w(\mathbf{y}) - u(\mathbf{x})$ is still a bit too large. With 50 iterations, we obtain the strategies

$$\mathbf{x}^{50} = (0.32,0.40,0.28,0)$$

and

$$\mathbf{y}^{50} = (0.32,0.02,0.28,0.38)$$

with security levels $u(\mathbf{x}) = 3.12$ and $w(\mathbf{y}) = 3.64$, respectively. The optimal strategies (obtained by using the simplex algorithm) are

$$\mathbf{x} = (0.25,0.39,0.36,0)$$

and

$$\mathbf{y} = (0.29,0,0.29,0.43)$$

with value $v = 3.43$, so we see that the approximation, while less than nearly perfect, is really not too bad.

This method is especially useful for the solution of very degenerate linear programs (reduced to games as explained in Section 5), for which approximate solutions are sufficient.

PROBLEMS ON SOLUTION OF GAMES BY FICTITIOUS PLAY

1.
$$\begin{pmatrix} 3 & 5 & 1 & 2 \\ 2 & 6 & 3 & 4 \\ 6 & 1 & 2 & 0 \\ 5 & 3 & 4 & 1 \end{pmatrix}$$

2.
$$\begin{pmatrix} 6 & 5 & 1 & 2 & 6 \\ 3 & 1 & 4 & 3 & 3 \\ 2 & 5 & 8 & 6 & 1 \\ 4 & 2 & 3 & 5 & 3 \end{pmatrix}$$

3.
$$\begin{pmatrix} 1 & 6 & 1 & 4 & 2 \\ 3 & 0 & 4 & 1 & 8 \\ 1 & 2 & 5 & 3 & 4 \\ 2 & 1 & 6 & 1 & 0 \end{pmatrix}$$

4. Show that the linear program

    Maximize $\mathbf{c} x^t$

    Subject to $A x^t \leq \mathbf{b}^t$

                               $x \geq 0$

where all the entries in the matrix A, and all components of the vectors $\mathbf{b}$ and $\mathbf{c}$, are positive, is equivalent to finding optimal strategies for P_2 in the game with matrix $A' = (a'_{ij})$, where

$$a'_{ij} = \frac{a_{ij}}{b_i c_j}$$

and that the value of the linear program is the reciprocal of the value of the game.

5. Obtain an approximate solution to the linear program

    Maximize

$$5x_1 + 3x_2 + x_3 + 4x_4 + 2x_5$$

    Subject to

$$\begin{aligned} 3x_1 + 2x_2 + 4x_3 + x_4 + x_5 &\leq 6 \\ x_1 + 3x_2 + x_3 + 2x_4 + 3x_5 &\leq 8 \\ 2x_1 + 4x_2 + 5x_3 + 2x_4 + x_5 &\leq 12 \\ x_1 + 3x_2 + 2x_3 + 4x_4 + x_5 &\leq 8 \\ x_i &\geq 0 \end{aligned}$$

300   8.   THE VON NEUMANN MODEL OF AN EXPANDING ECONOMY

We give here, as an application of game theoretic analysis (though not of game theory itself), a model, developed by von Neumann, of an expanding economy.

The von Neumann assumes that the economy produces n fundamental goods by means of m fundamental processes. The goods produced are in turn needed as inputs for the processes, so that the production at a future time period will depend on the production at previous periods. (More exactly, the inputs at the $(k+1)$th period are constrained to be no greater than the outputs at the kth period.)

The processes can be characterized by two $m \times n$ matrices, $A = (a_{ij})$ and $B = (b_{ij})$. The matrix A is the *input* matrix: the entry a_{ij} is the amount of product j necessary to operate the ith process at unit intensity. In turn, B is the *output* matrix: the entry b_{ij} is the amount of goods j produced by the ith process at unit intensity.

To explain this by an example, let us consider an underdeveloped country that is trying to "enter the twentieth century" by expanding its steel industry. To do this, it must construct additional *capacity* (i.e., more steel mills). This is complicated, however, because this capacity must itself be built of steel: thus, the new capacity produced is bounded by the previous capacity.

To simplify our model, we shall assume that labor, raw materials, and certain other industrial products (such as cement) are in plentiful supply; we shall also assume that no steel goes into the production of automobiles or other consumer goods (this would cause a permanent "leak" in our model). We have, then, an economy with two fundamental products: steel and steel mills, and two fundamental industries: steel and construction.

Let us assume, now, that 1 ton of steel is necessary to construct a mill of 12-ton capacity. We shall assume, also, that 1/6 ton of steel is consumed in the production of each new ton of steel. We obtain, then, the following input-output matrices:

$$
A = \begin{array}{l} \text{Construction} \\ \text{Steel} \end{array} \begin{pmatrix} \overset{\text{Capacity}}{0} & \overset{\text{Steel}}{1} \\ 1 & 1/6 \end{pmatrix}
$$

$$
B = \begin{array}{l} \text{Construction} \\ \text{Steel} \end{array} \begin{pmatrix} 12 & 0 \\ 1 & 1 \end{pmatrix}
$$

Note that in the second process (the steel industry) the capacity that goes in as input comes out once again as an output of the process: this represents the fact that the steel mill is capital: it is not consumed in the production of steel, and will still be there after the steel has been forged.

Let us see now how this economy will expand in practice. The principal question is whether such an economy can expand at a steady rate, without the difficulties (such as unemployment and goods shortages) that often characterize overly quick growth. Suppose that the country starts the first time period with 300 T. of steel, and the capacity for producing 600 T. In the first time period, 100 T. will be used for production, while the remaining 200 T. are used to build mills with a capacity of 2400 T. At the end of the first period, then, the country has an inventory of 600 T. of steel, and capacity to produce 3000 T. In the second period, 500 T. will be used for steel production, and the other 100 for construction, so that at the end of this period, the country will have 3000 T. in inventory and capacity to produce 4200 T. In the third period, 700 T. will be used for steel-making, and 2300 T. for construction, so that the country has 4200 T. in inventory and capacity for 31800 T. production. There is now not enough steel available to prime the mills: some capacity must go unused. (There is also a danger, at this point, of severe unemployment in the construction industry.)

On the other hand, suppose the country starts with 200 T. in inventory and a capacity to produce 600 T. Then, in the first period, 100 T. will be used to produce more steel, while the other 100 T. will be used for construction. Thus at the end of the first period, the country will have 600 T. in inventory, and a capacity to produce 1800 T. In the second period, 300 T. will be used for production, and 300 T. for construction, giving a total of 1800 T. in inventory, plus a capacity for 5400 T. We see that, in this case, the country's assets triple each time: it is clear that expansion can be continued indefinitely in this way (until, say, a shortage of other goods, presently in surplus, begins to make itself felt). The economy is then expanding *in equilibrium* with an *expansion factor* of 3.

In the general case, let us suppose that the ith process is carried out at intensity x_i. This will mean that it uses $a_{ij}x_i$ units of goods j. We can represent the operation of the entire economy by an *intensity vector* $\mathbf{x} = (x_1, \ldots, x_m)$; in this case the total amount of goods j needed as an input is given by

$$\sum_{i=1}^{m} a_{ij}x_i$$

and so the total input of the economy can be expressed by the vector $\mathbf{x}A$. In a similar manner, the total amount of goods j obtained as an output will be

$$\sum_{i=1}^{m} b_{ij}x_i$$

and so the total output is represented by the vector $\mathbf{x}B$.

Now, the input for one cycle must be taken from the output of the previous cycle. As we are interested in a steady expansion of the economy, we want the intensity for each cycle to be some scalar constant, α, times the intensity vector for the previous cycle. (In the foregoing example, $\alpha = 3$.) If the intensity for the last cycle was $\mathbf{x}$, the intensity for this cycle must be $\alpha\mathbf{x}$. The feasibility constraint can then be represented by the inequality

$$(\alpha\mathbf{x})A \leq \mathbf{x}B$$

or, equivalently,

6.8.1
$$\mathbf{x}(B - \alpha A) \geq 0$$

A related problem is that of finding the "natural" prices for the system. In fact, suppose that a ton of steel costs \$150, whereas the capacity to produce this steel costs \$50 (i.e., a mill with a capacity of t tons is valued at \$50t). Then operation of the construction industry at unit intensity requires an investment of \$150 and produces \$600. (Again, we are disregarding the price of labor and raw materials for the purpose of simplification; these will be introduced later on.) On the other hand, operation of the steel industry at unit intensity will require an investment of \$75 (\$50 for capacity, plus \$25 for steel) and give a return of \$200. We see that the investment in the steel industry is not quite tripled in one period, whereas investment in the construction industry is quadrupled in one period. This will cause greater investment in construction, while steel will be neglected — which in turn will cause severe economic dislocations.

Suppose, on the contrary, that a ton of steel costs \$180, while a 1-ton production capacity is valued at \$45. In this case, unit operation of the construction industry requires a \$180 investment and returns \$540, while unit operation of the steel industry requires a \$75 investment and returns \$225. In this case, both industries triple the invested money, so there is once again an equilibrium: neither is more profitable than the other. This factor of 3 is the "natural" interest rate as well: if the interest rate were lower, it would be "forced up" by people investing in the industries.

In the general case, we can let y_j be the price of one unit of the jth commodity. The vector $\mathbf{y} = (y_1, \ldots, y_n)$ is known as the *price vector*. Now, the cost of operating the ith industry at unit intensity is given by

$$\sum_{j=1}^{n} a_{ij} y_j$$

and the return from this industry is

$$\sum_{j=1}^{n} b_{ij} y_j$$

The two vectors, $A\mathbf{y}^t$ and $B\mathbf{y}^t$, represent the investments and returns from the several industries.

As was mentioned earlier, no industry can give a greater return than the prevailing interest rate (otherwise this rate will be forced up by people investing to seek the higher profit from such an industry). Thus, if β is the interest rate, we must have

$$B\mathbf{y}^t \leq \beta A\mathbf{y}^t$$

or equivalently,

6.8.2
$$(B - \beta A)\mathbf{y}^t \leq 0$$

We have considered an example, and seen that an equilibrium exists and that the interest and expansion rates are equal. The questions arise whether such equilibria always exist and whether, at equilibrium, the two rates, α and β, are always equal. Intuitively, this seems so, since an expansion rate α will normally "pull" β up to the same level or higher, i.e., $\alpha \leq \beta$. At the same time, investments will not be made unless the return equals the interest rate, i.e., $\alpha \geq \beta$.

We have the two conditions, (6.8.1) and (6.8.2), concerning $\mathbf{x}, \mathbf{y}, \alpha$, and β. Certain other conditions can also be obtained. It is a known economic fact that surplus goods become worthless. The jth goods is in surplus if more is produced during a cycle than is needed during the following cycle, i.e., if

$$(\mathbf{x}B)_j > \alpha(\mathbf{x}A)_j$$

Thus, we conclude that we must have $y_j = 0$ whenever $\mathbf{x}(B - \alpha A)_j > 0$. This gives us the condition

6.8.3
$$\mathbf{x}(B - \alpha A)\mathbf{y}^t = 0$$

An investment is *unprofitable* if its return is smaller than the going interest rate, β. It is natural that, in equilibrium, unprofitable investments are not made. We find that, if

$$(B\mathbf{y}^t)_i < \beta(A\mathbf{y}^t)_i$$

then the ith industry will not be operated; $x_i = 0$, which gives us

6.8.4
$$\mathbf{x}(B - \beta A)\mathbf{y}^t = 0$$

We shall put one final condition on $\mathbf{x}$ and $\mathbf{y}$. This is that the goods produced must have some value. But the total value of goods produced is $\mathbf{x}B\mathbf{y}^t$. Thus

6.8.5
$$\mathbf{x}B\mathbf{y}^t > 0$$

We have obtained five conditions for equilibrium, enough to prove the equality of the interest and expansion factors. In fact, (6.8.3) and (6.8.4) can be restated in the form

$$\mathbf{x}B\mathbf{y}^t = \alpha\mathbf{x}A\mathbf{y}^t$$
$$\mathbf{x}B\mathbf{y}^t = \beta\mathbf{x}A\mathbf{y}^t$$

Thus

$$\alpha\mathbf{x}A\mathbf{y}^t = \beta\mathbf{x}A\mathbf{y}^t$$

But, by (6.8.5), $\mathbf{x}B\mathbf{y}^t$ is not zero; hence $\mathbf{x}A\mathbf{y}^t$ is not zero, and this will mean

6.8.6 $$\alpha = \beta$$

We see, then, that interest and expansion rates must be equal. Thus (6.8.2) may be restated in the form

6.8.7 $$(B - \alpha A)\mathbf{y}^t \leq 0$$

Consider, now, the vectors $\mathbf{x}$ and $\mathbf{y}$. Since these are necessarily non-negative, and by (6.8.5), neither can be identically zero, each can be expressed as a positive scalar times a mixed-strategy vector. In our analysis, it is only the relative intensities of the industries and the relative prices of the products that matter. It follows, then, that we may assume both $\mathbf{x}$ and $\mathbf{y}$ are mixed-strategy vectors. The two constraints, (6.8.1) and (6.8.7), mean that the matrix $B - \alpha A$ (thought of as a game) has value

6.8.8 $$v(B - \alpha A) = 0$$

and that $\mathbf{x}$ and $\mathbf{y}$ are optimal strategies. Note, now, that (6.8.8) will guarantee the existence of $\mathbf{x}$ and $\mathbf{y}$ satisfying (6.8.1) and (6.8.7). These two conditions, in turn, will imply (6.8.3) and (6.8.4). Thus the question of existence of an equilibrium reduces to the two conditions:

$$v(B - \alpha A) = 0$$
$$\mathbf{x}B\mathbf{y}^t > 0$$

where $\mathbf{x}$ and $\mathbf{y}$ are optimal strategies for the game $B - \alpha A$.

9. EXISTENCE OF AN EQUILIBRIUM EXPANSION RATE

The question now reduces to the existence of an α such that (6.8.8) holds. So far, we have placed no conditions on the matrices

A and B, other than the obvious one of non-negativity. We want our process, however, to be realistic. This means that we cannot produce something out of nothing. Every process must use some goods. Therefore, each row in the matrix A has at least one positive entry. This will mean that

6.9.1 $$v(-A) < 0$$

Another condition for realism is that every product must be produced by some process; each column of B must have at least one positive entry. This means

6.9.2 $$v(B) > 0$$

Consider, now, the game $B - \alpha A$. As α increases (from 0 to ∞) it is clear that the entries in $B - \alpha A$ will either decrease or remain unchanged. In this way the value $v(B - \alpha A)$ will decrease steadily (or at least never increase); it is, moreover, a *continuous* function of α. For small values of α, the term αA is negligible, and so $v(B - \alpha A)$ will be close to $v(B)$, which is positive. When α is large, the term B becomes negligible, and so $v(B - \alpha A)$ will be near the value $v(-\alpha A)$, which is negative. By the continuity of v, it may be seen that, for some value of α, condition (6.8.8) must hold.

In our previous example, the matrix

$$B - \alpha A = \begin{pmatrix} 12 & -\alpha \\ 1 - \alpha & 1 - \frac{1}{6}\alpha \end{pmatrix}$$

has no saddle points for $\alpha \geq 0$. Its value may be computed by the methods of Section 5:

$$v(B - \alpha A) = \frac{12\left(1 - \frac{1}{6}\alpha\right) - (1 - \alpha)(-\alpha)}{12 + 1 - \frac{1}{6}\alpha - (1 - \alpha) - (-\alpha)}$$

or

$$v(B - \alpha A) = \frac{12 - \alpha - \alpha^2}{12 + \frac{11}{6}\alpha}$$

The graph of this function is seen in Figure VI.9.1. To make the value vanish, it is sufficient to let the numerator vanish, which gives us

$$12 - \alpha - \alpha^2 = 0$$

FINITE MATHEMATICS

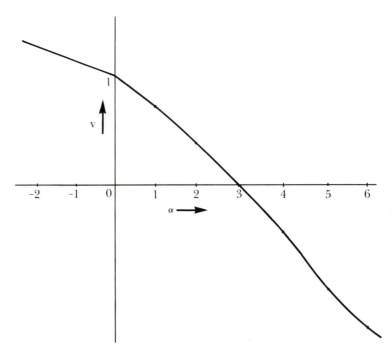

FIGURE VI.9.1 The value of the game, $B - \alpha A$, as a function of α. Note that to the left of the v-axis this graph is a straight line.

This equation has the solution $\alpha = 3$ and $\alpha = -4$; the negative solution is extraneous, and we see that this economy can have an equilibrium only at an expansion rate of 3. Then

$$B - 3A = \begin{pmatrix} 12 & -3 \\ -2 & \frac{1}{2} \end{pmatrix}$$

Optimal strategies for this game may be obtained: it will be seen that they correspond to the equilibrium intensity and price vectors already obtained.

One final question remains, and that is the existence of optimal strategies **x** and **y** for $B - \alpha A$ that satisfy (6.8.5). It can be shown (though we shall not attempt to do so) that one of two cases must hold. One possibility is that there is only one number α such that (6.8.8) holds. If so, then there are always optimal strategies satisfying (6.8.5). The other possibility is that there may be more than α for which (6.8.8) holds, i.e., there may be numbers, α and α', such that $\alpha < \alpha'$, and

$$v(B - \alpha A) = v(B - \alpha' A) = 0.$$

In this case, if α'' is any number satisfying $\alpha < \alpha'' < \alpha'$, it is not difficult to see that α'' also satisfies (6.8.8), since v either decreases steadily or remains constant as α increases. Thus there are an infinite number of possible expansion rates. It turns out, however, that most of these represent cases in which nothing of value is produced; i.e., (6.8.5) cannot be satisfied. Nevertheless, there are always some values of α (finitely many of them) that satisfy both (6.8.8) and (6.8.5). These include both the maximal and minimal values of α that satisfy (6.8.8). An example will perhaps make this most clear.

VI.9.1 *Example.* Consider a two-product economy with the products "wheat" and "diamonds," each produced by a process. The input-output matrices are:

$$A = \begin{array}{c} \\ W \\ D \end{array} \begin{array}{cc} W & D \\ \begin{pmatrix} 10 & 0 \\ 5 & 10 \end{pmatrix} \end{array}$$

$$B = \begin{array}{c} W \\ D \end{array} \begin{pmatrix} 40 & 0 \\ 0 & 20 \end{pmatrix}$$

We have then,

$$B - \alpha A = \begin{pmatrix} 40 - 10\alpha & 0 \\ -5\alpha & 20 - 10\alpha \end{pmatrix}$$

Suppose we let $\alpha = 4$. For this factor,

$$B - 4A = \begin{pmatrix} 0 & 0 \\ -20 & -20 \end{pmatrix}$$

It is easy to see that this game has value 0. P_1's unique optimal strategy is $\mathbf{x} = (1,0)$, but any strategy will be optimal for P_2. At this expansion rate, we see that only wheat is produced; diamonds must be neglected as this industry simply cannot expand so quickly. The prices are irrelevant since there is, in effect, only one commodity. The only constraint, now, is that

$$\mathbf{x}B\mathbf{y}^t = 40y_1$$

so that y_1 must be positive if (6.8.5) is to hold.

Suppose, next, we let $\alpha = 2$. We will have

$$B - 2A = \begin{pmatrix} 20 & 0 \\ -10 & 0 \end{pmatrix}$$

308

Again, we have a game with value zero. A strategy, x, will be optimal for P_1 if it satisfies

$$x_1 \geq \frac{1}{3}$$

P_2's unique optimal strategy is $(0,1)$. In this case, we find that wheat becomes worthless. The trouble here is that wheat production can expand more rapidly than the factor $\alpha = 2$, and so it becomes surplus goods whose production will get out of hand if it has any value. It may be seen that by choosing any x such that $x_1 \geq 1/3$ and, moreover,

$$x_2 > 0$$

the conditions will be satisfied.

Suppose, finally, we take an α such that $2 < \alpha < 4$. Say $\alpha = 3$. We have

$$B - 3A = \begin{pmatrix} 10 & 0 \\ -15 & -10 \end{pmatrix}$$

For this game, the value is still 0, but the unique optimal strategies are $x = (1,0)$ and $y = (0,1)$. But this would mean only wheat, which is worthless, will be produced, i.e., (6.8.5) does not hold. The trouble here is that diamonds, unable to expand so quickly, will not be produced; wheat can expand this quickly, but there is always a surplus. This same analysis will hold for any α such that $2 < \alpha < 4$.

We see from this example that a new difficulty arises in such cases. The possible expansion rate is not unique; moreover, a product that has value at one rate will become worthless at another expansion rate. What really happens is that the economy tends to split up into independent sub-economies. The thing to remember, however, is that these sub-economies can never exist in a vacuum: each one must pay wages to its workers, who proceed to link up the economy again by their demands. If, in the foregoing example, we can get the wheat industry workers to buy some diamonds (in the form, say, of jewelry for their wives) the economy will be put together again.

Let us introduce, therefore, the "wages" matrix, $W = (w_{ij})$. The entry w_{ij} means that, if process i is operating at unit intensity, then its workers must be paid an amount w_{ij} of commodity j. (We are assuming that the workers are paid in goods rather than money. In reality the situation is, of course, more complicated, since the workers are paid in money and then decide what to buy with it, but the two procedures are not too dissimilar in effect.)

Suppose we have a wage matrix:

$$W = \begin{pmatrix} 2 & 2 \\ 1 & 1 \end{pmatrix}$$

This means that, for the wheat industry to operate at unit intensity, it must hire workers, who demand 2 units of wheat and 2 units of diamonds. The diamond industry pays only half as much in wages, perhaps because it requires less labor.

It now turns out that the required input is no longer represented by the matrix A, but rather, by the sum of the two matrices, A and W. (In other words, the entrepreneur finds that he now needs the commodities, both as raw materials and as capital goods, and to pay wages. We need, now, a factor α such that

6.9.3 $$v(B - \alpha A - \alpha W) = 0$$

Expansion will then be less rapid (but more realistic).

For our example, we have

$$B - \alpha(A + W) = \begin{pmatrix} 40 - 12\alpha & -2\alpha \\ -6\alpha & 20 - 11\alpha \end{pmatrix}$$

It is not too difficult to check that there is no value of α that will make the game have a saddle point *and* have value zero. We must solve for α under the assumption that the game does not have a saddle point. In this case, the value is given by

$$v = \frac{(40 - 12\alpha)(20 - 11\alpha) - (6\alpha)(2\alpha)}{(40 - 12\alpha) + (20 - 11\alpha) + 6\alpha + 2\alpha}$$

We set the numerator of this expression equal to zero:

$$120\alpha^2 - 680\alpha + 800 = 0$$

which gives the solutions $\alpha = 4$, $\alpha = 5/3$. The larger value, $\alpha = 4$ gives us the matrix

$$B - 4(A + W) = \begin{pmatrix} -8 & -8 \\ -24 & 24 \end{pmatrix}$$

which clearly does not have value 0 (i.e., $\alpha = 4$ is an extraneous solution). The other solution, $\alpha = 5/3$, gives us the matrix

$$B - \frac{5}{3}(A + W) = \begin{pmatrix} 20 & -\dfrac{10}{3} \\ -10 & \dfrac{5}{3} \end{pmatrix}$$

310

which has the desired value and unique optimal strategies

$$x = \left(\frac{1}{3}, \frac{2}{3}\right)$$

$$y = \left(\frac{1}{7}, \frac{6}{7}\right)$$

Note how the price of diamonds is considerably greater than the price of wheat. This is due, of course, to many things, not least of which is their relative scarcity (they cannot be produced as quickly).

We see, here, how the influence of workers' demands ties up the economy, giving it a unique expansion factor and price system. It can be proved (though we shall not do so here) that the system will have a *unique* expansion factor if all the elements in the matrix $A + B + W$ are positive. Heuristically, this means that each process should be somehow related to each product, either using it as an input or producing it as an output, or else obtaining it to pay as wages to its employees. (This condition is sufficient, but not necessary, for the uniqueness of the expansion rate.)

PROBLEMS ON THE VON NEUMANN MODEL

1. Find the intensity and price vectors at equilibrium in the economies represented by the following pairs of matrices

(a)
$$A = \begin{pmatrix} 3 & 1 \\ 0 & 2 \end{pmatrix} \qquad B = \begin{pmatrix} 0 & 5 \\ 2 & 0 \end{pmatrix}$$

(b)
$$A = \begin{pmatrix} 1 & 4 \\ 2 & 1 \end{pmatrix} \qquad B = \begin{pmatrix} 5 & 2 \\ 1 & 4 \end{pmatrix}$$

(c)
$$A = \begin{pmatrix} 5 & 3 \\ 3 & 8 \end{pmatrix} \qquad B = \begin{pmatrix} 2 & 10 \\ 6 & 4 \end{pmatrix}$$

(d)
$$A = \begin{pmatrix} 2 & 1 \\ 1 & 3 \end{pmatrix} \qquad B = \begin{pmatrix} 0 & 10 \\ 5 & 0 \end{pmatrix}$$

2. Consider an economy consisting of two kinds of goods, fuel and transportation. (The raw materials are assumed in plentiful supply.) Fuel must be transported 100 miles from the mine to a processing plant so that each ton of fuel requires 100 T.-mi. of transportation. In turn, each ton-mile of transportation requires 4 lb. of fuel. Find the equilibrium price and intensity vectors for this economy. (1 T. = 2000 lb.)

3. Consider once again the economy of Problem 2, but assume this time that each ton of fuel requires wages of 100 lb. of fuel and 10 T.-mi. of transportation, while each ton-mile of transportation

requires wages of 1/2 lb. of fuel and 10 lb.-mi. of transportation. How does this change prices and the expansion factor?

4. In Problem 3, change the wages to 200 lb. of fuel and 5 T.-mi. of transportation per ton of fuel; and 1 lb. of fuel and 5 lb.-mi. of transportation for each ton-mile of transportation. How does this affect prices?

10. TWO-PERSON NON-ZERO-SUM GAMES

Up to now we have almost exclusively studied games in which the two players' interests were directly opposed. Though this theory has many obvious applications, it is clear that there are very many cases in which the two players of a game can both gain by certain courses of action. Unfortunately, there is no such satisfactory theory for these games as there is for the strictly competitive (zero-sum) games studied so far.

These *non-zero-sum games* can generally be represented by two $m \times n$ matrices, $A = (a_{ij})$ and $B = (b_{ij})$. In this case, a_{ij} and b_{ij} are the amounts won by P_1 and P_2, respectively, assuming that they choose the ith row strategy and jth column strategy, respectively. (In the zero-sum case, we would find that $B = -A$.) Thus P_1 wishes to maximize a_{ij}, while P_2 wishes to maximize b_{ij}. Such games are called *bimatrix* games.

We consider two possibilities: cooperative games, in which such things as preplay communications, binding contracts, and side payments are allowed, and non-cooperative games, in which these are forbidden (e.g., by antitrust laws).

In the non-cooperative case, the "obvious" idea is to look for equilibrium points. These are defined in a manner analogous to that for matrix games:

VI.10.1 Definition. Let (A, B) be a bimatrix game. The ith row strategy and jth column strategy are an *equilibrium pair* if, for every k,

$$a_{ij} \geq a_{kj}$$

and, for every l,

$$b_{ij} \geq b_{il}$$

Definition VI.10.1 states that a_{ij} is the largest element in its column, while b_{ij} is the largest in its row. Neither player can gain by a unilateral change in strategy. (Remember that, in this case, P_2 wishes to *maximize* b_{ij}, not to minimize it.)

The principal question, now, is whether bimatrix games will

have equilibrium pairs. The answer is exactly the same as for matrix (zero-sum) games: there need be no equilibrium pair of pure strategies, but there will always be an equilibrium pair of mixed strategies. (The proof of this statement uses some very advanced mathematics and must therefore be omitted here). Unfortunately, the behavior of equilibrium points for these games can be quite singular, as the following examples show.

VI.10.2 Example. A company, C_1, has a factory that may be used to process either an expensive blend or a cheap blend of coffee. A second company, C_2, can produce either an expensive percolator or an inexpensive coffee urn. The expensive coffee can be made in either the percolator or the urn, but the cheap coffee can only be made in the percolator. If C_1 makes the expensive blend, and C_2 the urn, then C_1 makes a large profit and C_2 a small profit. If C_1 makes the cheap blend and C_2 the percolator, then it is C_1 that obtains the small profit, and C_2 the large profit. If both make the expensive products, then sales are very small and not enough to cover expenses. If both make the cheap products, there are no sales.

The matrices for this game might reasonably be assumed to look like this:

$$A = \begin{pmatrix} 4 & -1 \\ -1 & 1 \end{pmatrix} \qquad B = \begin{pmatrix} 1 & -1 \\ -1 & 4 \end{pmatrix}$$

This game has two equilibrium pairs in pure strategies, corresponding to one company making its cheap product, while the other makes its expensive product. (There is a third equilibrium pair, using mixed strategies, but we can safely ignore it.) It may be seen, however, that the behavior of the equilibrium pairs is not nearly as satisfactory as for matrix games. First of all, it may be seen that C_1 prefers one equilibrium pair, while C_2 prefers the other (i.e., the pairs are not equivalent; they do not give the same payoffs). For this game, disclosure as to one's intentions might actually be helpful. If C_1 can just make an announcement, "We will produce the expensive coffee, and that's that!" and make it sound convincing enough, C_2 might just be forced to follow along and produce the cheap urns. (Of course, C_2 might believe C_1 was only bluffing, which would lead to disaster for both of them.) Note also that, in the absence of preplay communications (possibly forbidden by antitrust laws), C_1 might aim for one equilibrium pair while C_2 aimed for the other, giving a non-equilibrium result. For this game, equilibrium pairs are not interchangeable.

Even in a game that has only one equilibrium pair, the results might be rather unsatisfactory.

VI.10.3 Example (The Price War). (This example is the work of A. W. Tucker.) Two companies produce an identical product, which may be sold at either of two prices, high or low. If both sell at high price, then both make a small profit; if at low price, both break even. If, however, one sells at a high price, and the other at a low price, then the company selling at the lower price makes a large profit, whereas the other gets no customers and takes a loss. The game may be represented by the two matrices

$$A = \begin{pmatrix} 2 & -1 \\ 3 & 0 \end{pmatrix} \qquad B = \begin{pmatrix} 2 & 3 \\ -1 & 0 \end{pmatrix}$$

It may be seen that the second row of A dominates the first, whereas the second column of B dominates the first. The only equilibrium pair is (low price, low price). This gives a payoff of 0 units to each player. Yet both players could do better if they were to use the *dominated* strategies (high price, high price), which gives them a profit of 2 units each. The difficulty is that neither can count on the other to maintain a high price, and price-fixing agreements are (by hypothesis) forbidden.

VI.10.4 Example (The Prisoner's Dilemma). An unscrupulous district attorney has two prisoners, whom he suspects of engineering a big robbery. Unfortunately, he has no evidence, so that his only hope of obtaining a conviction is to have one of them confess.

Isolating the two prisoners, the D.A. talks to each one as follows: "If neither of you two confesses, then I will see that you are both sent to jail on some minor, trumped-up charges. If you both confess, then you will both receive moderate sentences. But if one of you confesses, then he will receive a very light sentence, whereas the other will receive a long prison term. You had better confess before your partner does so."

We can reasonably assign utilities of 0 for the very light sentence, -1 for the sentence on trumped-up charges, -5 for the moderate sentence, and -10 for the long prison term. In this case, each prisoner has two strategies, "do not confess," and "confess." The game matrices are:

$$A = \begin{pmatrix} -1 & -10 \\ 0 & -5 \end{pmatrix} \qquad B = \begin{pmatrix} -1 & 0 \\ -10 & -5 \end{pmatrix}$$

It may be seen that the second row dominates the first row, while the second column dominates the first column. Hence the only equilibrium point corresponds to the second row, second column: both players should confess. Yet, if they both confess, they both do worse than if neither confesses. As in the previous example, we see it may be better for both to act "irrationally."

Let us now consider cooperative games. Generally speaking, the problem here is one of bargaining. If they can reach an agreement, the two players can obtain a certain amount of money or utility; it is then merely a question of deciding how to divide the spoils.

To take a trivial example, suppose a benefactor offers to give two players $100, to divide any way they wish, if only they can agree ahead of time on this division. There is, strictly speaking, no single natural outcome for this game. The "even split," $50 each, seems fair, but there is no way in which it can be enforced if, say, one of the players insists on $75. In other words, every division of the money is, in some sense, possible: if one of the two players needs $20 very badly, he may be forced to accept $20 rather than nothing at all.

The search in these games is for a "fair" solution — an outcome that will adequately represent the players' bargaining *positions*, but not their bargaining *abilities*. It might be thought of as an "arbitrated" outcome: the outcome that a fair and impartial arbitrator might give.

The *arbitrated threat solution* is obtained in the following way: given the two matrices, A and B, find the *threat value*, which is merely the value of the matrix game $A - B$. Let this number be v. This represents more or less, the relative advantage which P_1 is able to obtain over P_2. Then the arbitrated threat solution divides the maximal possible utility that can be obtained by the two players in such a way that P_1 receives v more units than P_2. In other words, the solution divides the possible gains in such a way as to maintain the relative advantage v.

VI.10.5 Example. Two companies, C_1 and C_2 produce equivalent automobiles. Neither can produce enough to saturate the market, but, if both produce as much as possible, they can oversaturate the market.

If one company produces its maximum while the other produces only a small amount, then the one that produces a lot makes a large profit, while the other makes a small profit. If both overproduce, there is a glut on the market, and both lose money. If both underproduce, then there is great danger of a recession, which will cause losses to both companies. The situation is somewhat asymmetrical because C_1 has recently bought some very expensive and efficient machinery that enables it to produce at a somewhat lower cost, but must, on the other hand, be amortized. Hence C_1 suffers a large loss if there is underproduction, but only a small loss if there is overproduction, while C_2's loss is large for overproduction and small for underproduction.

The game can be represented by the matrices

$$A = \begin{pmatrix} -1 & 4 \\ 1 & -4 \end{pmatrix} \qquad B = \begin{pmatrix} -4 & 1 \\ 4 & -1 \end{pmatrix}$$

where the first row or column represents high production; the second row or column, low production.

At first glance, the game seems to be entirely symmetric, which would suggest that both companies should produce a moderate amount, giving a payoff of 5/2 units to each. On the other hand, if we form the "threat matrix"

$$A - B = \begin{pmatrix} 3 & 3 \\ -3 & -3 \end{pmatrix}$$

we see that the *threat value* is $v = 3$. The maximal possible utility available to both companies is obtained by taking the matrix

$$A + B = \begin{pmatrix} -5 & 5 \\ 5 & -5 \end{pmatrix}$$

whose maximal entry is 5. There are 5 units of utility to be divided between C_1 and C_2, in such a way as to maintain C_1's threat advantage of 3 units. This gives us the *arbitrated threat solution* $(4,1)$: C_1 should produce as much as it can, while C_2 keeps its production level low.

PROBLEMS ON TWO-PERSON NON-ZERO-SUM GAMES

1. Find the equilibrium pairs for the following bimatrix games.

(a)
$$A = \begin{pmatrix} 3 & 5 \\ 2 & 1 \end{pmatrix} \qquad B = \begin{pmatrix} 1 & 4 \\ 8 & 2 \end{pmatrix}$$

(b)
$$A = \begin{pmatrix} 6 & 1 \\ 1 & 6 \end{pmatrix} \qquad B = \begin{pmatrix} 2 & 1 \\ 1 & 2 \end{pmatrix}$$

(c)
$$A = \begin{pmatrix} 2 & 3 \\ 5 & 1 \end{pmatrix} \qquad B = \begin{pmatrix} 4 & 3 \\ 1 & 5 \end{pmatrix}$$

(d)
$$A = \begin{pmatrix} 1 & 5 \\ 3 & 2 \end{pmatrix} \qquad B = \begin{pmatrix} 2 & 3 \\ 4 & 1 \end{pmatrix}$$

2. *The Inspector's Game* (This example is the work of M. Maschler.)

Player I, an *inspector*, tries to prevent E, an *evader*, from carrying out some forbidden action. E can carry out the action in either of two time periods or not at all. I can inspect in either of those two time periods, but he is only allowed to inspect once. If I inspects at the same time that E tries to act, E will certainly be caught. We assume a payoff to I of 1 if E does not act, 0 if E acts and is caught, -1 if E acts and escapes detection. The payoff to E is 1 if he acts successfully, 0 if he does not act, and -1 if he is caught.

(a) Show that this game can be represented by the two matrices

$$A = \begin{pmatrix} 0 & -1 \\ -1 & 1 \end{pmatrix} \qquad B = \begin{pmatrix} -1 & 1 \\ 1 & 0 \end{pmatrix}$$

(What are the players' strategies?)

(b) Show that $x = (1/3, 2/3)$ and $y = (2/3, 1/3)$ form an equilibrium pair. What are the expected payoffs in this case? (In fact, this is the only equilibrium pair.)

(c) Show that, if I "announces" a strategy $(1/3 + \epsilon, 2/3 - \epsilon)$, e.g., $(0.34, 0.66)$, E's best strategy is $(0,1)$. Show that this gives a better expectation to I than the equilibrium pair (x, y).

11. *n*-PERSON GAMES

We shall consider, now, games for more than two players. For these, the emphasis is not so much on the strategies as on the coalitional considerations; we are interested in knowing what coalitions will form among the players, and the way in which the members of these coalitions bargain among themselves. Generally speaking, each coalition is faced with the problem of paying its members enough so that they will not be tempted to defect.

We shall label the *players* $1, 2, \ldots, n$. Then $N = \{1, 2, \ldots, n\}$ is the *set of all players* (for a particular game). All the non-empty subsets $S \subset N$, including N itself, and all the single-player subsets (but not including the empty set) are called *coalitions*. As we are not generally interested in the strategies used by the players, we shall dispense with the *normal form* of the game and study, rather, the *characteristic function*. In essence, the characteristic function of a game gives the amount (of utility, money, and so on) that can be obtained by the members of a coalition when acting together (i.e., assuming that the coalition actually forms and uses its resources in optimal manner), whatever the remaining players may do.

VI.11.1 Definition. An *n*-person game in *characteristic function form* is a function, v, whose domain is the collection of all subsets of $N = \{1, 2, \ldots, n\}$ and whose range is in the real numbers, satisfying

6.11.1 $$v(\varnothing) = 0$$

6.11.2 $$v(S \cup T) \geq v(S) + v(T) \quad \text{if} \quad S \cap T = \varnothing$$

Essentially, $v(S)$ is the amount of utility that the coalition S can obtain for its members, to be divided among them in any way they see fit. Condition (6.11.1) then says simply that the empty set can obtain nothing for its (non-existent) members. Condition (6.11.2) is known as *super-additivity* and represents the fact that coalitions S and T, acting together, can obtain at least as much utility for their members as when acting separately.

VI.11.2 Example. Consider the following three-person game. Each of three players chooses either heads (H) or tails (T). If two players choose the same, but the remaining player chooses differently, then this last player must give 1 unit to each of the other players. In any other case, there is no payoff.

For this game, it is clear that

$$v(\varnothing) = 0$$

since the empty set can obtain nothing, and also,

$$v(\{1,2,3\}) = 0$$

since the game is zero-sum: the three players together can gain nothing since all payments must take place among them.

Suppose, next, that the coalition $\{2,3\}$ forms. It will, naturally, be opposed by the remaining player, giving what is essentially a two-person game between the two coalitions, $\{1\}$, and $\{2,3\}$. The coalition $\{1\}$ has two strategies, H and T, while $\{2,3\}$ has four strategies, HH, HT, TH, and TT, giving us a 2×4 matrix game

$$
\begin{array}{c c}
 & \begin{array}{cccc} HH & HT & TH & TT \end{array} \\
\begin{array}{c} H \\ T \end{array} & \left(\begin{array}{cccc} 0 & 1 & 1 & -2 \\ -2 & 1 & 1 & 0 \end{array} \right)
\end{array}
$$

For this matrix, the second and third columns are dominated, and it is easy to see that the value is -1.

$$v(\{1\}) = -1$$

and

$$v(\{2,3\}) = 1$$

By the symmetry of the game, we conclude that the characteristic function of this game is

$$
v(S) = \begin{cases} 0 & \text{if } |S| = 0 \text{ or } 3 \\ -1 & \text{if } |S| = 1 \\ 1 & \text{if } |S| = 2 \end{cases}
$$

where $|S|$ is the *number of elements* in the set S.

This example shows how the characteristic function of a game may be obtained from its normal form. Very often, however, the game is given directly in characteristic function form.

VI.11.3 Example. Let R and L be such that $R \cap L = \varnothing$ and $R \cup L = N$. The members of R manufacture right-hand gloves only, while the members of L manufacture left-hand gloves only; each player manufactures the same quantity of gloves and these are all identical except for the right-left differences.

Let us assume that the total number of pairs that can be produced by one right-hand and one left-hand player together are valued at 1 unit, but that right-hand or left-hand gloves alone are useless. Then it is clear that

$$v(S) = \min\left(|S \cap R|, |S \cap L|\right)$$

(where once again $|S \cap R|$ is the number of players in $S \cap R$) for any coalition S. In particular, we see that $v(\{i\}) = 0$ for any single player: it is only by forming coalitions that any gains can be made.

Let us suppose, now, that a game, v, is played. What payoffs can the players expect from this game? This is not an easy question to answer. We may assume, however, that no player will accept less than he could obtain for himself. This gives us the concept of an imputation.

VI.11.4 Definition. An *imputation* of the n-person game v is a vector $\mathbf{x} = (x_1, \ldots, x_n)$ satisfying

6.11.3
$$\sum_{i=1}^{n} x_i = v(N)$$

6.11.4
$$x_i \geq v(\{i\})$$

Conditions (6.11.3) and (6.11.4) are known respectively as *group rationality* and *individual rationality*. In essence, an imputation is one of the possible distributions of profits (payoffs) that might be obtained if the game v is played. Unfortunately, most games have an infinity of imputations; it is generally desired to distinguish one, or at least a small subset, of these as being most "reasonable." Several "solution concepts" have been suggested for n-person games: we shall consider only two.

VI.11.5 Definition. The *core* of an n-person game, v, is the set of all imputations $\mathbf{x} = (x_1, \ldots, x_n)$ such that, for every coalition S,

$$\sum_{i \in S} x_i \geq v(S)$$

If an imputation belongs to the core, this means that every coalition is obtaining as much as possible for its members by forming; thus a core imputation exhibits a good deal of stability. If an imputation does not belong to the core, there is a coalition S whose members are not getting as much as they could: there therefore tends to be some pressure for a change on the part of the members of S.

Let us look for the cores of the games just described. For Example VI.11.3, the core is empty. Suppose x is in the core. As x is an imputation, we must have

$$x_1 + x_2 \geq 1$$

and

$$x_1 + x_2 + x_3 = 0$$

But this means $x_3 \leq -1$. Similarly, it may be seen that $x_2 \leq -1$ and $x_3 \leq -1$. But this contradicts the hypothesis that

$$x_1 + x_2 + x_3 = 0$$

There are no imputations in the core. This may be interpreted as meaning that the game is unstable; it may also be interpreted as meaning that the core is not an entirely satisfactory concept.

Let us consider now the game of Example VI.11.4. The core condition, (6.11.5), tells us that each right-hand-glove maker and left-hand-glove maker, together, must receive at least 1 unit. Let us assume that there are as many right-handers as left-handers, say $|R| = |L| = m$. Then there are m units to be divided, and it is not difficult to see that the core consists of all the vectors $x(t)$, for $0 \leq t \leq 1$, such that

$$x_i = \begin{cases} t & \text{if } i \in R \\ 1-t & \text{if } i \in L \end{cases}$$

Thus, an imputation is in the core if it corresponds to some "market price" t for right-hand gloves: each right-hand-glove maker receives t units, and each left-hand maker, $1 - t$ units. The parameter t, thought of as a market price, is arbitrary except that it must satisfy $0 \leq t \leq 1$.

Suppose, next, that there are fewer right-hand-glove makers than left-hand: $|R| < |L|$. In this case, the only imputation in the core is x, given by

$$x_i = \begin{cases} 1 & \text{if } i \in R \\ 0 & \text{if } i \in L \end{cases}$$

In other words, the surplus of left-hand gloves makes them worthless.

If $|R| > |L|$, the opposite result holds: the right-hand gloves become worthless, the members of R receive 0, and the members of L, 1.

The core is a very useful solution concept: it gives us results that generally coincide with those of classic economic analysis, as is shown by our treatment of Example VI.11.4. Unfortunately, many interesting games have empty cores, as is shown by Example VI.11.3.

12. THE SHAPLEY VALUE

The second solution concept that we will study is the *power index* of L. S. Shapley, also known as the Shapley value.

The Shapley value should be thought of as a function, ϕ, which assigns to each n-person game, v, an n-vector $\phi[v]$. This vector represents a "reasonable" *a priori* expectation for the payoff of the game; it assumes that the game is adequately represented by the characteristic function alone. It also disregards such intangibles as personal affinities among the players and differences in bargaining ability.

The Shapley value is developed axiomatically, as follows.

Let W be any coalition. Consider the game w, defined by

$$6.12.1 \qquad w(S) = \begin{cases} 0 & \text{if } W \not\subset S \\ 1 & \text{if } W \subset S \end{cases}$$

For this game, it is clear that the only question of importance is whether the members of the coalition W can all agree: these, acting together, can obtain 1 unit. The other players are *dummies*: they contribute nothing to any coalition. A "reasonable" expectation for this game is $\phi[w]$, defined by

$$6.12.2 \qquad \phi_i[w] = \begin{cases} 0 & \text{if } i \notin W \\ \dfrac{1}{|W|} & \text{if } i \in W \end{cases}$$

i.e., the 1 unit should be divided evenly among the members of W. Other players get nothing.

Suppose, next, that u and v are n-person games in characteristic function form. If α and β are positive numbers, it is easy to see that the function $\alpha u + \beta v$ is also the characteristic function of an n-person game. (We need only verify that it satisfies (6.11.1) and (6.11.2).) Then it seems reasonable that

$$6.12.3 \qquad \phi[\alpha u + \beta v] = \alpha \phi[u] + \beta \phi[v]$$

Condition (6.12.3) states that the expected payoff from playing the two games, αu and βv, simultaneously, should be the sum of the expected payoffs from each game. This corresponds, more or less, to the additive properties of expectations (Theorem V.9.1).

These two conditions, (6.12.2) and (6.12.3), are reasonable enough on the surface, so that they may be treated as axioms. Now it may be shown that every n-person game can be written as the sum and difference of games such as w (defined by (6.12.1)) multiplied by constants. For these games w, (6.12.2) defines a value $\phi[w]$; condition (6.12.3) then is used to extend this definition to all n-person games.

Consider the following probabilistic model: take all the $n!$ possible orderings (permutations) of the players of an n-person game, and assign the probability $1/n!$ to each. For each ordering form a payoff vector in the following way: if player i is preceded in the ordering by the members of the coalition S, and no others, he receives the amount

$$v(S \cup \{i\}) - v(S)$$

Then the expected value of this payoff is exactly equal to the Shapley value. In fact, it is trivial to verify (6.12.2); (6.12.3) follows from Theorem V.9.1.

(We *do not* suggest that this probabilistic model is a realistic model of the way payoffs would actually be made; we introduce it only as an aid in computing the Shapley value.)

Using the probabilistic model, we can obtain a closed formula for the power index. In fact, we can compute the probability that, in a random ordering, player i will be preceded by the members of S and no others. If S has s elements, the members of S can be ordered in $s!$ ways, and the members of $N - S - \{i\}$ can be ordered in $(n-s-1)!$ ways. There are $s!(n-s-1)!$ permutations of N in which i is preceded by the members of S and followed by all others. The probability of such an ordering, then, is

6.12.4 $$\gamma_{n,s} = \frac{s!(n-s-1)!}{n!}$$

Thus, the Shapley value is given by the formula

6.12.5 $$\phi_i[v] = \sum_{\substack{S \subset N \\ i \notin S}} \gamma_{n,s}[v(S \cup \{i\}) - v(S)]$$

where the sum is taken over all coalitions that *do not* include player i. The value is *always* an imputation.

Formula (6.12.5) is complicated, and, for large games, it can be quite difficult to apply. For $n=3$ or $n=4$, it is sometimes easiest to take all the permutations of the players: since $3!=6$ and $4!=24$, the

games are still small enough to do this. For larger games, it is sometimes possible to take advantage of some special feature in the
structure of the game to simplify the calculations.

Let us consider the game of Example VI.11.3, with $|R|=1$
and $|L|=2$. In our model, the right-hand-glove manufacturer re-
ceives 1 unit if he is preceded in the ordering by at least one of the
other players. There are six possible orderings:

$$1,2,3 \qquad 2,1,3 \qquad 3,1,2$$
$$1,3,2 \qquad 2,3,1 \qquad 3,2,1$$

If player 1 is the right-hand manufacturer, we see that, in four of the
six orderings, he obtains 1 unit, and for the game,

$$\phi_1 = \frac{2}{3}$$

Let us look, now, at player 2. He receives 1 unit if he is preceded in
the ordering by player 1 but not by player 3. There is only 1 ordering
that does this.

$$\phi_2 = \frac{1}{6}$$

In a similar way, we find that

$$\phi_3 = \frac{1}{6}$$

In the general case, suppose there are r right-hand-glove manufacturers, and l left-handers. A member of R receives 1 unit if he is
preceded by more elements of L than of R. Through mathematics
more advanced than can be considered here, it can be shown that,
for $r > l$, the value to a player $i \in R$ is

$$\phi_i = \frac{1}{2} - \frac{r-l}{2r} \sum_{k=0}^{l} \frac{r!l!}{(r+k)!(l-k)!}$$

and that for a player $j \in L$ is

$$\phi_j = \frac{1}{2} + \frac{r-l}{2l} \sum_{k=1}^{l} \frac{r!l!}{(r+k)!(l-k)!}$$

For small values of r and l, these values can be calculated directly,
and are given in Table VI.12.1. (The entries in the table represent the
value to a member of R for each value of r and l.)

TABLE VI.12.1

r	0	1	2	3	4	5	6	7	8
				l					
1	0	0.500	0.667	0.750	0.800	0.833	0.857	0.875	0.889
2	0	0.167	0.500	0.650	0.733	0.786	0.822	0.847	0.867
3	0	0.083	0.233	0.500	0.638	0.720	0.774	0.811	0.838
4	0	0.050	0.133	0.272	0.500	0.629	0.710	0.764	0.802
5	0	0.033	0.086	0.168	0.297	0.500	0.622	0.701	0.755
6	0	0.024	0.060	0.113	0.194	0.315	0.500	0.616	0.693
7	0	0.018	0.044	0.081	0.135	0.214	0.330	0.500	0.610
8	0	0.014	0.033	0.061	0.099	0.153	0.230	0.341	0.500

As may be seen, the "short" end of the market has a definite competitive advantage. With three right-hand and five left-hand manufacturers, the members of R receive 0.72 units each, or 2.16 units altogether. The members of L receive only 0.168 units each, or 0.84 units altogether. In a strictly competitive situation, however, as represented by the core, the members of L would receive nothing. The reason that the Shapley value allows them to receive something is that it takes account of the possibility that the members of L might form a coalition among themselves and limit production so as to raise the price of their product.

VI.12.1   Example.   A landlord owns a large piece of ground, which can be worked by $n-1$ peasants. It is assumed that if k peasants work the field, the production is $\sqrt{k}$ units.

This is an n-person game in which player 1 is the landlord and the other players are the peasants. Assuming the landlord does not work, the characteristic function is

$$v(S) = \begin{cases} 0 & \text{if } 1 \notin S \\ \sqrt{|S|-1} & \text{if } 1 \in S \end{cases}$$

where, as usual, $|S|$ is the number of elements in S. We may compute the value to the first player in the following way: since the game treats the other players equally, we need not concern ourselves with *what* players precede him in a random ordering, but only with *how many* precede him. If he is preceded by k players, he receives $\sqrt{k}$ units according to our model. For any k, the probability that he is preceded by k players is $1/n$. Thus,

$$1 = \frac{1}{n} \sum_{k=0}^{n-1} \sqrt{k}$$

For small values of n, this sum may be computed exactly. For large n, it may be shown that this is approximately $2\sqrt{n}/3$.

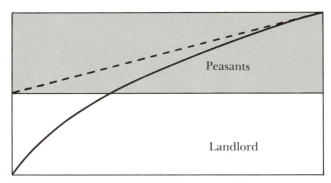

FIGURE VI.12.1   Competitive "market" solution of Example VI.12.1.

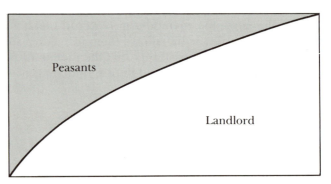

FIGURE VI.12.2   Shapley value solution of Example VI.12.1.

For the remaining players, the value can be computed by remembering that all receive the same value, and that the value is an imputation. Since there are $\sqrt{n-1}$ units to be divided, all told, we have

$$\phi_2 = \phi_3 = \ldots = \phi_n = \frac{\sqrt{n-1} - \phi_1}{n-1}$$

For large values of n, this would be approximately

$$\phi_2 = \phi_3 = \ldots = \phi_n = \frac{1}{3\sqrt{n}}$$

We may compare this with the "competitive" solution, according to which each peasant would receive an amount equal to the marginal productivity of one peasant, or $\sqrt{n} - \sqrt{n-1}$. For large n, this is approximately equal to $1/2\sqrt{n}$. This is the "natural" wage. Note that we obtain very different results according to the two schemes.

13. VOTING STRUCTURES

We mentioned that the formula (6.12.5) for the Shapley value is normally difficult to apply unless the game has some special structure. A type of game for which these calculations are considerably simplified is the simple game.

VI.13.1 Definition. An n-person game, v, is said to be *simple* if, for every coalition S, $v(S)$ is either 0 or 1.

For a simple game, each coalition is either *winning* (which means, roughly, that it can obtain any imputation on which its members agree) or *losing* (which means that it can get its members nothing at all). If S is winning, then $v(S) = 1$, while if S loses, then $v(S) = 0$.

For simple games, (6.12.5) is considerably simplified by the observation that many of the terms

$$v(S \cup \{i\}) - v(S)$$

that appear in the sum will vanish. In fact, each of $v(S)$ and $v(S \cup \{i\})$ must be either 0 or 1. The difference will be 0 if both S and $S \cup \{i\}$ are winning coalitions, or if both are losing coalitions. It will be 1 if S loses but $S \cup \{i\}$ wins. (The remaining case, S winning but $S \cup \{i\}$ losing, is not possible by the super-additivity condition.) The formula will reduce to

6.13.1
$$\phi_i = \sum_S \gamma_{n,s}$$

where the sum is taken over all S such that S is losing but $S \cup \{i\}$ is winning.

An alternative method is to consider our probabilistic model once again. In an ordering of the players, we shall say i is the *pivot* if the players preceding i form a losing coalition S, but $S \cup \{i\}$ is winning. In other words, i is the *pivot*, or *pivotal player*, if he completes a winning coalition. Then the power index of player i is the probability that he will be pivotal (assuming that each ordering has the probability $1/n!$).

Of great importance, among simple games, are *voting games*. In these, each of the players has a certain number of votes; a coalition wins if it has at least q votes, where q is often (but not always) equal to one more than half the total number of votes.

VI.13.2 *Example.* Consider a corporation with four stockholders, having, respectively, 1, 2, 3, and 4 shares of stock. Decisions can be made by any coalition having a simple majority of the total stock.

Since there are 10 shares of stock, 6 shares constitute a majority. In other words, a winning coalition is any coalition with 6 or more votes. To compute the value, let us take all 4! orderings of the players:

1 2 3* 4	2 1 3* 4	3 1 2* 4	4 1 2* 3
1 2 4* 3	2 1 4* 3	3 1 4* 2	4 1 3* 2
1 3 2* 4	2 3 1* 4	3 2 1* 4	4 2* 1 3
1 3 4* 2	2 3 4* 1	3 2 4* 1	4 2* 3 1
1 4 2* 3	2 4* 1 3	3 4* 1 2	4 3* 1 2
1 4 3* 2	2 4* 3 1	3 4* 2 1	4 3* 2 1

In each ordering, we have designated the pivot by an asterisk. We see that, of the 24 orderings, player 1 is the pivot in 2 cases; players 2 and 3, in 6 cases each, and player 4, in 10 cases. Thus, the value vector is

$$\phi = \left(\frac{1}{12}, \frac{1}{4}, \frac{1}{4}, \frac{5}{12}\right)$$

Note that the power indices of 1 and 3 are slightly less than their shares of stock, while players 2 and 4 have slightly more power than their shares of stock. Note also that 2 and 3 have the same power, although they have different amounts of stock.

VI.13.3 *Example.* The security council of the United Nations consists of 15 members, 5 of which are permanent and have "veto" powers. A motion is carried by vote of 9 members, except that opposition by any one of the permanent members defeats a motion.

Disregarding the possibility of abstention (used by the permanent members when they wish to show partial opposition to a motion), the winning coalitions for this game are any coalitions with nine or more members *including* the five permanent members. We shall compute the power index by considering one of the non-permanent members.

If i is a non-permanent member, the sets S such that S is losing but $S \cup \{i\}$ is winning must consist of exactly three non-permanent members and five permanent members. By (6.13.1), we must add the coefficients $\gamma_{n,s}$ of all these sets. In this case, $s = 8$ and $n = 14$, so

$$\gamma_{n,s} = \frac{8!5!}{14!} = \frac{1}{18018}$$

We must multiply this by the number of such sets. The five permanent members can be chosen in only one way, but the three non-permanent members can be chosen from among the nine remaining members (not counting i) in $\binom{9}{3} = 84$ different ways. Thus

$$\phi_i = \frac{84}{18018} = \frac{2}{429} = 0.00466$$

Each non-permanent member has approximately 1/2 of 1 per cent voting power. This means the non-permanent members have 4.66 per cent of the voting power. The five permanent members have 95.34 per cent of the power, or 19.07 per cent each. We see that the veto gives the permanent members almost complete control over the security council.

VI.13.4 Example. A committee consists of four members, each with one vote. The chairman of the committee is given the power to break ties.

It is not difficult to see that the chairman will be the pivotal player if he appears in second or third position. The probability of this is 2/4, or 1/2. Thus

$$\phi_1 = \frac{1}{2}$$

The remaining 1/2 unit of voting power is divided evenly among the other three committee members. Thus

$$\phi_2 = \phi_3 = \phi_4 = \frac{1}{6}$$

PROBLEMS ON N-PERSON GAMES

1. A committee consists of a chairman with three votes, two senior members with two votes each, and five junior members with one vote each. What is the distribution of voting power?

2. In a committee with four members the chairman is given the power to break ties. What is his voting power? What would it be if he were given veto power?

3. In a committee of five members, a simple majority of three is needed to carry a motion. Two members agree to join forces and vote together at all times. Does this strengthen or weaken them?

4. In a committee of five members, a unanimous vote is necessary

328

to carry a motion. Two members agree to join forces and vote together at all times. Does this strengthen or weaken them?

5. Consider a three-person simple game, v, in which the one-person coalitions lose, but the two- and three-person coalitions win, i.e., $v(S) = 0$ if S has one member, but $v(S) = 1$ if S has one or two members. Show that the three *imputations*

$$V = \left\{ \begin{array}{c} \left(\frac{1}{2}, \frac{1}{2}, 0\right) \\[2mm] \left(\frac{1}{2}, 0, \frac{1}{2}\right) \\[2mm] \left(0, \frac{1}{2}, \frac{1}{2}\right) \end{array} \right\}$$

form a *stable set* in the following sense:

(a) Given any two imputations in this set, there is no coalition all of whose members prefer one to the other, and can actually enforce it, (i.e., a winning coalition).

(b) Given any other imputation z (i.e., any other possible payoff vector) there is a winning coalition S, and an imputation $x \in V$, such that the members of S all prefer x to z. (Any set with these two properties is known as a von Neumann-Morgenstern solution. Are there any other such sets for this game?)

6. Show that, for the game of Problem 5, the imputation (1/2, 1/2, 0) is *stable for the coalition* {1,2} in the following sense: if either member of this coalition wishes to obtain more, and defects to join player 3, then the remaining player can "protect his share" of 1/2 while outbidding his former partner for player 3's services.

Show also that (1/2, 1/2, 0) is the only imputation stable for the coalition {1,2}. Show, e.g., that the imputation (0.51, 0.49, 0) is not stable, because player 2 can form a coalition with 3, with payoffs (0, 0.5, 0.5), and player 1 is then unable to "protect" his share of 0.51.)

(Payoff vectors that are stable in this manner are said to belong to the *bargaining set*.)

CHAPTER VII

DYNAMIC
PROGRAMMING

1. THE PRINCIPLE OF MAXIMALITY

Let us suppose that a firm has a finite amount of resources, X, which is to be invested in several enterprises. It is assumed that these enterprises are all independent, so that the return from any one of them depends only on the resources invested. If the return from an investment of x_i units in the ith enterprise is $g_i(x_i)$, the firm is faced with the problem of finding the function

7.1.1
$$F(X) = \max \sum_{i=1}^{n} g_i(x_i)$$

where the maximum is taken over all vectors $(x_1, \ldots, x_n)$ satisfying

7.1.2
$$\sum_{i=1}^{n} x_i = X$$

7.1.3
$$x_i \geq 0$$

Our desire, then, is to maximize the function

7.1.4
$$G(x_1, \ldots, x_n) = \sum_{i=1}^{n} g_i(x_i)$$

subject to the constraints (7.1.2) and (7.1.3). We shall do this by a technique that, while quite straightforward, yields very valuable (and unintuitive) results.

The *principle of maximality*, as it is called, is, on the face of it, nothing more than common sense. Yet even common sense often needs to be well formulated before it can be profitably used; the formulation of this principle must be recognized as an important step. It is generally attributed to R. Bellman, though a very similar technique was independently developed, more or less at the same time, by R. Isaacs.

To obtain a better understanding of the principle, we might think of our hypothetical firm's problem as a *sequential decision* problem. Let us assume that the firm must first decide how much of its resources to devote to the first enterprise. After this decision is made, it must decide how much of the *remaining* resources is to be spent on the second enterprise; then, how much of the remainder to spend on the third enterprise, and so on. This is probably easiest to visualize if we assume that each of these decisions must be made in a different period of time. For instance, a firm might wish to decide how to distribute its advertising budget over the 12 months of the year.

We shall call any distribution of resources over the n enterprises, a *policy*. It will be *optimal* if it maximizes the function (7.1.4) subject to the constraints (7.1.2) and (7.1.3). By a *sub-policy*, we shall mean a distribution of the resources left after k decisions, among the remaining $n - k$ enterprises. The principle of maximality can then be expressed in the intuitive form

7.1.5. *Every optimal policy contains only optimal sub-policies.*

In effect, (7.1.5) says that, whatever resources may have been spent in the first k enterprises, the remaining resources must be optimally distributed among the remaining $n - k$ enterprises. As such, its truth is obvious. Its importance, however, lies in the fact that the decision made at any one stage need consider only what can be done in subsequent stages with the amount of resources left; it need not, in other words, be concerned with the allocation of the resources already spent.

To see just what this means, let us imagine that a firm with a fixed advertising budget must decide how much to spend for this purpose in each of the 12 months of the year. When December comes, there will be no decision to make since all remaining monies are to be spent then: there is no question of saving any resources for future use. In November the advertising director must make a decision, but this will be simple enough, as it is only a question of deciding how much to spend then and how much to save for December, and this process involves only maximizing a function of one variable. Then the October problem reduces to deciding how much should be spent then and how much saved for the combined November-December advertising. We find that the advertising director has to make 11 simple (1-variable) decisions instead of 1 extremely complicated (11-variable) decision.

Let us give this a mathematical treatment. We shall use the notation $F_k(y)$ to denote the return that can be obtained by an optimal distribution of y units of resource among the kth, $(k + 1)$th, ..., and nth enterprises, i.e., among all enterprises from the kth to the nth inclusive.

Now the principle of maximality tells us that, whatever amount, x, may be invested in the kth enterprise, the remaining amount,

$y - x$, must be invested optimally among the remaining $n - k$ enterprises. Thus,

7.1.6 $$F_k(y) = \max_x g_k(x) + F_{k+1}(y-x)$$

This will allow us to compute F_k if we know both the functions g_k and F_{k+1}. For the final decision, we have

7.1.7 $$F_n(y) = g_n(y)$$

The two equations, (7.1.6) and (7.1.7), represent the solution to our problem. In fact, (7.1.7) gives us the function F_n. This allows us to compute F_{n-1}, which in turn gives us F_{n-2}. It is then merely a question of tabulating the functions $F_1, F_2, \ldots, F_n$; this tabulation can also include the optimal decision to be made at each stage for each level of resources that may then remain.

A question now arises about tabulating the functions F_k. In some cases, these will be elementary functions, say, powers, exponentials, logarithms, and such. Such cases are, unfortunately, quite rare; even if the functions g_i are elementary functions, it does not follow that the F_k will be so. Rather, the F_k are generally made up of pieces from several elementary functions and can usually only be obtained in tabular form (see Figure VII.1.1). Since an approximate solution is usually possible, this is not a great hardship; approximations can be made to as great a degree of accuracy as may be desired. In some cases, the resources available may consist of indivisible units, i.e., the x_i have to be integers. In this case, the dynamic programming technique is extremely powerful.

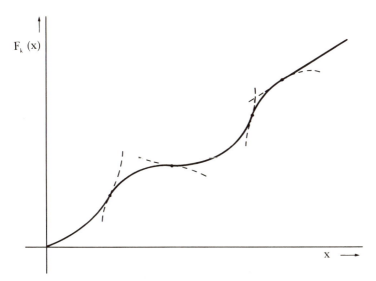

FIGURE VII.1.1 The curves $F_k(x)$ generally consist of pieces from several elementary curves.

332

VII.1.1   Example.   A company has 14 men whom it can assign to four projects. It estimates that the return on the ith project, given that x men are assigned to it, is the function $g_i(x)$ given in Table VII.1.1.

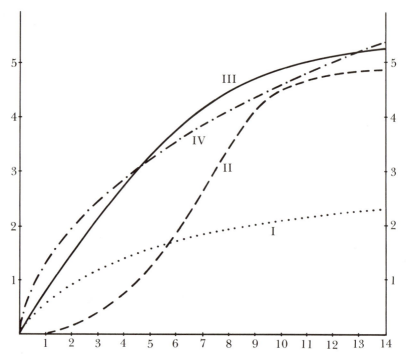

FIGURE VII.1.2   Returns on four jobs as a function of men assigned to each (Example VII.1.1).

TABLE VII.1.1

x	\multicolumn{4}{c}{i}			
	1	2	3	4
0	0	0	0	0
1	0.53	0.04	.75	1.35
2	0.88	0.18	1.50	1.98
3	1.15	0.40	2.15	2.48
4	1.37	0.75	2.75	2.88
5	1.58	1.25	3.30	3.22
6	1.75	1.90	3.75	3.56
7	1.86	2.64	4.15	3.87
8	1.95	3.43	4.48	4.13
9	2.04	4.12	4.73	4.37
10	2.13	4.50	4.90	4.62
11	2.20	4.70	5.05	4.85
12	2.25	4.82	5.15	5.05
13	2.29	4.89	5.22	5.25
14	2.33	4.90	5.26	5.45

In view of the nature of the problem, we can assume that only integral values are permissible for the variables. We must, now, compute and tabulate the functions F_k. We already know F_4: it is equal to g_4. To compute F_3, we have the definition

$$F_3(y) = \max_{0 \le x \le y} \{g_3(x) + F_4(y - x)\}$$

We will simultaneously tabulate the function $\hat{x}_3(y)$, defined by

$$F_3(y) = g_3(\hat{x}_3) + F_4(y - \hat{x}_3)$$

i.e., $\hat{x}_3(y)$ is the number of men that should be assigned to the third project, assuming that there are y men available for the third and fourth projects together.

To compute $F_3(14)$, say, we shall consider all the sums $g_3(x) + F_4(14 - x)$, for $x = 0, 1, 2, \ldots, 14$. These are:

$$
\begin{aligned}
g_3(0) &+ F_4(14) = 0 &+ 5.45 &= 5.45 \\
g_3(1) &+ F_4(13) = 0.75 &+ 5.25 &= 6.00 \\
g_3(2) &+ F_4(12) = 1.50 &+ 5.05 &= 6.55 \\
g_3(3) &+ F_4(11) = 2.15 &+ 4.85 &= 7.00 \\
g_3(4) &+ F_4(10) = 2.75 &+ 4.62 &= 7.37 \\
g_3(5) &+ F_4(9) &= 3.30 + 5.37 &= 7.67 \\
g_3(6) &+ F_4(8) &= 3.75 + 4.13 &= 7.88 \\
g_3(7) &+ F_4(7) &= 4.15 + 3.87 &= 8.02 \\
g_3(8) &+ F_4(6) &= 4.48 + 3.56 &= 8.04 \\
g_3(9) &+ F_4(5) &= 4.73 + 3.22 &= 7.95 \\
g_3(10) &+ F_4(4) &= 4.90 + 2.88 &= 7.78 \\
g_3(11) &+ F_4(3) &= 5.05 + 2.48 &= 7.53 \\
g_3(12) &+ F_4(2) &= 5.15 + 1.98 &= 7.13 \\
g_3(13) &+ F_4(1) &= 5.22 + 1.35 &= 6.57 \\
g_3(14) &+ F_4(0) &= 5.26 + 0 &= 5.26
\end{aligned}
$$

The largest of these sums is 8.04, obtained by letting $x_3 = 8$. We thus have

$$\hat{x}_3(14) = 8 \qquad F_3(14) = 8.04$$

which means that, if there are 14 men available for projects 3 and 4 only, 8 should be assigned to project 3 and 6 to project 4, giving a total return of 8.04 units.

To determine $F_3(13)$, we consider all the sums $g_3(x) + F_4(13-x)$, as x goes from 0 to 13. These are, respectively, 5.25, 5.80, 6.35, 6.77, 7.12, 7.43, 7.62, 7.71, 7.70, 7.61, 7.38, 7.03, 6.30, 5.22. The largest of these is 7.71, obtained for $x_3 = 8$. Therefore,

$$\hat{x}_3(13) = 8 \qquad F_3(13) = 7.70$$

In a similar way, we complete the tabulation of $F_3(y)$, obtaining

TABLE VII.1.2

y	$\hat{x}_3(y)$	$F_3(y)$
0	0	0
1	0	1.35
2	1	2.10
3	2	2.85
4	3	3.50
5	3	4.13
6	4	4.73
7	5	5.28
8	6	5.73
9	6	6.23
10	6	6.63
11	7	7.03
12	7	7.37
13	8	7.70
14	8	8.04

We are now in a position to compute $F_2(y)$, using the formula

$$F_2(y) = \max_{0 \le x \le y} \{g_2(x) + F_3(y - x)\}$$

and $\hat{x}_2(y)$, defined by

$$F_2(y) = g_2(\hat{x}_2) + F_3(y - \hat{x}_2)$$

Proceeding as before, we compute the sums $g_2(x) + F_3(14 - x)$, as x goes from 0 to 14. These are, respectively, 8.04, 7.74, 7.55, 7.43, 7.38, 7.48, 7.63, 7.92, 8.16, 8.25, 8.00, 7.55, 6.92, 6.24, 4.90. The largest of these is 8.25, obtained with $x_2 = 9$. Thus,

$$\hat{x}_2(14) = 9; \qquad F_2(14) = 8.25$$

Continuing in this manner, we obtain the functions F_2 and $\hat{x}_2$, given in

TABLE VII.1.3

y	$\hat{x}_2(y)$	$F_2(y)$
0	0	0
1	0	1.35
2	0	2.10
3	0	2.85
4	0	3.50
5	0	4.13
6	0	4.73
7	0	5.28
8	0	5.73
9	0	6.23
10	0	6.63
11	0	7.03
12	0	7.37
13	0	7.70
14	9	8.25

This says that no men should be assigned to project 2 unless there are 14 men available for projects 2, 3, and 4 together. In this case, 9 men should be assigned to project 2.

We are now in a position to solve the original problem. In fact, we are looking for the two functions, $F_1(14)$ and $\hat{x}_1(14)$. Again, we compute the sums $g_1(x) + F_2(14-x)$, and find that these are, respectively, 8.25, 8.23, 8.25, 8.18, 8.00, 7.81, 7.48, 7.14, 6.68, 6.17, 5.63, 5.05, 4.35, 3.64, 2.33. Here there is a tie for maximum, with $x_1 = 0$ and $x_1 = 2$ both giving a total return of 8.25, and there will be two optimal policies. One is obtained by choosing $x_1 = 0$. There are then 14 men available for projects 2, 3, and 4. We see that $\hat{x}_2(14) = 9$, so we assign 9 men to project 2. This leaves 5 men for projects 3 and 4; we have $\hat{x}_3(5) = 3$, and so 3 are assigned to project 3, and 2 are assigned to project 4.

The other optimal policy is to assign $x_1 = 2$; then $\hat{x}_2(12) = 0$, and $\hat{x}_3(12) = 7$. This assigns 2 men to project 1, 7 to project 3, and 5 to project 4. It may be checked that these two policies,

$$(0,9,3,2)$$

and

$$(2,0,7,5)$$

each give a total return of 8.25.

Example VII.1.1 shows the general procedure to be followed for dynamic programming problems. The computation gives us, not only the functions $F_k(y)$, but also the assignment functions, $\hat{x}_k(y)$, which tell us the amount of resources to be assigned to the kth enterprise, assuming there are still y units left at the time, i.e., $\hat{x}_k$ is defined mathematically by

7.1.8 $$F_k(y) = g_k(\hat{x}_k) + F_{k+1}(y - \hat{x}_k)$$

It is of interest to see how this technique compares with the crude evaluation of all possible assignments of the 14 men to four jobs. In the dynamic programming method we notice that 15 additions were necessary to compute $F_3(14)$, 14 additions were necessary to compute $F_3(13)$, and so on. We needed $1+2+\ldots+15 = 120$ additions to compute the entire function F_3. Another 120 were necessary to compute F_2, and 15 were necessary to obtain $F_1(14)$. In all, 255 additions were necessary. Against this, the total number of possible assignments is equal to the binomial coefficient $\binom{17}{3}$ or 680. A very real saving has been effected. More important, however, is the fact that, if a fifth project were available, the number of dynamic programming additions would increase only by another 120, to 375, but the total number of possible assignments would increase to $\binom{18}{4}$, or 3060. Clearly, the larger the problem, the greater the saving effected by dynamic programming techniques will be. Although the dynamic program increases in an arithmetic progression, the number of possible assignments increases much more rapidly.

2. THE FIXED-CHARGE TRANSPORTATION PROBLEM

We consider here a problem similar to those treated in Chapter III, Section 12: again, a company must deliver goods from its warehouses to its distributors; the difference lies in the fact that the cost function is not entirely linear. We shall assume that the cost of delivering x units from warehouse i to distributor j is $c_{ij}x$ (a linear cost), *plus* an additional amount, k_{ij}, which must be paid if *anything at all* is shipped from W_i to D_j. (For example, k_{ij} might be the cost of having a road repaired so that the company's trucks can use it. This is then the *fixed charge*; it is the same whether 1 unit or 100 units are sent along this route.

Because the form of the cost function is not linear, it follows that we cannot solve this problem by the linear programming techniques studied earlier. However, if there are only *two* warehouses (or only two destinations), the dynamic programming technique can be used.

VII.2.1 Example. To see how this is done let us assume that the availabilities at the two warehouses are, respectively, 9 and 13 units.

We shall assume six destinations, with requirement vector

$$\mathbf{b} = (4, 3, 5, 4, 2, 4)$$

and cost matrices

$$C = \begin{pmatrix} 7 & 6 & 7 & 9 & 1 & 4 \\ 7 & 10 & 3 & 4 & 5 & 4 \end{pmatrix}$$

and

$$K = \begin{pmatrix} 7 & 7 & 1 & 0 & 4 & 9 \\ 3 & 1 & 6 & 6 & 0 & 4 \end{pmatrix}$$

We repeat that C is the matrix of linear costs, K that of fixed charges, i.e., the cost of shipping x units on the route (i,j) is $c_{ij}x + k_{ij}$ if x is positive, but 0 if $x = 0$.

Let us assume that x units of the goods are shipped from the first warehouse to the jth destination. Since there is a total requirement of b_j units here, the remaining $b_j - x$ units must be sent from the second warehouse. The total cost of servicing this destination will be

$$7.2.1 \quad g_j(x) = \begin{cases} c_{2j}b_j + k_{2j} & \text{if } x = 0 \\ c_{1j}b_j + k_{1j} & \text{if } x = b_j \\ c_{1j}x + c_{2j}(b_j - x) + k_{1j} + k_{2j} & \text{if } 0 < x < b_j \end{cases}$$

The function $g_j(x)$ can be thought of as undefined, or (equivalently) very large for $x < 0$ or $x > b_j$. The problem now becomes one of finding the minimum of

$$\sum_{j=1}^{n} g_j(x_j)$$

subject to the constraints that the x_j must be non-negative, and their sum is equal to the total amount available at W_1. There are additional constraints, namely, that the x_j cannot exceed the requirements b_j. We shall see, however, that these constraints present no problem at all: we simply ignore other values of x_j in computing the functions F_k.

Since this is a minimization problem, we will have the equations

$$F_k(y) = \min \{ g_k(x) + F_{k+1}(y - x) \}$$

where the minimum is taken over all non-negative x satisfying $x \leq y$ and $x \leq b_j$.

The first problem is to tabulate all the functions $g_j(x)$. Strictly speaking, this is not necessary, but it helps to see things at a glance. It can be proved (though we shall not do so here) that the minimum for a problem such as this will always be obtained at an extreme point of the constraint set. We saw in Chapter III that for these problems, the extreme points have only integral values for the variables (assuming, of course, that the requirements and availabilities are themselves integers). Thus, we need only tabulate the functions, given by (7.2.1), for integral values of x. These are given in

TABLE VII.2.1

x			j			
	1	2	3	4	5	6
0	31	31	21	22	10	20
1	38	34	26	27	10	29
2	38	30	30	32	6	29
3	38	25	34	37	—	29
4	35	—	38	36	—	25
5	—	—	36	—	—	—

As may be seen, the functions are simply not defined for $x > b_j$.

As before, we have $F_6 = g_6$. This allows us to compute $F_5(y)$, as well as $\hat{x}_5(y)$, for y between 0 and 6 (Table VII.2.2).

TABLE VII.2.2

y	$\hat{x}_5(y)$	$F_5(y)$
0	0	30
1	1	30
2	2	26
3	2	35
4	2	35
5	2	35
6	2	31

We compute, next, $\hat{x}_4(y)$ and $F_4(y)$, for y ranging between 0 and 9 (Table VII.2.3). (Note that 9 is the total availability at W_1; hence we will not consider larger values of y.)

TABLE VII.2.3

y	$\hat{x}_4(y)$	$F_4(y)$
0	0	52
1	0	52
2	0	48
3	1	53
4	0	57
5	0	57
6	0	53
7	1	58
8	2	63
9	3	67

We obtain, next, $\hat{x}_3$ and F_3 (Table VII.2.4).

TABLE VII.2.4

y	$\hat{x}_3(y)$	$F_3(y)$
0	0	73
1	0	73
2	0	69
3	1	74
4	2	78
5	0	78
6	0	74
7	0	79
8	2	83
9	3	87

Next, we have $\hat{x}_2$ and F_2 (Table VII.2.5).

TABLE VII.2.5

y	$\hat{x}_2(y)$	$F_2(y)$
0	0	104
1	0	104
2	0	100
3	3	98
4	3	98
5	3	94
6	3	99
7	3	103
8	3	103
9	3	99

Finally, using this, we find

$$\hat{x}_1(9) = 4; \; F_1(9) = 129$$

This gives us the solution; we must let $x_1 = 4$. This leaves 5 units at W_1; we see next that $\hat{x}_2(5) = 3$. Then, $\hat{x}_3(2) = 0$; $\hat{x}_4(2) = 0$, and $\hat{x}_5(2) = 2$. Thus the amounts shipped from W_1 are given by the vector

$$(4,3,0,0,2,0)$$

The total shipping schedule is obtained by including, also, the amount shipped from W_2. This will give us the matrix

$$\begin{pmatrix} 4 & 3 & 0 & 0 & 2 & 0 \\ 0 & 0 & 5 & 4 & 0 & 4 \end{pmatrix}$$

It may be checked that this will indeed give a total cost of 129 units. Note also that the program is degenerate; this degeneracy causes no trouble at all in our computations.

340   PROBLEMS ON DYNAMIC PROGRAMMING

1. A company has 10 men which it must assign to four different projects. The revenue expected from assigning j men to the ith project is given in the table.

Men assigned	Job			
	1	2	3	4
0	0	0	0	0
1	13	8	15	10
2	23	17	22	20
3	29	29	31	30
4	37	31	39	35
5	42	40	44	40
6	51	47	51	50
7	64	54	62	65
8	80	67	70	78
9	88	80	80	88
10	95	86	91	99

How many men should be assigned to each job?

2. A company wishes to send a cargo of gadgets, widgets, and beepers to a distributor; it has a truck with a capacity of 25,000 lb. Each gadget weighs 3000 lb., each widget 5000 lb., and each beeper, 4000 lb. Each gadget yields a profit of $500, each widget $900, and each beeper, $650. The demand for these items is a random variable; the probability that there be a demand for k items of a given type is given in the following table.

	Gadgets	Widgets	Beepers
0	0.05	0.05	0
1	0.15	0.10	0.05
2	0.15	0.15	0.08
3	0.15	0.20	0.11
4	0.15	0.20	0.20
5	0.10	0.15	0.15
6	0.10	0.10	0.13
7	0.10	0.05	0.10
8 or more	0.05	0	0.18

How many items of each type should be sent to maximize expected profits? It may be assumed that leftover items can be kept at very little cost, so they represent no loss.

3. These are fixed-cost transportation problems. In each case, **a** is the vector of availabilities at the two sources, and **b** is the vector of requirements at the n destinations. The matrices C and K represent, respectively, the linear costs per unit between the sources and destinations, and the fixed costs of keeping the corresponding routes open.

(a) $\qquad$ **a** $= (8, 13)$ $\qquad\qquad$ **b** $= (3, 3, 5, 4, 6)$

$$C = \begin{pmatrix} 3 & 5 & 2 & 6 & 1 \\ 1 & 3 & 1 & 8 & 2 \end{pmatrix} \qquad K = \begin{pmatrix} 1 & 4 & 8 & 3 & 4 \\ 4 & 5 & 6 & 2 & 3 \end{pmatrix}$$

(b) $\qquad$ **a** $= (13, 17)$ $\qquad\qquad$ **b** $= (5, 7, 8, 3, 2, 5)$

$$C = \begin{pmatrix} 6 & 1 & 3 & 5 & 8 & 2 \\ 2 & 2 & 4 & 6 & 3 & 1 \end{pmatrix} \qquad K = \begin{pmatrix} 1 & 4 & 2 & 3 & 0 & 5 \\ 5 & 3 & 2 & 8 & 3 & 1 \end{pmatrix}$$

(c) $\qquad$ **a** $= (9, 6)$ $\qquad\qquad$ **b** $= (3, 4, 5, 3)$

$$C = \begin{pmatrix} 1 & 2 & 6 & 4 \\ 2 & 0 & 4 & 4 \end{pmatrix} \qquad K = \begin{pmatrix} 2 & 3 & 1 & 6 \\ 0 & 2 & 5 & 3 \end{pmatrix}$$

3. INVENTORIES

One of the more common applications of mathematical analysis to industrial situations today is in the control of inventory size. Generally speaking, an inventory is an amount of goods that is kept in storage to service future demand. The decision-maker will, in normal situations, wish to maintain a certain amount of inventory, in part because of uncertainties in demand and in part because the "setting-up" costs connected with keeping supply exactly equal to demand at all times are generally prohibitive. On the other hand, too large an inventory ties up working capital and must, moreover, be stored (for a price) in a warehouse. An optimal policy must be sought that reconciles these two extremes.

Let us see how this can best be done. We shall assume that the demands for each month, over a period extending n months into the future, are known. Let d_j be the amount needed at the end of the jth month, x_j the number of units produced during the jth month, and y_j the number of units in inventory at the *beginning* of the jth month. The x_j, y_j, and d_j are connected by the relation

7.3.1 $\qquad\qquad\qquad y_{j+1} = y_j + x_j - d_j$

and we have the additional constraints

342

7.3.2 $$x_j, y_j \geq 0$$

The total costs connected with the system include setting-up costs, production costs, and inventory storage costs. We shall simplify the model by assuming that production runs can only be made once a month (though there is no need to make a run every month). The cost of producing x units in a month will be a function, $c_j(x)$, while the cost of keeping y units in inventory will be another function, $f_j(y)$. The total costs connected with the jth month will be

7.3.3 $$g_j(x_j, y_j) = K\delta_j + c_j(x_j) + f_j(y_j)$$

where K is the setting-up cost, and $\delta_j = 0$ if $x_j = 0$, but is equal to 1 if $x_j > 0$. Then the problem is to minimize

$$\sum_{j=1}^{n} g_j(x_j, y_j)$$

where g_j is, as in (7.3.3), subject to the constraints (7.3.1) and (7.3.2). In form, this is slightly different from the problems studied in the first two sections of this chapter. Essentially, however, the problem is the same and can be solved by the methods of dynamic programming.

Let us, as before, write

7.3.4 $$F_k(\xi) = \min \sum_{j=k}^{n} g_j(x_j, y_j)$$

where ξ represents inventory on hand at the beginning of the kth period. The minimum is taken over all x_j, y_j satisfying (7.3.1) and (7.3.2) and such, moreover, that $y_k = \xi$.

If an amount x_k is produced during the kth period, the inventory on hand at the end of this period will be $\xi + x_k - d_k$. Applying the recurrence equation of the principle of maximality, we have

7.3.5 $$F_k(\xi) = \min_{x \geq 0} \{g_k(x, \xi) + F_{k+1}(\xi + x - d_k)\}$$

We shall use (7.3.5) to compute the functions F_k. It very often happens that all the functions c_j and f_j are equal, which simplifies the calculations considerably. This is even more so when the c_j and f_j are linear functions: thus, if

$$c_j(x) = cx \qquad \text{for all } j, x$$
$$f_j(y) = hy \qquad \text{for all } j, y$$

then

$$g_j(x,y) = \begin{cases} hy & \text{if } x = 0 \\ K + cx + hy & \text{if } x > 0 \end{cases}$$

VII.3.1 Example. A tool company has orders for wrenches to be delivered over the course of a year. The total number to be delivered at the end of each month is shown in

TABLE VII.3.1

Month	Requirement
January	300
February	100
March	200
April	200
May	400
June	300
July	500
August	700
September	400
October	700
November	200
December	600

We shall assume that each wrench costs $3.00 to produce, plus a setting-up cost of $100 per run. Inventory costs are 10c per wrench per month.

The fact that the variable costs of production (as opposed to setting-up costs) are linear means, more or less, that they can be disregarded. There are 4600 wrenches to be produced; these will cost $13,800, regardless of when they are produced. (This is, of course, a considerable simplification; in general, interest charges mean that future costs can be discounted, and there is also the possibility that excessive production in any time period requires overtime labor, with consequently higher costs.) For this problem, therefore, we shall consider only the setting-up and inventory costs.

We proceed, now, to calculate the functions g_j and F_k. Disregarding the $3.00 production costs, we have

$$g_j(x,y) = \begin{cases} 0.1y & \text{if } x = 0 \\ 0.1y + 100 & \text{if } x > 0 \end{cases}$$

for each j. Suppose, now, that we wish to finish the year with no inventory on hand, i.e., $y_{13} = 0$. The month of December can be started with any amount up to 600 wrenches on hand. Of necessity we will have

$$x_{12} = 600 - y_{12}$$

and so we can form a table (Table VII.3.2) of the functions $F_{12}(\xi)$, $\hat{x}_{12}(\xi)$,

TABLE VII.3.2

ξ	$\hat{x}_{12}(\xi)$	$F_{12}(\xi)$
0	600	100
100	500	110
200	400	120
300	300	130
400	200	140
500	100	150
600	0	60

where, as before, $\hat{x}_{12}(\xi)$ is the amount to be produced in December assuming an inventory level ξ at the beginning of that month.

We proceed, next, to calculate F_{11}. For example, we know that

$$F_{11}(0) = \min_{x} \{g_{11}(x,0) + F_{12}(x - 200)\}$$

since the requirement for November is 200 wrenches. It is easy to see that x must be at least 200, and that the desired value $F_{11}(0)$ is then the smallest of the numbers 200, 210, 220, 230, 240, 250, 160. The smallest number, 160, is obtained by letting $x = 800$. Thus

$$\hat{x}_{11}(0) = 800 \qquad F_{11}(0) = 160$$

We proceed similarly, to compute all the functions $\hat{x}_k(\xi)$, $F_k(\xi)$, shown in Table VII.3.3.

TABLE VII.3.3

ξ	F_2	F_3	F_4	F_5	F_6	F_7	F_8	F_9	F_{10}	F_{11}	F_{12}
0	750	710	650	590	510	460	360	310	220	160	100
100	720	720	660	600	520	470	370	320	230	170	110
200	740	670	610	610	530	480	380	330	240	120	120
300	700	690	630	620	490	490	390	340	250	140	130
400	730	650	650	550	510	500	400	260	260	160	140
500	700	680	670	570	530	410	410	280	270	180	150
600	740	710	610	590	550	430	420	300	280	200	60
700	780	740	640	560	570	450	380	320	230	220	–
800	820	690	670	590	490	470	400	340	250	140	–
900	770	730	650	620	520	490	420	360	210	–	–
1000	830	770	690	650	550	510	440	380	240	–	–

ξ	$\hat{x}_2$	$\hat{x}_3$	$\hat{x}_4$	$\hat{x}_5$	$\hat{x}_6$	$\hat{x}_7$	$\hat{x}_8$	$\hat{x}_9$	$\hat{x}_{10}$	$\hat{x}_{11}$	$\hat{x}_{12}$
0	500	400	600	700	800	500	1100	1300	900	800	600
100	0	300	500	600	700	400	1000	1200	800	700	500
200	0	0	0	500	600	300	900	1100	700	0	400
300	0	0	0	400	0	200	800	1000	600	0	300
400	0	0	0	0	0	100	700	0	500	0	200
500	0	0	0	0	0	0	600	0	400	0	100
600	0	0	0	0	0	0	500	0	300	0	0
700	0	0	0	0	0	0	0	0	0	0	–
800	0	0	0	0	0	0	0	0	0	0	–
900	0	0	0	0	0	0	0	0	0	–	–
1000	0	0	0	0	0	0	0	0	0	–	–

It is now possible to choose the optimal policy. In fact, the table allows us to compute $\hat{x}_1(0)$ and $F_1(0)$:

$$\hat{x}_1(0) = 600 \qquad F_1(0) = 800$$

The optimal policy is now given by the recursive definition

$$y_1 = 0$$
$$x_k = \hat{x}_k(y_k)$$
$$y_{k+1} = y_k + x_k - d_k$$

and Table VII.3.4.

TABLE VII.3.4

k	1	2	3	4	5	6	7	8	9	10	11	12
y_k	0	300	200	0	400	0	500	0	400	0	200	0
x_k	600	0	0	600	0	800	0	1100	0	900	0	600

We find then, that six production runs are scheduled. Most of these are for two months' requirements, but the first is for three months, and last for only one month. Note that no production is ever scheduled until inventory has vanished: this is a common property of systems such as this (though it need not be if the cost functions are more complicated). The reason is that it is clearly better to plan production to avoid such leftovers, which contribute nothing but expenses.

It can be checked that this policy gives a total cost of $200 for inventory, plus $600 setting-up cost. Added to the $13,800 linear costs, the total expenses are $14,600.

VII.3.2 *Example.* A vulcanizing plant must produce 30 tires per month, during five months, to meet a contract. The costs of production and storage are independent of time, and given in

TABLE VII.3.5

x(Tires)	Production cost	Inventory cost (per month)
10	170	15
20	300	30
30	430	45
40	550	75
50	670	105
60	790	150
70	900	
80	1000	
90	1090	

We shall assume that there are no tires on hand at the beginning of the five months, and none on hand at the end. Proceeding as before, we obtain the functions $\hat{x}_5(\xi)$ and $F_5(\xi)$ shown in

TABLE VII.3.6

ξ	$\hat{x}_5(\xi)$	$F_5(\xi)$
0	30	430
10	20	315
20	10	200
30	0	45

and so on. We finally obtain Table VII.3.7 for $\hat{x}_k$ and F_k:

TABLE VII.3.7

ξ	$\hat{x}_1$	$\hat{x}_2$	$\hat{x}_3$	$\hat{x}_4$	$\hat{x}_5$	F_1	F_2	F_3	F_4	F_5
0	30	60	30	60	30	2100	1670	1265	835	430
10	–	50	20	50	20	–	1565	1150	730	315
20	–	40	10	40	10	–	1460	1005	625	200
30	–	0	0	0	0	–	1310	880	475	45
40	–	0	0	0	–	–	1225	805	390	–
50	–	0	0	0	–	–	1110	730	420	–
60	–	0	0	0	–	–	1030	625	195	–

We conclude from this that the optimal policy is to produce 30 tires the first month and 60 each in the second and fourth months. The total cost is $2100.

PROBLEMS ON INVENTORIES

1. A company manufactures bolts; there is a steady demand of 160 T. of bolts per month. Production costs are $1250 per ton, plus $3000 in setting-up costs for each production run. Inventory storage costs are $200 per ton, per year. How frequently should production runs be scheduled to minimize total costs over the next eight months?

2. A canned food company has a steady demand for 5000 cans of beans per month. Production costs are 12¢ per can, plus $15 setting-up costs for each run. Inventory storage costs are 5¢ per can per year. How many cans should be made at each production run to minimize costs over the next six months?

3. A miller in a resort town has a contract to supply 400 lb. of flour per week during the five-week season. Production and storage costs (per week) for the flour are given in the following table:

Lb.	Production	Storage
100	13	2
200	17	4
300	21	6
400	25	8
500	28.5	9.5
600	32	11
700	35.5	12
800	39	13
900	42	14
1000	45	14.5.
1100	47.5	15
1200	50	15.5

How should the miller schedule his production?

4. A tool company has a contract to deliver 300 screwdrivers in August, 500 in September, 200 in October, 400 in November and 600 in December. Production costs are 75¢ per screwdriver, plus $25 setting-up costs for each run. Inventory costs are 3c per screwdriver per month. How should production be scheduled?

4. STOCHASTIC INVENTORY SYSTEMS

Throughout the third section of this chapter, we have assumed that the demands on the system are deterministic, i.e., not subject to random fluctuations. Planning can be done well in advance, and there is never any danger that the system will be unable to meet the demands. In practice, demands are often known only probabilistically; they are subject to random fluctuations. There is a very real danger that the stock will run out. As a result, we shall see, such systems must almost always be maintained at a "safe" level: more stock is kept than one expects to use.

We shall, in what follows, make the assumption that a production run can be ordered at the beginning of any time period, but at no other time. The amount produced in this period is immediately available, i.e., it can be used to meet demands received during this period. (This does not mean, of course, that production is instantaneous, but rather, that there are substantially equal lags in the production and retail processes, i.e., customers generally must wait a while before receiving their orders.) We shall, moreover, assume that the demand during each period is a random variable with a known probability distribution.

In general, suppose that we have an inventory y_j on hand at the beginning of the jth period. If we then order x_j units, we will have $x_j + y_j$ units available for this period. If the demand is Z_j units, we will be able to deliver either Z_j or $x_j + y_j$, whichever is smaller. Thus we will have

7.4.1
$$y_{j+1} = \begin{cases} x_j + y_j - Z_j & \text{if } Z_j \leq x_j + y_j \\ 0 & \text{if } Z_j > x_j + y_j \end{cases}$$

Let us consider, next, the costs incurred in the jth period. There is, first of all, the cost $f_j(y_j)$ of holding y_j units in inventory. Next, there are production costs $c_j(x_j)$, plus a setting-up cost $A\delta_j$, where $\delta_j = 0$ if $x_j = 0$, but $\delta_j = 1$ if $x_j > 0$. Finally, a penalty cost is incurred whenever $Z_j > x_j + y_j$. We shall, as before, assume that the charges (except for the setting-up cost) are linear functions of their variables.

7.4.2 $\quad g_j(x_j, y_j, Z_j) = \begin{cases} hy_j + cx_j + A\delta_j & \text{if } Z_j \leq x_j + y_j \\ hy_j + cx_j + A\delta_j + p(Z_j - x_j - y_j) & \text{if } Z_j > x_j + y_j \end{cases}$

The problem, now, is to choose the variables, x_j, in such a way as to minimize the *expected* sum of the costs g_j over the n periods. We will have

$$F_k(\xi) = \min_{x_k,\dots,x_n} \mathbf{E}\left[\sum_{j=k}^{n} g_j(x_j,y_j,Z_j)\right]$$

where the minimum is taken over all non-negative $x_k,\dots,x_n$, with y_j and g_j given by (7.4.1) and (7.4.2), and $y_k = \xi$. We will then have

7.4.3 $F_k(\xi) = \min_x \mathbf{E}\left[g_k(x,\xi,Z_k) + F_{k+1}(y_{k+1})\right]$

where y_{k+1} is given by (7.4.1). The usual dynamic programming technique is to be used, but with expected values replacing the previous deterministic costs.

VII.4.1 Example. A butcher must decide how many turkeys to keep in stock during a three-day period. He considers that he might sell as many as four turkeys each day, with the probability distributions (for each day) given in

TABLE VII.4.1

Number of turkeys	Day		
	Thursday	*Friday*	*Saturday*
0	0.3	0.2	0.1
1	0.4	0.3	0.1
2	0.2	0.3	0.3
3	0.1	0.1	0.4
4	0	0.1	0.1

Each turkey costs $3, but there is a charge of $2 for each delivery from the farm (whatever size the delivery may be). The storage charge for leftover turkeys is 50¢ per bird per night. Additionally, there is a "penalty" of $7 (in lost business) for each customer who must be sent away for lack of a turkey.

Once again, the problem is attacked by the "backward" technique. Suppose that the stock level on Saturday (including any deliveries made on Saturday morning) is three turkeys. In that case, there is a 0.1 probability of three leftover turkeys, causing costs of $1.50; a 0.1 probability of two leftover turkeys, causing costs of $1; a 0.3 probability of one left-over turkey, with a cost of 50¢, and a 0.1 probability that there will be a shortage of one turkey, causing a penalty of $7. The expected costs for the day, *exclusive of order costs*, amount to

$$(0.1)(1.5) + (0.1)(1) + (0.3)(0.5) + (0.1)(7) = 1.10$$

FINITE MATHEMATICS

350

or $1.10. For other possible levels of supply $y_3 + x_3$, the expected inventory and penalty charges are computed and given in

TABLE VII.4.2

$y_3 + x_3$	Costs
0	16.10
1	9.85
2	4.35
3	1.10
4	0.85
5	1.35
6	1.85
7	2.35
8	2.85
9	3.35
10	3.85

It is clear from this that costs will be least on Saturday if we have four turkeys available that day. Suppose, however, that the stock on Saturday morning is less than four turkeys. Then we must decide whether it is worth-while to order any, remembering that order costs of $3 per bird, plus a $2 delivery charge, must then be paid. It may be seen, for example, that if $y_3 = 1$, then it is best to order two turkeys, incurring further charges of $8, but lowering the total expected costs to only $9.10. If, on the other hand, $y_3 = 2$, then no order should be made. In general, if $y_3 = 0$ or 1, an order should be made to increase the supply to three. If $y_3 \geq 2$, however, no order should be made. We obtain, in this way, the functions $\hat{x}_3(\xi)$ and $F_3(\xi)$ as shown in

TABLE VII.4.3

ξ	$\hat{x}_3(\xi)$	$F_3(\xi)$
0	3	12.10
1	2	9.10
2	0	4.35
3	0	1.10
4	0	.85
5	0	1.35
6	0	1.85
7	0	2.35
8	0	2.85
9	0	3.35
10	0	3.85

We repeat the procedure. Suppose we have no turkeys available on Friday. Then there is an expected shortage of 1.6 turkeys, with a corresponding expected penalty of \$11.20, plus the prospect of having no birds Saturday morning, which, as we have just calculated, represents expected costs of \$12.10. The total costs will be \$23.30. In general, for various levels of supply, $y_2 + x_2$, the expected future costs are given by

TABLE VII.4.4

$y_2 + x_2$	Costs
0	23.30
1	17.20
2	12.10
3	8.12
4	5.12
5	3.90
6	3.77
7	4.32
8	5.25
9	6.25
10	7.25

As before, we can compute the optimal policies for each level of supply on Friday morning: it may be seen that, if $y_2 = 0$ or 1, then an order should be made, raising the level to 3 or 4 (there is a tie here). Otherwise no order should be made. We have

TABLE VII.4.5

ξ	$\hat{x}_2(\xi)$	$F_2(\xi)$
0	3 or 4	19.12
1	2 or 3	16.12
2	0	12.20
3	0	8.12
4	0	5.12
5	0	3.90
6	0	3.77
7	0	4.32
8	0	5.25
9	0	6.25
10	0	7.25

Finally, we compute the expected costs for each available supply $y_1 + x_1$ on Thursday. These are given in Table VII.4.6.

352

TABLE VII.4.6

$y_1 + x_1$	Costs
0	27.82
1	21.17
2	17.04
3	13.40
4	10.29
5	7.01
6	6.97
7	7.05
8	9.00
9	10.20
10	10.68

from which we can find the optimal policy for Thursday: if $y_1 = 0$ or 1, we place an order to raise the level to five birds; if $y_1 \geq 2$, we place no order. The functions $\hat{x}_1(\xi)$ and $F_1(\xi)$ are given in

TABLE VII.4.7

ξ	$\hat{x}_1(\xi)$	$F_1(\xi)$
0	5	24.01
1	4	21.01
2	0	17.04
3	0	13.40
4	0	10.29
5	0	7.01
6	0	6.97
7	0	7.05
8	0	9.00
9	0	10.20
10	0	10.68

We have thus determined our optimal policy for this problem. According to this policy, there are two numbers, s_k and S_k, assigned to each period (these numbers are $(1,5)$, $(1,4)$, and $(1,3)$, respectively, for the problem). If at any period, the supply $y_k \leq s_k$, then an order should be made to raise supply to S_k. If, however, $y_k > s_k$, no order should be made. Many inventory problems exhibit this sort of solution. A sufficient condition for this type of behavior is that the costs (production costs, storage costs, and shortage penalties) all be linear functions of the relevant variables.

PROBLEMS ON STOCHASTIC INVENTORY SYSTEMS

1. An automobile dealer makes a profit of $800 on each automobile that he sells; inventory carrying cost is $65 per car per month.

The demand for this car is a random variable with distribution given by the following table.

Demand	Probability
0	0.05
1	0.15
2	0.23
3	0.23
4	0.17
5	0.11
6	0.04
7 or more	0.02

Assuming that his shortage penalty is simply the $800 (lost profit), and that he can order cars only at the beginning of each month, how many automobiles should he keep in stock?

2. A television dealer sells color television sets for $400. He pays storage costs of $10 per month per set; his supply costs are $300 per set plus $50 delivery charge for each shipment of whatever size. For the last three months of the model year, the demand is expected to have the probability distribution shown in the table:

	July	August	September
0	0.05	0.05	0.10
1	0.15	0.20	0.30
2	0.20	0.30	0.25
3	0.30	0.20	0.20
4	0.20	0.15	0.10
5	0.10	0.10	0.05

Sets can be ordered at the beginning of each month; if demand is higher than supply, the excess demand is simply lost. What is the best ordering policy?

3. An airplane company has noticed that, in any month, the demand for its "Family Jet" model is a random variable with probability distribution $P(0) = 0.1$; $P(1) = 0.6$; $P(2) = 0.3$. Production costs are $50,000 per plane, plus a $10,000 setting-up cost for each production run; storage costs are $2000 per plane per month. Each plane can be sold for $70,000, but any left over at the end of the model year must be sold below cost, at $40,000. Assuming that at most one production run can be held each month, how should the company schedule production?

CHAPTER VIII

GRAPHS AND NETWORKS

1. INTRODUCTION

In this chapter we shall study problems dealing with mathematical systems known as *networks.* One special problem of this type, the *transportation* problem, was studied in some detail in Chapter III. Here we shall be concerned with more general networks.

The theory of networks is an old and honorable branch of applied mathematics, dating back to the physicists James Clerk Maxwell (1831-1879) and Gustav Kirchhoff (1824-1887), who were, however, interested in the theory of electrical networks. The economically oriented theory of networks dates to the 1940's and 1950's. Because network problems can generally be treated by linear or dynamic programming methods, the authors mentioned in those chapters have obviously contributed to this work. Starting in the late 1950's, however, some authors, notably L. R. Ford, Jr., and D. R. Fulkerson, have raised network theory to the level of a nearly independent branch of mathematics.

In the context of this chapter (as in Chapter III, Section 12), a graph will be a collection of points called *nodes,* together with lines, called *arcs,* that join pairs of these nodes; we shall assume that there is at most one arc joining any two nodes. The graph is connected if, given any two nodes, A and B, it is possible to go from A to B along the arcs of the graph. The graph in Figure VIII.1.1 is connected, whereas Figure VIII.1.2 shows a *disconnected* graph: it is not connected because vertices A and G cannot be joined by arcs of the graph.

A *network,* as we shall use the word here, is a connected graph. Normally, each of the arcs in a network has a number assigned to it, corresponding, perhaps, to the capacity of the arc or to the distance

354

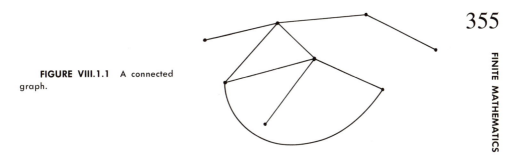

FIGURE VIII.1.1 A connected graph.

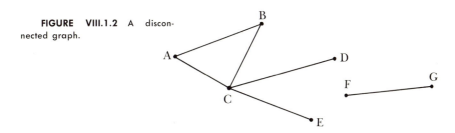

FIGURE VIII.1.2 A disconnected graph.

between the two nodes that it joins. The network may be *oriented*, which means that each arc has a definite direction (say, from A to B but not from B to A) or that the capacities or distances along the arc are different in the two directions, or it may be *unoriented*. We shall find that it does not generally matter, for many applications, whether the graph is oriented or not. If it does matter, we shall definitely say so.

2. CRITICAL PATH ANALYSIS

The method of *critical path analysis* is another example of the ways in which "plain common sense" can be used for rather complex problems, if only it is correctly formulated. Using no advanced mathematics (though its problems could be reformulated as linear programs), for *small* problems, critical path analysis is an obvious way to do things; many people have used it for years without knowing of it. For larger problems, its use is not so common, mainly because people do not take time to formulate problems in these terms. The strict mathematical formulation of this method is principally the work of J. E. Kelley, Jr., and M. R. Walker, and dates to the late 1950's.

In general, critical path analysis is used to find the minimal time necessary to complete a project. It also shows which of the several parts of the project actually affect the overall completion time.

Generally speaking we define a *project* as a set of *activities,* together with an *order* relation among these activities; we say that the *i*th activity *precedes* the *j*th activity if the *i*th activity must be complete before the *j*th activity can begin. There is, furthermore, a nonnegative number, *t*, assigned to each activity, representing the time it takes to carry out this activity, the *time of performance* of the activity.

A project can be represented by a directed graph, whose *nodes* represent the several activities. There will be an *arc* from node *i* to node *j* if the activity *i precedes* activity *j*. (Actually, it is not necessary to draw an arc from *i* to *j* unless *i* is an *immediate* predecessor of *j*.)

VIII.2.1 Example. Let us consider the project of repairing a broken door for an automobile. The door must be removed, hammered back into shape, and welded; some parts must also be ordered. We have a table:

TABLE VIII.2.1

Activity	Time (days)	Immediate predecessors
a. Start	0	none
b. Remove door	1	*a*
c. Order parts	3	*a*
d. Weld and paint door	2	*b,c*
e. Reinstall window	1	*b,c*
f. Replace door	1	*d,e*
g. Finish	0	*f*

The project can be represented by the network shown in Figure VIII.2.1. As may be seen, we assign to each arc of the network a number corresponding to the time of performance of the activity that is at the *beginning* of the arc. We call this the *length* of the arc. The

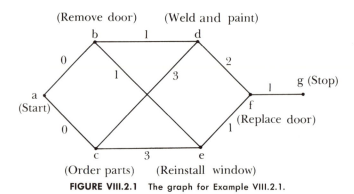

FIGURE VIII.2.1   The graph for Example VIII.2.1.

network diagram is especially useful in displaying the order of work; one does not need to know much about automobile repairs in order to understand it.

Suppose we wish to know the earliest possible finishing time for this project. This can be done in either of two ways. One is to treat the problem as a linear program and minimize the finishing time, subject to the obvious constraints (e.g., that activity d cannot begin until after activity b is finished). The other is to think of this as a dynamic program and look for the longest path through the network, from start to finish. We shall, in essence, do this.

Let us start at node a, the "start" of the network. Starting here, we assign times to each node and arc, according to the following rules.

1. Node a is assigned the time 0.
2. Each arc is assigned a time equal to the sum of its *length* and the time assigned to the node at which the arc begins.
3. Each node is assigned a time equal to the *maximum* of the times assigned to the arcs that *end* at this node.

Let us see, now, exactly what we have done. The time that has been assigned to each node is the earliest possible time at which the corresponding activity can be started, and the number assigned to an arc is the earliest time at which such an activity can be finished. Rule 1 tells us that we start at time 0; rule 2 tells us that the finish time for an activity is equal to its starting time plus the time that it takes (the "length" of the arc); rule 3 tells us that an activity can be started as soon as its predecessors have been completed, but no sooner. The times obtained in this manner are known as the *early start time* (e.s.t.) and *early finish time* (e.f.t.) of the various activities. The e.s.t. for the "finish" node is the earliest time at which the project can be completed and is known as the *target time* of the project.

Figure VIII.2.2 shows our network with e.s.t. and e.f.t. assigned to each of the activities. As may be seen, the target time is 6.

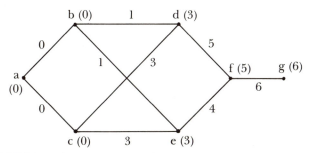

FIGURE VIII.2.2 Early start times (at nodes) and early finish times (on arcs) for Example VIII.2.1.

Suppose, next, that for some reason one of the activities is delayed. We would like to know whether this will affect the target time of the project.

What we actually wish to know, in this case, is the latest time at which an activity can be started without affecting the target date of the project. We shall call this the *late start time* (l.s.t.) of the activity. Working backward from the project's finish, we find the l.s.t. of the several activities by assigning times to the nodes and arcs, according to the following rules:

1. Assign the *target time* to the finish node.
2. To each arc, assign the time of the node at which the arc *ends*, *minus* the length of the arc.
3. To each node, assign a time equal to the minimum of the times assigned to the arcs that *begin* at that node.

The times assigned to the nodes in this manner are the l.s.t.'s. Figure VIII.2.3 shows the network of Example VIII.2.1 with l.s.t.'s included.

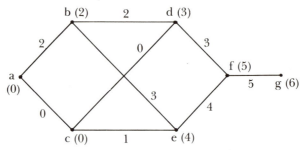

FIGURE VIII.2.3 Late start times (at nodes) for Example VIII.2.1.

It is clear that the l.s.t. can never be smaller than the e.s.t. On the other hand, the e.s.t. might be smaller than the l.s.t. We shall call the difference between these two the *slack time* for the activity involved. Thus,

8.2.1   Slack time = late start time − early start time.

We define an activity as *critical* if its slack time is 0. In the example, the critical activities are a, c, d, f, and g. Of these, a and g ("start" and "finish") are obviously critical; they are critical for all projects, but of course they are not really activities in the strict sense of the word.

In Figure VIII.2.4, the critical activities for this project are shown; the arcs joining these nodes are represented by heavier lines. It may be seen that these arcs, each of which starts and ends at a critical activity, form a connected chain from "start" to "finish," i.e., they form a *path* through the network. This path is known as the *critical path* of the project; it is not difficult to see that the length of this path, i.e., the sum of the lengths of the arcs that make up the path, is precisely equal to the target time.

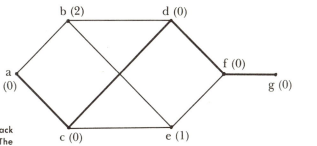

FIGURE VIII.2.4 Slack times for Example VIII.2.1. The heavy line shows the critical path.

The critical activities are important in the following sense. Suppose it is decided that the project has to be completed in less than six days. This can be done only if some of the activities are carried out on a "crash" basis. Since this usually represents additional expenditures, it follows that only a few or, if possible, only one of the activities should be expedited in this manner. The problem is to determine which one. Clearly, it does no good to put activities b or e, the non-critical activities, on a special basis. Not lying on the critical path, they cannot help to shorten it. Additional expenditures, if any, should always be made on the critical activities: *if the target time is to be decreased, it is always necessary to decrease the performance time for one or more of the critical activities.*

In the general case, similar considerations always hold, the only difference being that there may be more than one critical path. To prove this, we use the following result:

VIII.2.2 Lemma. In any project a critical activity other than "start" or "finish" always has at least one critical activity among its immediate successors, and at least one among its immediate predecessors. The activity "start" has at least one critical activity among its immediate successors, while "finish" has at least one among its immediate predecessors.

To prove the lemma, we note that an activity is critical if its e.s.t. and l.s.t. are equal. Suppose that activity j is critical. Then the e.s.t. of each successor of j is at least equal to the e.s.t. of j, plus j's time of performance. On the other hand, the l.s.t. of j is equal to the l.s.t. of some immediate successor of j, which we shall call k, minus j's time of performance. We thus obtain the relations:

$$\text{e.s.t. of } k \geq \text{e.s.t. of } j + \text{t.o.p. of } j$$
$$\text{l.s.t. of } j = \text{l.s.t. of } k - \text{t.o.p. of } j$$

which will give us

$$\text{e.s.t. of } k \geq \text{l.s.t. of } k$$

or

$$\text{slack time of } k \leq 0$$

Since the slack time is never negative, we conclude that k has slack time 0, i.e., k is *critical*, and j has a critical immediate successor. Similar considerations may be used to prove the rest of the lemma.

Let us see just what the lemma means. If we consider only those arcs of the network that have critical activities at both ends, the lemma states that each such arc, unless it ends at the node "finish," must be followed by another; unless it begins at the node "start," it must be preceded by another. Each of these arcs is on a path, made up entirely of such arcs, going from "start" to "finish." This path is known as the *critical path* of the network. (There may, however, be more than one critical path.) The critical path is always the longest path through the network, and it is easy to see that, to decrease the target time of the project, it is necessary to decrease the time of completion of at least one activity on the critical path (or on each of the critical paths, if there is more than one).

VIII.2.3 Example. Consider the project of Example VIII.2.1. Suppose that the time of performance for activity d (Remove and paint door) is decreased from two days to one day. Since this is a critical activity, the target time of the project is decreased from six to five days. The new project network, together with e.s.t.'s and l.s.t.'s is shown in Figure VIII.2.5. It may be seen that there are now *two* critical paths: (a,c,d,f,g) and (a,c,e,f,g). To decrease the target time further, we can decrease the time of performance of activities c or f, which lie on both these paths. On the other hand, it will not be sufficient to decrease the time of performance of activity d, even though it is critical, because activity d lies on only one of the two paths. We may, however, decrease the target time by decreasing the times of performance for both activity d and activity e (this gives us one activity on each of the critical paths).

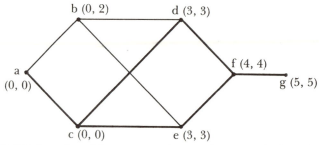

FIGURE VIII.2.5   Early and late start times for Example VIII.2.3. The heavy lines show the critical paths.

VIII.2.4 Example. A contractor makes plans for building a house as shown in

TABLE VIII.2.2

Activity	Time of completion	Immediate predecessors
a. Start	0	none
b. Lay foundation	3	*a*
c. Construct basement	2	*b*
d. Underground wiring	1	*b*
e. Underground pipes	1	*b*
f. Construct main floor	4	*c,d,e*
g. Install fixtures	1	*f*
h. Connect plumbing	2	*g*
i. Connect wiring	1	*g*
j. Place roof	2	*f*
k. Roof drains	2	*h,j*
l. Paint house	1	*h,i*
m. Stop	0	*k,l*

Figure VIII.2.6 shows the network for this project; the target date is 14 weeks, and, as may be seen, the critical path passes through nodes (a,b,c,f,g,h,k,m). The other nodes all have one or two units slack.

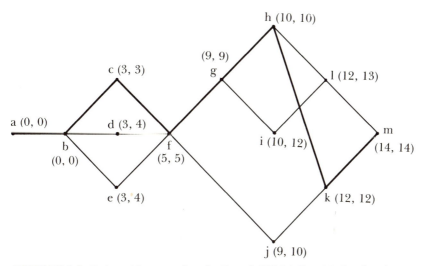

FIGURE VIII.2.6 Early and late start times for Example VIII.2.4. The critical path is shown.

362   In practice, the times of completion of the several activities in a project are generally uncertain. In fact, there are always unforeseen delays connected with any such activity. A project manager normally asks an expert for the estimated time of completion of each activity; the expert will be the first to admit the existence of these uncertainties. When the possibilities of such delays are not negligible, it is sometimes necessary to ask the expert for a description of the probability distribution of each completion time, which is treated as a random variable. The means and variances of each of these random variables are then obtained. (It must, of course, be remembered that in the great majority of cases these are, at best, informed estimates and, at worst, wild guesses.) The critical path method is then used with the *means* of the times of completion as the lengths of the arcs.

If the non-critical activities have large enough slack time, and if the *variances* of the times of completion are small enough, then the critical path obtained in this manner will, with very high probability, be the critical path for the *actual* times of completion. (There is, unfortunately, always the probability that a non-critical activity will suffer an unexpected delay that will use up all its slack time and more, but this has low probability.) Then the actual target time will be a random variable, equal to the sum of the times of completion of the activities on the critical path. In turn, these can generally be considered as independent random variables. The mean of their sum will be the sum of their means, while the variance of their sum will be the sum of their variances.

VIII.2.5 *Example.* Consider once again the project of Example VIII.2.1, but suppose a mechanic tells us that there is a "likely deviation" of 1 day in the time of completion of activity *c* (ordering parts) and of 1/4 day in the time of completion of each of the other activities (excepting, of course, "start" and "finish"). We interpret this to mean that each time of completion is a random variable with a mean equal to the value shown and a *standard deviation* of 1 day for activity *c* and 1/4 day for the other activities.

With this interpretation, the critical path is unchanged; the target time will now be a random variable whose mean is 6 units. To find its variance, we add the variances of jobs, *c*, *d*, and *f*, which are the squares of 1/4, 1, and 1/4, respectively. Thus,

$$\sigma^2 = \frac{1}{16} + 1 + \frac{1}{16} = \frac{18}{16}$$

and

$$\sigma = \frac{3\sqrt{2}}{4} = 1.06$$

We conclude that the target time is a random variable with mean 6 days and a standard deviation of slightly over 1 day.

PROBLEMS ON CRITICAL PATH ANALYSIS

1. The project of installing new light cables for an auditorium is described in the following table:

Activity	Time (days)	Immediate predecessors
a. Start	0	none
b. Obtain specifications	4	a
c. Make duplicates	3	b
d. Order cable	1	b
e. Order clamps	3	c
f. Order cutter	3	b
g. Thread cable	1	c,d
h. Cut cable	2	f,g
i. Install cable	3	e,h
j. Stop	0	i

(a) Find the e.s.t. and l.s.t. for each activity in this project.

(b) What is the target time?

(c) Find the critical path.

2. In preparing a shop for rental, a contractor must perform the following tasks:

Activity	Time (weeks)	Immediate predecessors
a. Start	0	none
b. Draw blueprints	2	a
c. Hire mechanics	1	a
d. Order parts	4	b
e. Duplicate blueprints	1	b
f. Adjust equipment	2	c,d,e
g. Hire sub-contractors	3	e
h. Contract utilities	2	e
i. Install equipment	5	g,f
j. Landscape	2	c,g
k. Stop	0	h,i,j

(a) How long will it take to complete this project?

(b) Find the critical path.

3. In critical path analysis, the *free slack* of an activity is the amount of time that an activity can be delayed without delaying any

of the other activities. The free slack of i is equal to the minimum of the *early start times* of the successors of i, minus the *early finish time* of i. Find the free slack of the several activities in the project of Problem 2.

4. The *independent* slack of activity i is the amount

$$A - B - C$$

where A is the minimum of the early start times of the successors of i, B is the time of performance of i, and C is the maximum of the late finish times of the predecessors of i. (Late finish time = l.s.t. + t.o.p.) The independent slack is the amount of leeway available in scheduling activity i, without in any way affecting the scheduling of other activities. Find the independent slack of the several activities involved in the project of Problem 2. (In general, very few activities will have any independent slack.)

3. THE SHORTEST PATH THROUGH A NETWORK

In the previous section, we studied a network problem and saw that it reduces, in essence, to finding the longest path through the network. We are often faced, however, with the problem of finding the *shortest* path between two given nodes of a network. When we consider that the length of an arc can represent not only physical distance, but also time required or costs, we see that this is a very common problem. We shall approach it through what is, essentially, a dynamic programming technique.

In general, the problem of finding a shortest path is not substantially different from that of finding the critical path. One principal difference is, perhaps, the fact that the new problem does not require the network to be oriented; there may be "two-way" arcs between the nodes. We shall assume this to be the case throughout: all arcs are "two-way." Practically, this may mean that the distance between two points is the same in either direction (which is a well-known fact), or it may mean that transportation costs between two places are the same in either direction (which is not so obvious). At any rate, the generalization to oriented graphs is obvious and does not require different treatment.

The method we shall use for the solution of this problem depends on the fact that, if a network contains n nodes, then the shortest path through the network (between any two nodes) will contain at most $n-1$ arcs (since more arcs would imply passing through the same node twice). The principle of maximality then states that, if a path is minimal, any sub-path will also be minimal.

Mathematically, the procedure is as follows. From the graph, a table of distances, d_{ij}, is formed, in which d_{ij} is the length of the arc between nodes i and j, if such an arc exists, or ∞, if such an arc does not exist. Let us now use the notation

$$F_m(i,j)$$

to denote the length of the shortest path from node i to node j, *using at most m arcs*. It is easy to see that the functions F_m satisfy the equation

8.3.1 $$F_m(i,j) = \min_k \{F_{m-1}(i,k) + d_{kj}\}$$

which is simply the principle of optimality; we also have

8.3.2 $$F_1(i,j) = d_{ij}$$

Using this formulation, it is trivial to apply the dynamic programming technique to the solution of these problems.

VIII.3.1 Example. A motorist wishes to travel from one city to another by the shortest possible route. A road map shows the network of routes that is reproduced schematically in Figure VIII.3.1. The motorist starts from city 1 and wishes to reach city 11.

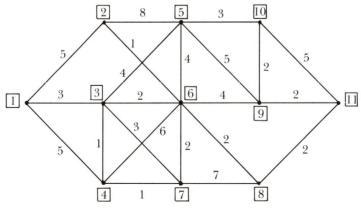

FIGURE VIII.3.1 Example VIII.3.1.

366

We tabulate the distances as follows:

TABLE VIII.3.1

	1	2	3	4	5	6	7	8	9	10	11
1	0	5	3	5	—	—	—	—	—	—	—
2	5	0	—	—	8	1	—	—	—	—	—
3	3	—	0	1	4	2	3	—	—	—	—
4	5	—	1	0	—	6	1	—	—	—	—
5	—	8	4	—	0	4	—	—	5	3	—
6	—	1	2	6	4	0	2	2	4	—	—
7	—	—	3	1	—	2	0	7	—	—	—
8	—	—	—	—	—	2	7	0	—	—	2
9	—	—	—	—	5	4	—	—	0	2	2
10	—	—	—	—	3	—	—	—	2	0	5
11	—	—	—	—	—	—	—	2	2	5	0

where the entries — mean that there is no direct-line connection between the corresponding cities. The number that we are looking for is, of course, $F_{10}(1,11)$. We shall compute the several functions $F_k(1,j)$ by equations (8.3.1) and (8.3.2).

The function $F_1(i,j)$ is given by (8.3.2); it is, of course, given by the table. We compute the functions $F_2(1,j)$ and $\hat{k}_1(1,j)$ by

$$F_2(1,j) = \min_k \{F_1(1,k) + d_{kj}\}$$

and

$$F_2(1,j) = F_1(1,\hat{k}_1) + F_1(\hat{k}_1,j)$$

In other words, $\hat{k}_{m-1}(1,j)$ is defined by saying that, to go from node 1 to node j in m steps, the $(m-1)$th step should pass through node k_{m-1}. We have, then,

TABLE VIII.3.2

j	$F_2(1,j)$	$k_1(1,j)$
1	0	1
2	5	2
3	3	3
4	4	3
5	7	3
6	5	3
7	6	3

This says, for instance, that the shortest two-step route from node 1 to node 6 is five miles long, and passes through node 3. The two-step distances are not given for the other nodes, mainly because these cannot be reached in two steps.

Proceeding in the usual manner, we construct a table of the several functions $F_m(1,j)$ and $\hat{k}_{m-1}(1,j)$, given in Table VIII.3.3.

TABLE VIII.3.3

j	$F_2(1,j)$	$F_3(1,j)$	$F_4(1,j)$	$\hat{k}_1(1,j)$	$\hat{k}_2(1,j)$	$\hat{k}_3(1,j)$
1	0	0	0	1	1	1
2	5	5	5	2	2	2
3	3	3	3	3	3	3
4	4	4	4	3	3	3
5	7	7	7	3	3	3
6	5	5	5	3	3	3
7	6	5	5	3	4	4
8	–	7	7	–	6	6
9	–	9	9	–	6	6
10	–	10	10	–	5	5
11	–	–	9	–	–	8

Values F_m can be computed for $m = 5,6$, and so on, but these are not necessary. In fact, if we compute them, we shall see that distances are not decreased: the minimal path between node 1 and any other node in this network always takes 4 steps or less. The solution is obtained, as usual, by working backward. We have $\hat{k}_3(1,11) = 8$; $\hat{k}_2(1,8) = 6$ and $\hat{k}_1(1,6) = 3$. Thus, the shortest path through this network, from node 1 to node 11, is 9 miles long, passing through nodes 3, 6, and 8 on the way. Figure VIII.3.2 shows the *minimal path* in the network.

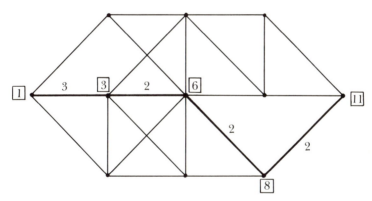

FIGURE VIII.3.2 Minimal path for Example VIII.3.1.

As can be seen from the example, our technique is strictly a dynamic programming technique. Several types of modification are possible. We may, for example, find the distances $F_m(i,11)$, rather than $F_m(1,j)$; this would tell us how to proceed so as to reach the destination node 11 from any starting point. The technique we used tells us, instead, how to reach any destination from the fixed starting point, node 1. Depending on the actual purpose of the decision maker, one technique may be more useful than the other. There is not, generally, any difficulty in effecting a variation.

368 PROBLEMS ON SHORTEST PATHS THROUGH NETWORKS

1. A traveler wishes to fly from Central City to Capital City, some 1700 miles away. Unfortunately, he has only a very small plane, with a flying limit of 800 miles. He must, therefore, make several stops along the way. There are twelve intermediate cities with airports. The distances among these (with 0 for Central City and 13 for Capital City) are given in the following table. (A blank space means that the distance is greater than 800 miles and thus unmanageable for the plane.)

	0	1	2	3	4	5	6	7	8	9	10	11	12	13
0	0	513	582	597	611	—	—	—	—	—	—	—	—	—
1	513	0	243	158	211	561	550	520	663	610	—	—	—	—
2	582	243	0	172	204	520	565	671	605	543	—	—	—	—
3	597	158	172	0	169	508	580	553	525	498	—	—	—	—
4	611	211	204	169	0	515	522	603	583	485	—	—	—	—
5	—	561	520	508	515	0	216	242	185	206	520	585	602	—
6	—	550	565	580	522	216	0	223	191	300	564	520	511	—
7	—	520	671	553	603	242	223	0	216	220	591	616	567	—
8	—	663	605	525	583	185	191	216	0	265	548	561	522	—
9	—	610	543	498	485	206	300	220	265	0	517	542	568	—
10	—	—	—	—	—	520	564	591	548	517	0	321	214	427
11	—	—	—	—	—	585	520	616	561	542	321	0	265	442
12	—	—	—	—	—	602	511	567	522	568	214	265	0	507
13	—	—	—	—	—	—	—	—	—	—	427	442	507	0

What is the shortest route available to the traveler?

2. Find the shortest path through the graph in Figure VIII.3.3, from node 0 to node 11.

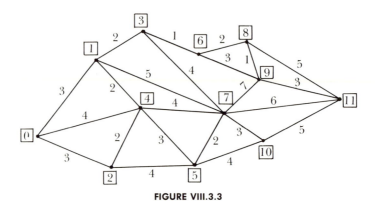

FIGURE VIII.3.3

4. MINIMAL TREES

We consider here a slightly different problem: that of finding a *minimal tree* in a network.

To understand what this means, let us suppose that a utility company must service each of several customers. It must then construct lines that connect it with each of the customers, so as to deliver its product. It is not necessary to draw independent lines from the utility to each customer; a single line, passing by each customer, will do; so will several lines, each servicing a small number of customers. In general, it is necessary only that the lines all be connected and reach all the customers. In other words, the customers must be the nodes of a *connected* graph; the utility lines must be the arcs of this graph. The company will naturally wish to build the lines with as little expense as possible, and so we look for the minimal cost graph that can be drawn subject to these restrictions. In general, we start with a network showing all possible distances between nodes (i.e., between pairs of customers, or between the company and a customer); the problem is to choose the arcs that will give a minimal total distance. It is not too difficult to see that these lines never form a loop; we talk, then, about the *minimal tree* that spans the network (i.e., passes through all the nodes).

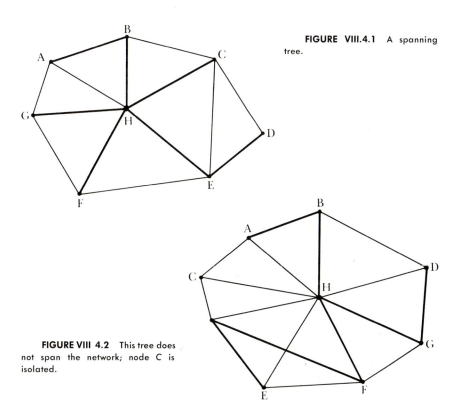

FIGURE VIII.4.1 A spanning tree.

FIGURE VIII 4.2 This tree does not span the network; node C is isolated.

The method we shall use for solving this problem is an extremely simple one, known as *Kruskal's algorithm*. Kruskal's algorithm tells us to build the tree by adding arcs, one at a time, according to the rule: *of all the arcs that are not yet part of the tree, and that do not form a loop when added to those already in the tree, choose the shortest one.* (In case of ties, any one of those tying for shortest may be added to the tree.)

We shall now prove that Kruskal's algorithm does indeed give us the minimal spanning tree. We shall prove this by making the "non-degeneracy" assumption that no two arcs have equal lengths. If, in fact, this is not so, we can proceed, as we did with linear programs, by a *perturbation* technique: when two or more arcs have equal lengths, the lengths are altered by small amounts ϵ, 2ϵ, 3ϵ, and so on. If ϵ is small enough, the minimal tree for the perturbed network will still be minimal for the original, unperturbed problem.

VIII.4.1 Proof of Kruskal's Algorithm. To see that the algorithm will give us a spanning tree, note that, so long as the graph is dis-connected, there are still arcs that can be added. The algorithm will not stop until a connected graph is obtained. Clearly, this graph can have no loops, and so it must be a spanning tree. Let us call this tree H. (We think of H as a collection of arcs and nodes.)

Suppose, now, that H is not the minimal spanning tree. This means that there is a shorter one, W. The two trees, W and H, have the same number of arcs, namely, one less than the number of nodes. Since they do not have the same arcs (this would make them identical), it follows that H has at least one arc that does not belong to W. Let (i,j) be the shortest arc that is in H but not in W. Then $W \cup \{(i,j)\}$ is a graph with n nodes and n arcs, and therefore contains a loop. In this loop, there is an arc (k,l) that belongs to W but not to H (since, after all, H does not contain any loops). Consider then the graph

$$F = W \cup \{(i,j)\} - \{(k,l)\}$$

The graph of F is connected (since removing an arc from a loop does not disconnect a graph) and has n nodes and $n-1$ arcs. Therefore F is a spanning tree. Now the length of F is equal to the length of W, plus the length (i,j), minus the length (k,l).

Suppose, now, that (k,l) were shorter than (i,j). We know that all the arcs in H that precede (i,j) are also in W. Thus, (k,l) cannot form a loop when added to these arcs. But this would mean that (i,j) was added to H against the algorithm's rules (i.e., it was not the shortest available), and (k,l) is not shorter than (i,j). By the non-degeneracy assumption, (i,j) is shorter than (k,l). This means that F is shorter than W, contradicting the minimality of W. The contradiction tells us that H is the desired minimal tree.

We conclude this section with an example:

VIII.4.2   Example.   Find the minimal trees in the networks shown In Figures VIII.4.3 and VIII.4.4.

For Figure VIII.4.3, we add the following arcs: *EH, FG, BE, DG, EG, AB, AC*. For Figure VIII.4.4, we take the arcs *AB, FG, BH, BF, CG, CD, EF*. Figure VIII.4.5 and VIII.4.6 show the minimal trees on these networks.

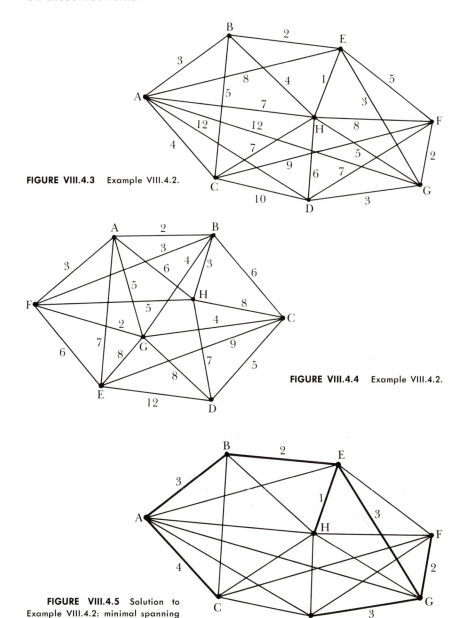

FIGURE VIII.4.3   Example VIII.4.2.

FIGURE VIII.4.4   Example VIII.4.2.

FIGURE VIII.4.5   Solution to Example VIII.4.2: minimal spanning tree for Figure VIII.4.3.

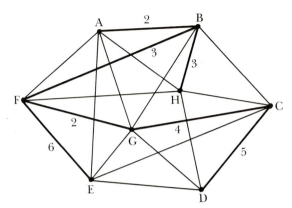

FIGURE VIII.4.6   Solution to Example VIII.4.2: minimal spanning tree for Figure VIII.4.4.

PROBLEMS ON MINIMAL TREES

1. A company wishes to connect its 10 branches by means of an intercommunication system. The distances between branches are given in the table:

	1	2	3	4	5	6	7	8	9	10
1	0	23	42	31	27	38	29	33	41	25
2	23	0	29	36	25	34	44	27	31	29
3	42	29	0	31	38	27	25	41	43	28
4	31	36	31	0	26	36	33	45	37	26
5	27	25	38	26	0	31	37	32	41	28
6	38	34	27	36	31	0	25	24	38	33
7	29	44	25	33	37	25	0	31	36	29
8	33	27	41	45	32	24	31	0	41	27
9	41	31	43	37	41	38	36	41	0	32
10	25	29	28	26	28	33	29	27	32	0

Assume that the cost of connecting two branches directly is proportional to the distance between them, and that messages from one branch to another can be relayed with no loss in efficiency. What is the most efficient way to connect the system?

2. Find the minimal spanning trees in Figures VIII.4.7, VIII.4.8, VIII.4.9, and VIII.4.10.

FIGURE VIII.4.7

FIGURE VIII.4.8

FIGURE VIII.4.9

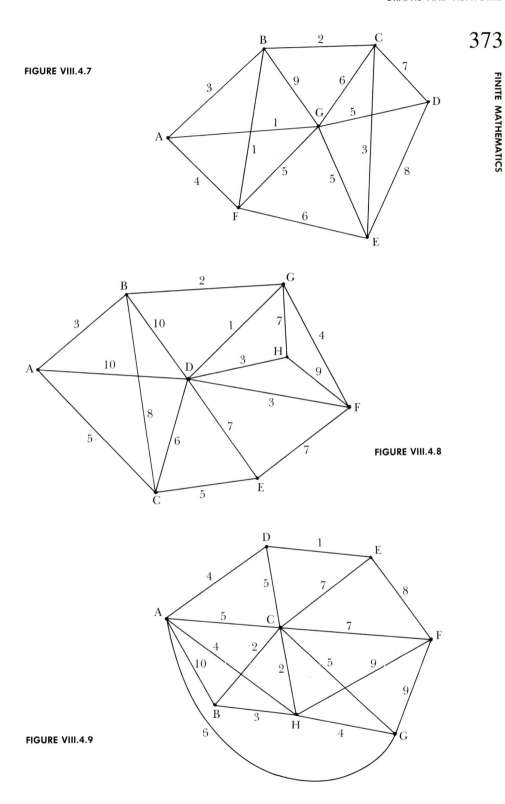

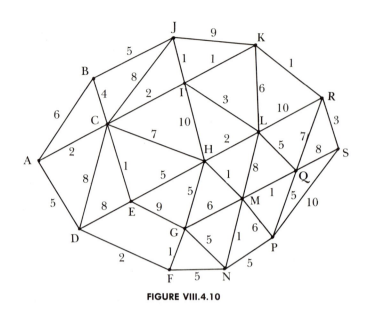

FIGURE VIII.4.10

5. THE MAXIMAL FLOW THROUGH A NETWORK

In this section, we shall study a considerably more complicated problem, that of finding the maximal flow through a network.

In essence, we have here a network with two distinguished nodes called a *source* and a *sink*. The numbers assigned to the arcs in this case represent, not costs or distances, but capacities. We assume that some material is to be shipped from the source to the sink over the arcs of the network; the problem is to find the maximal amount that can be shipped, subject to the constraint that the volume over any arc must be no greater than the capacity of the arc.

Mathematically, we shall formulate this problem as a linear program. The source and sink will be called nodes 0 and n, respectively; the *flow* (i.e., the amount shipped) from node i to node j will be denoted by x_{ij}, and must be smaller than the arc capacity, c_{ij}. The problem is then to

8.5.1 Maximize

$$\sum_{j=1}^{n} x_{0j} - \sum_{j=1}^{n} x_{j0} = w$$

Subject to

8.5.2 $$\sum_{j=0}^{n} x_{ij} - \sum_{j=0}^{n} x_{ji} = 0 \quad \text{for } i = 1, \ldots, n-1$$

8.5.3 $$x_{ij} \leq c_{ij} \quad \text{for all } i,j$$

8.5.4 $$x_{ij} \geq 0 \quad \text{for all } i,j$$

Let us see just what this means. The constraints (8.5.2) state that, for any node other than the source or the sink, the amount flowing into the node is equal to the amount flowing out. The other constraints are merely the capacity and non-negativity constraints. The objective function (8.5.1) is the amount flowing out of the source; constraints (8.5.2) guarantee that nothing is lost on the way, so that this same amount flows into the sink.

Generally speaking, if the network is not very complex, it is possible to obtain the maximal flow by inspection. In the network of Figure VIII.5.1 (a), the maximal flow is easily seen to be 6 units, as shown in Figure VIII.5.1 (b). It is clear that this flow is maximal since the nodes leading from the source have no more capacity.

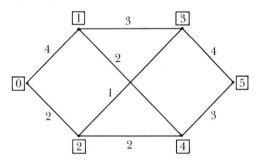

(a)

FIGURE VIII.5.1 Capacities (a) and maximal flow (b) in a network.

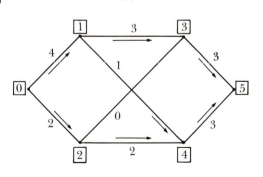

(b)

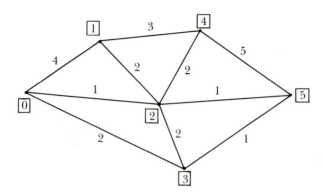

FIGURE VIII.5.2 A flow network with capacities.

Figure VIII.5.2 shows a slightly more complicated network, but even so, it is not difficult to construct the maximal flow by inspection. The principal problem lies in getting 5 units to node 4. A reasonable method would be to take, first, the maximal flow along the top route; this gives us a starting flow of 3 units. These capacities can be subtracted from the network, leaving a smaller network, as shown in Figure VIII.5.3. Again, we take the top route of this new network, which has a capacity of 1 unit, and subtract these capacities to obtain the network shown in Figure VIII.5.4. We proceed to do this several more times, until finally the maximal flow, shown in Figure VIII.5.5, is obtained.

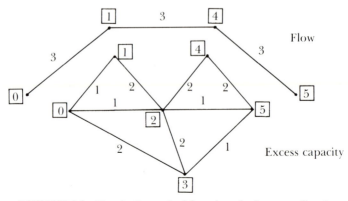

FIGURE VIII.5.3 We take the maximal flow along the "uppermost" route.

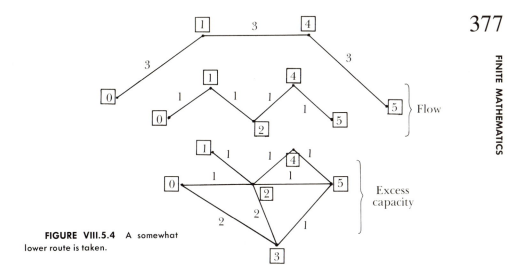

FIGURE VIII.5.4   A somewhat lower route is taken.

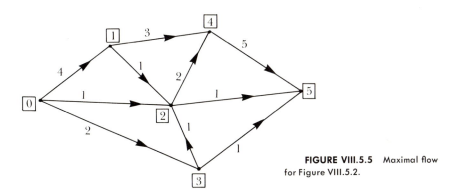

FIGURE VIII.5.5   Maximal flow for Figure VIII.5.2.

The method used for this example is quite straightforward and intuitive; for simple networks it is probably the quickest to use. Unfortunately, with complicated networks, it soon becomes impossible to keep track of all the possible paths without some bookkeeping techniques. This is especially so if the network is non-planar, i.e., if some of the arcs cross each other without actually meeting at a node.

We could, of course, solve this problem by using the simplex algorithm. Once again, however, we find that there are shorter methods, thanks to the special form of the problem. The particular method that we shall use, developed by Ford and Fulkerson, is known as the labeling technique.

We give first two definitions.

VIII.5.1   Definition.   We shall say that the arc (i,j) is *saturated* if $x_{ij} = c_{ij}$. It is *empty* if $x_{ij} = 0$.

VIII.5.2 Definition. We shall say that the arc (i,j) has *positive excess capacity* if either (a) the arc (i,j) is *unsaturated* or (b) the opposite arc (j,i) is *non-empty*.

In case (a), the *excess* capacity is

8.5.5 $$g_{ij} = c_{ij} - x_{ij}$$

and, in case (b),

8.5.6 $$g_{ij} = x_{ji}$$

In essence, the excess capacity of an arc is the amount that it can carry, over and above what it is already carrying. The flow along that arc may be increased by that amount. In the second case, (b), in which the flow is in the opposite direction, this reverse flow, from j to i, may be decreased by the amount of excess capacity, which, as far as we are concerned, has the same effect as increasing the flow from i to j by that amount.

An alternate definition of excess capacity could be

8.5.7 $$g_{ij} = c_{ij} - x_{ij} + x_{ji}$$

We now describe the labeling procedure, giving it as a set of rules. It is, first of all, assumed that a feasible flow (x_{ij}) has been given.

VIII.5.3 Rules for the Labeling Procedure

1. Start at the source (node 0). If any of the arcs $(0,j)$ have positive excess capacity, assign the label (t_j, k_j), where

$$t_j = g_{oj}$$
$$k_j = 0$$

to the corresponding vertices, j.

2. (General step.) From among the labeled nodes, let i be the smallest index that has not yet been treated in this manner. If there are any arcs (i,j) with positive excess capacity starting at this node, and such that their end vertices j *have not yet been labeled*, then assign the label (t_j, k_j), where

8.5.8 $$t_j = g_{ij}$$

8.5.9 $$k_j = i$$

to the corresponding vertices, j.

The rules VIII.5.3 tell us how to carry out the labeling procedure. In effect, if the node j is labeled (t_j, k_j), this means that it is possible to send an additional flow from the source (node 0) to node j. This flow is along a certain sequence of arcs; the last arc in this sequence starts at node k_j.

Let us suppose that the rules VIII.5.3 have been carried out. One of two things will happen: (a) the sink, node n, is eventually labeled; or (b) the sink cannot be labeled. This will happen in at most n steps, since each of the steps described treats one of the n nodes (other than the sink) in the network. We shall show that, in case (a), the objective function (8.5.1) can be increased. In case (b), the given flow (x_{ij}) is maximal.

Assume, then, that case (a) holds: the sink can be labeled. It is clear that a new flow can be sent from the source to the sink. The labels will tell us the route along which it must come. In fact, k_n will be the next to last node on this route, k_{k_n} will be the one before this, and so on. Let us say that the route of new flow, then, is given by the arcs $(0, j_1), (j_1, j_2), \ldots, (j_m, n)$. The flow can be increased along each of them by as much as the minimum of the excess capacities along these arcs. But these excess capacities are given by the labels t_j. Thus, the smallest of the t_j on this route is the amount of additional flow that can be put into the network.

Suppose that the capacities on the arcs are all integers. (This is not a very strong restriction, inasmuch as the network is not essentially changed if all capacities are multiplied by the same positive number. If the capacities are rational numbers, they can be multiplied by their least common denominator and become integers. If they are not rational, they can be approximated as well as desired by rational numbers.) In this case, the flows so obtained will always be integers (since there will be an integral flow along each arc, and the excess capacities will always be integers), so that, every time that the sink is labeled, the flow can be increased by at least one unit. Since the flow is clearly bounded above—i.e., the linear program (8.5.1) to (8.5.4) is bounded—it follows that the rules VIII.5.3 can only be followed a finite number of times; eventually a flow will be obtained for which the sink cannot be labeled.

We must now show that, if the sink cannot be labeled, then the flow in the network is maximal. To do this, we will define a *cut* in the network.

VIII.5.4 Definition. By a *cut* in a network is meant a partition of the set of all nodes in a network into two sets, U and L, such that

(a) $$0 \in U$$

(b) $$n \in L$$

(c) $$U \cap L = \varnothing$$

VIII.5.5 Definition. Let (U,L) be a cut in a network. Then by the value of the cut is meant

8.5.10 $$v(U,L) = \sum_{i \in U} \sum_{j \in L} c_{ij}$$

In other words, the value of a cut is the total capacity of the arcs leading from the set U into the set L. Intuitively, it is clear that no flow can ever be greater than the value of a cut, since any amount flowing from source to sink must pass through one of the arcs on the right side of (8.5.10). The net flow is equal to the flow across any cut, which must, by constraints (8.5.3), be less than the value of the cut. Mathematically, we can prove this by defining the *flow across a cut* (U,L) as

8.5.11 $$f(U,L) = \sum_{\substack{i \in U \\ j \in L}} x_{ij} - \sum_{\substack{i \in U \\ j \in L}} x_{ji}$$

Then, by constraints (8.5.3) and (8.5.4), it is easy to see that $f(U,L) \leq v(U,L)$. By constraints (8.5.2), it may be seen that $w = f(U,L)$ for any cut (U,L). Thus,

$$w \leq v(U,L)$$

Suppose, then, that we have a flow (x_{ij}), such that the sink cannot be labeled. We shall let U be the set of all nodes that can be labeled (including the source, which we shall think of as being labeled) and L be the set of unlabeled nodes. Then (U,L) is a cut in the network. Now, the rules VIII.5.3 for the labeling procedure are such that, if i is labeled and (i,j) has positive excess capacity, then j must also be labeled. It follows that the arcs (i,j) from U to L all have zero excess capacity. In other words, all arcs from U to L are saturated, while all arcs from L to U are empty.

Thus,

$$\sum_{\substack{i \in U \\ j \in L}} x_{ij} = \sum_{\substack{i \in U \\ j \in L}} c_{ij} = v(U,L)$$

and

$$\sum_{\substack{i \in U \\ j \in L}} x_{ji} = 0$$

so that

$$f(U,L) = v(U,L)$$

But this means that $w = v(U,L)$: the flow in the network is equal to the value of the cut (U,L). Since the flow can never exceed the value of a cut, we conclude that w is the maximal flow in the network.

We note that, in passing, we have also proved an important theorem of networks:

VIII.5.6 Theorem (The Max-Flow Min-Cut Theorem). In any network, the maximum possible flow is equal to the value of the minimum cut.

We point out that VIII.5.6 is a special case of the duality theorem for linear programs. Let us illustrate the labeling technique by an example.

VIII.5.7 Example. Find the maximal flow in the following network. The flow is to be in the direction of the arrows only.

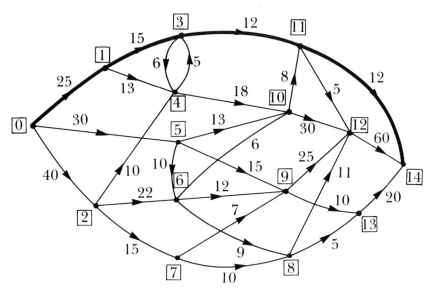

FIGURE VIII.5.6 Capacities for Example VIII.5.7. The flow can be increased along the route marked by the heavy line.

We will start with a null flow, i.e., set $x_{ij} = 0$ for all i,j. The excess capacities g_{ij}, are equal to the capacities, c_{ij}. We see then, that the source can be joined to nodes 1, 2, and 5 by arcs with excess capacity 25, 40, and 30 respectively, which gives us

$$t_1 = 25 \qquad t_2 = 40 \qquad t_5 = 30$$
$$k_1 = 0 \qquad k_2 = 0 \qquad k_5 = 0$$

Now node 1 has been labeled, so we consider it next. It is joined to nodes 3 and 4 by arcs with excess capacities 15 and 13, and

$$t_3 = 15 \qquad t_4 = 13$$
$$k_3 = 1 \qquad k_4 = 1$$

We consider next node 2. This can be joined to nodes 4, 6, and 7. Node 4 has already been labeled, so we simply write

$$t_6 = 22 \qquad t_7 = 15$$
$$k_6 = 2 \qquad k_7 = 2$$

Next, node 3 can be joined to node 4, which is already labeled, and to node 11. This gives us

$$t_{11} = 12$$
$$k_{11} = 3$$

We proceed in this manner, taking the labeled nodes one at a time, until finally we obtain

TABLE VIII.5.1

j	1	2	3	4	5	6	7	8	9	10	11	12	13	14
t_j	25	40	15	13	30	22	15	10	15	18	12	11	5	12
k_j	0	0	1	1	0	2	2	7	5	4	3	8	8	11

We see, here, that the sink (node 14) has been labeled. We have

$$k_{14} = 11 \qquad k_{11} = 3 \qquad k_3 = 1 \qquad k_1 = 0$$

which tells us that a flow can be sent from source to sink, passing through the nodes 1, 3, 11. The value of this flow is given by the minimum of

$$t_{14} = 12 \qquad t_{11} = 12 \qquad t_3 = 15 \qquad t_1 = 25$$

or 12. We add this flow to what is already in the network (nothing, in this case) and obtain the flow shown in Figure VIII.5.6. At each arc, the first number represents the capacity; the second number, the flow.

Treating Figure VIII.5.7 just as we did Figure VIII.5.6, we eventually obtain

FINITE MATHEMATICS

TABLE VIII.5.2

j	1	2	3	4	5	6	7	8	9	10	11	12	13	14
t_j	13	40	5	13	30	22	15	9	15	18	8	11	5	60
k_j	0	0	4	1	0	2	2	6	5	4	10	8	8	12

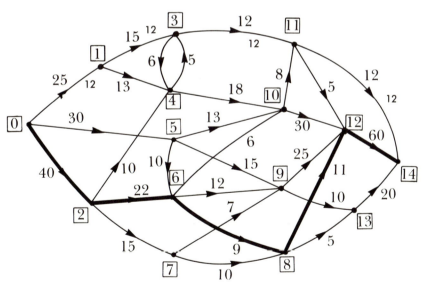

FIGURE VIII.5.7 Capacities (in large type) and flow (in smaller type).
W = 12.

384

Once again, we find that the sink is labeled; we have

$$k_{14} = 12, \; k_{12} = 8, \; k_8 = 6, \; k_6 = 2, \; k_2 = 0$$

and

$$t_{14} = 60, \; t_{12} = 11, \; t_8 = 9, \; t_6 = 22, \; t_2 = 40$$

which means that 9 units may be sent from source to sink through the nodes $2, 6, 8, 12$. This gives us Figure VIII.5.8.

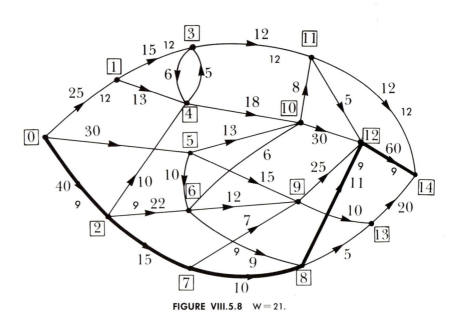

FIGURE VIII.5.8 $W = 21$.

For Figure VIII.5.8, we will have

TABLE VIII.5.3

j	1	2	3	4	5	6	7	8	9	10	11	12	13	14
t_j	13	31	3	13	30	13	15	10	15	18	8	2	5	51
k_j	0	0	1	1	0	2	2	7	5	4	10	8	8	12

We have, thus,

$$k_{14} = 12 \qquad k_{12} = 8 \qquad k_8 = 7 \qquad k_7 = 2 \qquad k_2 = 0$$
$$t_{14} = 51 \qquad t_{12} = 2 \qquad t_8 = 10 \qquad t_7 = 15 \qquad t_2 = 31$$

so that 2 units can be sent along the route 2,7,8,12. This gives us Figure VIII.5.9, for which we have Table VIII.5.4.

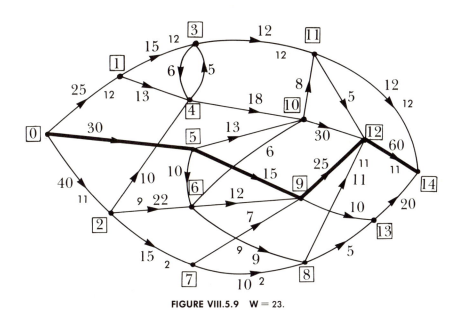

FIGURE VIII.5.9   W = 23.

TABLE VIII.5.4

j	1	2	3	4	5	6	7	8	9	10	11	12	13	14
t_j	13	29	3	13	30	13	13	8	15	18	8	25	5	49
k_j	0	0	1	1	0	2	2	7	5	4	11	9	8	12

which gives us

$$k_{14} = 12 \qquad k_{12} = 9 \qquad k_9 = 5 \qquad k_5 = 0$$
$$t_{14} = 49 \qquad t_{12} = 25 \qquad t_9 = 15 \qquad t_5 = 30$$

so that 15 more units can be sent along the route 5,9,12. This gives us the flow of Figure VIII.5.10.

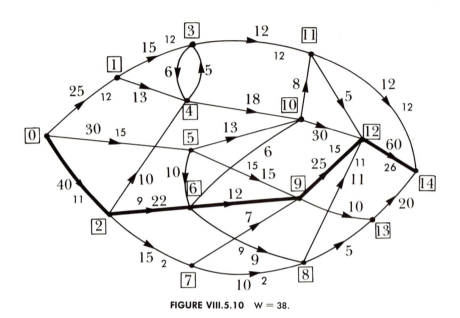

FIGURE VIII.5.10 W = 38.

For this figure, we have

TABLE VIII.5.5

j	1	2	3	4	5	6	7	8	9	10	11	12	13	14
t_j	13	29	3	13	15	13	13	8	12	18	8	10	5	34
k_j	0	0	1	1	0	2	2	7	6	4	10	9	8	12

and we see that 10 more units can be sent along the route 2,6,9,12. This will give us Figure VIII.5.11 and Table VIII.5.6,

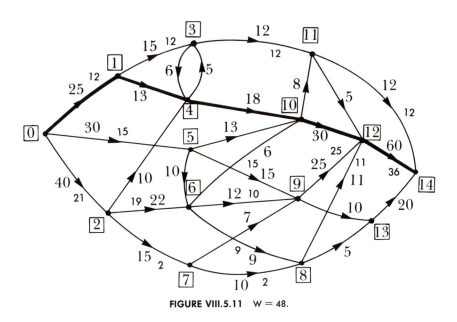

FIGURE VIII.5.11   W = 48.

TABLE VIII.5.6

j	1	2	3	4	5	6	7	8	9	10	11	12	13	14
t_j	13	19	3	13	15	3	13	8	2	18	8	30	5	24
k_j	0	0	1	1	0	2	2	7	6	4	10	10	8	12

so that 13 units can be sent along 1,4,10,12. This gives us Figure VIII.5.12.

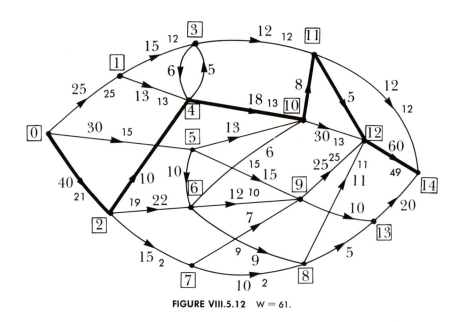

FIGURE VIII.5.12   W = 61.

TABLE VIII.5.7

j	1	2	3	4	5	6	7	8	9	10	11	12	13	14
t_j	13	19	5	10	15	3	13	8	2	5	8	5	5	11
k_j	4	0	4	2	0	2	2	7	6	4	10	11	8	12

We continue in this manner, obtaining the figures numbered VIII.5.12 through VIII.5.18 together with the corresponding tables (VIII.5.7 through VIII.5.12).

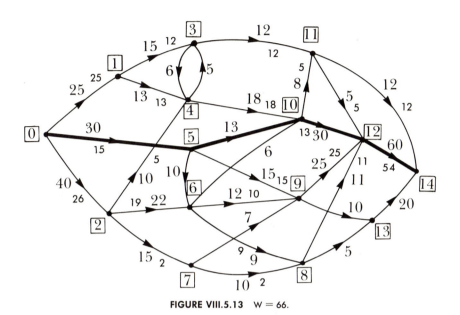

FIGURE VIII.5.13   W = 66.

TABLE VIII.5.8

j	1	2	3	4	5	6	7	8	9	10	11	12	13	14
t_j	13	14	5	5	15	3	13	8	2	13	3	17	5	6
k_j	4	0	4	2	0	2	2	7	6	5	10	10	8	12

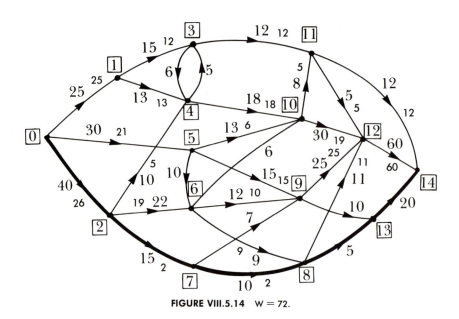

FIGURE VIII.5.14   W = 72.

TABLE VIII.5.9

j	1	2	3	4	5	6	7	8	9	10	11	12	13	14
t_j	13	14	5	5	9	3	13	8	2	7	3	11	5	20
k_j	4	0	4	2	0	2	2	7	6	5	10	10	8	13

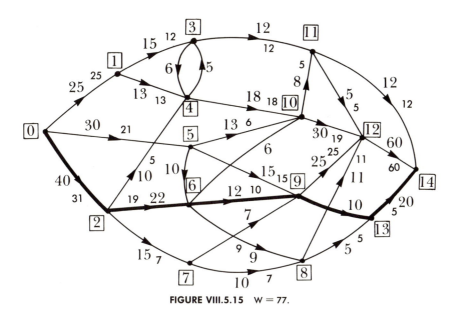

FIGURE VIII.5.15 W = 77.

TABLE VIII.5.10

j	1	2	3	4	5	6	7	8	9	10	11	12	13	14
t_j	13	9	5	5	9	3	8	3	2	7	3	11	10	15
k_j	4	0	4	2	0	2	2	7	6	5	10	10	9	13

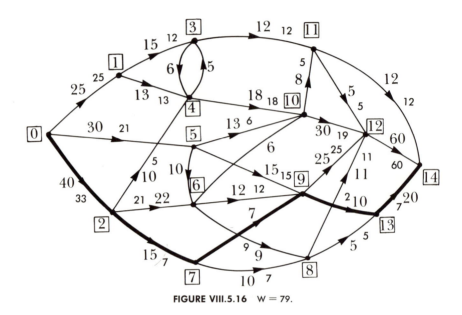

FIGURE VIII.5.16   W = 79.

TABLE VIII.5.11

j	1	2	3	4	5	6	7	8	9	10	11	12	13	14
t_j	13	7	5	5	9	1	8	3	7	7	3	11	8	13
k_j	4	0	4	2	0	2	2	7	7	5	10	10	9	13

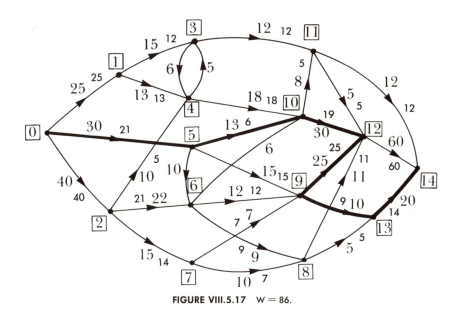

FIGURE VIII.5.17   W = 86.

TABLE VIII.5.12

j	1	2	3	4	5	6	7	8	9	10	11	12	13	14
t_j	13	21	5	5	9	10	1	3	25	7	3	11	1	6
k_j	4	6	4	2	0	5	2	7	12	5	10	10	9	13

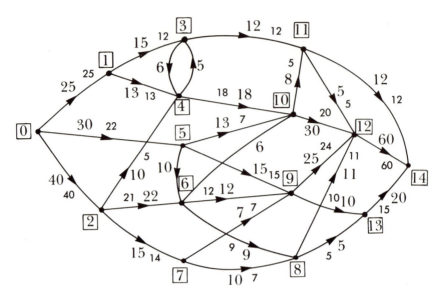

FIGURE VIII.5.18 Solution (maximal flow) for Example VIII.5.7. W = 87.

This last figure gives us the solution; the sink cannot be labeled. This may be checked directly by carrying out the labeling procedure. A shorter way of checking, however, is by noticing that the flow in the network is now 87 units. Consider, then, the cut

$$U = \{0,1,2,\ldots,12\} \quad L = \{13,14\}$$

There are 4 arcs leading from U into L; these are $(8,13), (9,13), (11,14),$ and $(12,14)$. Now,

$$c_{8,13} + c_{9,13} + c_{11,14} + c_{12,14} = 5 + 10 + 12 + 60 = 87$$

and so $v(U,L) = w$. By the max-flow min-cut theorem, we conclude that this is the maximum possible flow.

We have, throughout this chapter, relied considerably on the graphic technique. This is generally a good idea, inasmuch as the graph of a network is easily understood and helps to give a quick picture of the situation. On the other hand, graphs are not necessary. It is generally possible to replace a graph by a matrix, and, for purposes of storing the data (using a computer, which cannot appreciate diagrams) this may actually be necessary. The network of Example VIII.5.7 can be represented by the following matrix, which gives the capacities of the several arcs and also serves to give the excess capacities g_{ij} when there is no flow in the network.

	1	2	3	4	5	6	7	8	9	10	11	12	13	14
0	25	40	0	0	30	0	0	0	0	0	0	0	0	0
1	0	0	15	13	0	0	0	0	0	0	0	0	0	0
2	0	0	0	10	0	22	15	0	0	0	0	0	0	0
3	0	0	0	6	0	0	0	0	0	0	12	0	0	0
4	0	0	5	0	0	0	0	0	0	18	0	0	0	0
5	0	0	0	0	0	10	0	0	15	13	0	0	0	0
6	0	0	0	0	0	0	0	9	12	6	0	0	0	0
7	0	0	0	0	0	0	0	10	7	0	0	0	0	0
8	0	0	0	0	0	0	0	0	0	0	0	11	5	0
9	0	0	0	0	0	0	0	0	0	0	0	25	10	0
10	0	0	0	0	0	0	0	0	0	0	8	30	0	0
11	0	0	0	0	0	0	0	0	0	0	0	5	0	12
12	0	0	0	0	0	0	0	0	0	0	0	0	0	60
13	0	0	0	0	0	0	0	0	0	0	0	0	0	20

It is possible, from this matrix, to carry out the labeling procedure without needing to look at a graph. Two additional rows of the matrix would be sufficient for this purpose. After the additional flow has been decided on, a new matrix of excess capacities can be formed, according to the relation

$$g'_{ij} = g_{ij} - \Delta_{ij} + \Delta_{ji}$$

where g'_{ij} and g_{ij} are the new and old excess capacities, respectively, while Δ_{ij} and Δ_{ji} are the additional flows from i to j and from j to i, respectively. Corresponding to the network of Figure VIII.5.6, we have the matrix

	1	2	3	4	5	6	7	8	9	10	11	12	13	14
0	13	40	0	0	30	0	0	0	0	0	0	0	0	0
1	0	0	3	13	0	0	0	0	0	0	0	0	0	0
2	0	0	0	10	0	22	15	0	0	0	0	0	0	0
3	12	0	0	6	0	0	0	0	0	0	0	0	0	0
4	0	0	5	0	0	0	0	0	0	18	0	0	0	0
5	0	0	0	0	0	10	0	0	15	13	0	0	0	0
6	0	0	0	0	0	0	0	9	12	6	0	0	0	0
7	0	0	0	0	0	0	0	10	7	0	0	0	0	0
8	0	0	0	0	0	0	0	0	0	0	0	11	5	0
9	0	0	0	0	0	0	0	0	0	0	0	25	10	0
10	0	0	0	0	0	0	0	0	0	0	8	30	0	0
11	0	0	12	0	0	0	0	0	0	0	0	5	0	0
12	0	0	0	0	0	0	0	0	0	0	0	0	0	60
13	0	0	0	0	0	0	0	0	0	0	0	0	0	20

The labeling procedure can be continued, using matrices only, in this manner. There is generally no difference, conceptually, between the matric and the graphic procedures.

The reader is invited to construct the matrices corresponding to Figures VIII.5.7 to VIII.5.18.

PROBLEMS ON MAXIMAL FLOWS THROUGH NETWORKS

1. Find the maximal flows in the networks in Figures VIII.5.19 to VIII.5.23. What are the minimal cuts in each case?

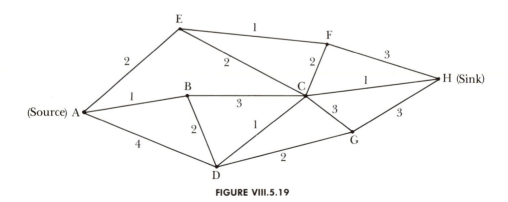

FIGURE VIII.5.19

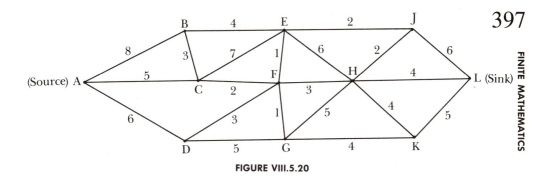

FIGURE VIII.5.20

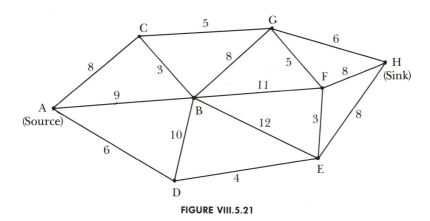

FIGURE VIII.5.21

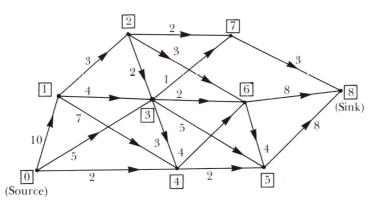

FIGURE VIII.5.22 Flow is allowed only in the direction of the arrows.

398

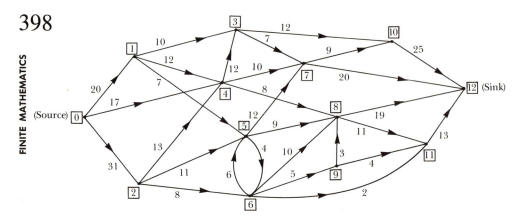

FIGURE VIII.5.23 Flow is allowed only in the direction of the arrows.

2. A steel company wishes to transport iron ore from the mine to its refining plant. It can do this by shipping from the mine to any one of several intermediate cities, among these cities, and on to the plant. The available space (in tons of ore) on the carriers between cities is given by the following table (with 0 for the mine, 13 for the plant).

	0	1	2	3	4	5	6	7	8	9	10	11	12	13
0	—	30	40	25	0	0	0	0	0	0	0	0	0	0
1	10	—	0	0	12	17	0	0	0	0	0	0	0	0
2	8	0	—	0	22	15	8	0	0	0	0	0	0	0
3	12	0	0	—	0	17	12	0	0	0	0	0	0	0
4	0	0	2	0	—	0	0	15	27	0	0	0	0	0
5	0	3	0	5	0	—	15	12	0	10	8	0	0	0
6	0	0	4	0	0	0	—	16	0	4	10	0	0	0
7	0	0	0	0	0	8	0	—	12	8	0	0	0	2
8	0	0	0	0	6	0	0	0	—	0	0	25	0	0
9	0	0	0	0	0	1	10	5	0	—	0	6	14	0
10	0	0	0	0	0	0	3	0	0	0	—	0	15	0
11	0	0	0	0	0	0	0	0	0	8	0	—	0	30
12	0	0	0	0	0	0	0	0	0	0	6	0	—	40
13	0	0	0	0	0	0	0	0	0	0	0	5	8	—

What is the maximal amount of ore that can be transported?

APPENDIX

1. THE SOLUTION OF EQUATIONS

From the beginning of this book, we have been dealing with the solution of equations, mainly linear in nature, but some also of higher degree—and some, indeed, transcendental (i.e., not algebraic). However, our principal purpose (in Chapter I, especially) was to show how *systems of simultaneous equations* may be solved; it was generally assumed that the student would know how to solve single equations with one unknown. This is usually the content (or part of the content) of high school algebra courses. We give here, for reference, the principal rules for the solution of such equations.

The general rule for equations is, briefly:

A.1.1 If an operation is performed on one side of an equation, it must also be performed on the other. This will preserve the equation if the operation is uniquely defined.

In effect, rule (A.1.1) states that the same quantity may be added or subtracted from both sides of an equation. Both sides may be multiplied by the same number or raised to the same power. On the other hand, some care must be exercised when dividing both sides: we must remember that division is not defined when the divisor is zero. Similarly, taking square roots is not permissible, unless some care is exercised, since a positive number has two square roots, (and a negative number has none, at least among the real numbers).

An important question is whether, after transforming an equation according to (A.1.1), the solutions of the transformed equation are also solutions of the original equation. The answer is that this will be so if the operations performed can be uniquely inverted (reversed). Addition and subtraction can be uniquely reversed, but multiplication cannot always be reversed (it might be multiplication by 0), and taking squares cannot be uniquely reversed. Any time that such an

399

operation is performed, there is a definite danger that so-called *extraneous roots* (roots that were not solutions of the original equation) will be introduced.

In brief, if an operation is well defined and has a well-defined inverse, there is no problem. If it is well defined, but not invertible, it may introduce extraneous roots. If it is not well defined, some solutions may be lost.

As an example of these difficulties, consider the problem of dividing by zero: Suppose $x = y$. Then:

(1) $$x = y$$

(2) $$x^2 = xy$$

(3) $$x^2 - y^2 = xy - y^2$$

(4) $$(x + y)(x - y) = y(x - y)$$

(5) $$x + y = y$$

(6) $$2y = y$$

(7) $$y = 0$$

We pass from equation (1), here, to equation (7), by a sequence of steps that seem valid. Note, however, that the step from (4) to (5) is not a well-defined operation: it involves division by zero. As a result, equation (7) has only one root, whereas equation (1) has an infinity of roots, most of which have been lost.

As an example of the introduction of extraneous roots, consider the following: Suppose $x^2 = -1$. Then

(1) $$x^2 = -1$$

(2) $$(x^2)^2 = (-1)^2$$

(3) $$x^4 = 1$$

(4) $$x = \sqrt[4]{1}$$

(5) $$x = \pm 1$$

Here, the step from equation (1) to equation (2) is valid, but is not invertible: the square root operation in not well defined. As a result no roots are lost (there were none to begin with), but new, extraneous roots have been introduced. Whenever something of this sort is done, it is always necessary to check each of the solutions obtained by introducing them into the original equation; this rids us of the extraneous solutions.

Let us return, now, to the solution of equations. The simplest is the general *linear* equation:

A.1.2 $$ax + b = c$$

To solve this, we subtract b from both sides:

$$ax + b - b = c - b$$

or

$$ax = c - b$$

and divide both sides by a:

$$\frac{ax}{a} = \frac{c - b}{a}$$

or

A.1.3
$$x = \frac{c - b}{a}$$

Note that this last step is permissible only when a is different from 0.

The next simplest equation is the *quadratic,* or second-degree equation,

A.1.4
$$ax^2 + bx + c = 0$$

in which $a \neq 0$.

This equation is solved by the method of *completing the square.* We divide both sides by a, to obtain

$$x^2 + \frac{b}{a}x + \frac{c}{a} = 0$$

and subtract c/a:

$$x^2 + \frac{b}{a}x = -\frac{c}{a}$$

Next, we add the term $b^2/4a^2$ to both sides:

$$x^2 + \frac{b}{a}x + \frac{b^2}{4a^2} = \frac{b^2}{4a^2} - \frac{c}{a}$$

This last equation may be rewritten in the form

$$\left(x + \frac{b}{2a}\right)^2 = \frac{b^2 - 4ac}{4a^2}$$

and we can take the square roots of both sides, being careful not to lose any roots:

$$x + \frac{b}{2a} = \pm\sqrt{\frac{b^2 - 4ac}{4a^2}}$$

or

$$x + \frac{b}{2a} = \frac{\pm\sqrt{b^2 - 4ac}}{2a}$$

We finally subtract $b/2a$ from both sides of this equation, obtaining the well-known quadratic formula

A.1.5
$$x = \frac{-b \pm \sqrt{b^2 - 4ac}}{2a}$$

Note that this equation has two solutions, corresponding to the positive and negative signs in front of the radical. There are three cases, depending on the sign of the *discriminant*

$$D = b^2 - 4ac$$

If $D > 0$, it has a positive and a negative square root, and so (A.1.5) gives two *real and distinct* values for x. If $D = 0$ then (A.1.5) gives only one value for x; we say that (A.1.4) has *real and coincident* roots. If $D < 0$, its square root is imaginary; in this case, (A.1.4) has two *conjugate-imaginary* roots; the roots are *not* real.

The next equations to be considered should logically, be the *cubic*, or third-degree, followed by the *biquadratic*, or fourth-degree equation. In actuality, the solution of these equations, though technically possible, is seldom treated because of its difficulty. We give brief outlines of the methods.

For the general cubic

A.1.6
$$x^3 + ax^2 + bx + c = 0$$

the change of variables

A.1.7
$$x = y - \frac{a}{3}$$

will give a *deficient* cubic

A.1.8
$$y^3 + py + q = 0$$

(where p, q, are constants depending on a, b, and c). The further change of variables

A.1.9
$$y = z - \frac{p}{3z}$$

gives the equation

$$z^3 + q - \frac{p^3}{27z^3} = 0$$

and multiplication by z^3 gives us

A.1.10
$$z^6 + qz^3 - \frac{p^3}{27} = 0$$

This can then be treated as a quadratic equation in $w = z^3$, which can be solved by using (A.1.5). Knowledge of w will give us z, and, in turn, y and x. Unfortunately, this method has the disadvantage that the solutions of (A.1.10) may all be imaginary. More precisely, (A.1.5) has three solutions, of which one at least is real. The other two may be either real or imaginary. In the case in which (A.1.5) has three real solutions, it happens that (A.1.10) has only imaginary solutions. For the reader who is willing to work with imaginary numbers, this is not a great drawback. Most people, however, prefer to avoid this.

For the general biquadratic

A.1.11
$$x^4 + ax^3 + bx^2 + cx + d = 0$$

the change of variables

A.1.12
$$x = y - \frac{a}{4}$$

will give a *deficient* biquadratic

A.1.13
$$y^4 + py^2 + qy + r = 0$$

This can be rewritten

$$y^4 + 2\lambda y^2 + \lambda^2 + (p - 2\lambda) y^2 + qy + (r - \lambda^2) = 0$$

Subtracting the last three terms from both sides, we obtain

A.1.14
$$(y^2 + \lambda)^2 = (2\lambda - p) y^2 - qy + (\lambda^2 - r)$$

It is now necessary to choose the parameter λ so that the right side of this equation is a perfect square. This will happen if λ satisfies the equation

A.1.15
$$4 (2\lambda - p)(\lambda^2 - r) = q^2$$

The equation (A.1.15), which is known as the *resolvent* of (A.1.13), is a cubic, or third-degree equation. It can be solved by the method just considered. If and when this has been done, its solution (any one of its solutions) can be introduced into (A.1.14) to give something of the form

$$(y^2 + \lambda)^2 = (Ay + B)^2$$

or

A.1.16
$$y^2 + \lambda = \pm (Ay + B)$$

This gives two quadratic equations, each of which can be solved by formula (A.1.5). Note that, in all, there are four solutions, two from each quadratic.

We have seen that the cubic equation can be solved by an auxiliary quadratic equation, and that the biquadratic can be solved through an auxiliary cubic. The question naturally arises whether the fifth-degree equation cannot, perhaps, be solved through a fourth-degree resolvent, and so on. In fact, this question was pursued for several centuries by mathematicians—a pursuit that came to an unsuccessful end when Abel proved that *the general equation of the fifth-degree (or higher) cannot be solved by means of radicals alone.*

In general, it is best to solve such equations by approximation. The usual method for solving, say

A.1.17
$$P(x) = 0$$

where $P(x)$ is a polynomial, is to find two values, a and b, such that $P(a) < 0$ but $P(b) > 0$. It will follow that a root of (A.1.17) lies between a and b. A third value, c, is taken, between a and b. If $P(c) > 0$, then a root will lie between a and c, while if $P(c) < 0$, then the root will lie between c and b. In this manner, smaller and smaller intervals are found, so that the root may be obtained with any degree of accuracy desired. Such methods are known as *numerical methods.*

Next in order of difficulty, after the polynomial equations, are the *rational* equations. A *rational function* is defined as the quotient of two polynomials; a rational equation is one in which only rational functions appear. By addition, subtraction, and simplification, such an equation can always be written in the form

A.1.18
$$\frac{P(x)}{Q(x)} = 0$$

Multiplication by $Q(x)$ will reduce this to the polynomial equation $P(x) = 0$, so that the solution of (A.1.18) depends on that of (A.1.17). It should be pointed out, however, that multiplication by $Q(x)$ might introduce extraneous roots. The general rule is that *the solutions of (A.1.18) are all those solutions of (A.1.17) that are not solutions of* $Q(x) = 0$.

As an example, consider the equation

$$\frac{x^2 - 1}{x - 1} = 0$$

Multiplication by the denominator gives

$$x^2 - 1 = 0$$

which has the solutions

$$x = \pm 1$$

Now, if we substitute the value $x = -1$, in the original equation, we see that we have a solution. If, however, we substitute $x = 1$, the left side of the original equation takes the form $0/0$, which is meaningless. Thus, $x = 1$ is an extraneous root.

We consider, next, equations with *surds*, or radicals. The usual approach to these equations consists in raising them to powers in order to eliminate the radicals. In so doing, one must guard against extraneous roots. Consider the equation

$$\sqrt{2x + 1} = x - 7$$

in which the radical is taken to represent the *positive* square root of $2x + 1$. Squaring both sides, we obtain

$$2x + 1 = x^2 - 14x + 49$$

which, after adding and subtracting some terms, gives the quadratic

$$x^2 - 16x + 48 = 0$$

The use of (A.1.5) will give the two solutions $x = 4$ and $x = 12$. Substitution of $x = 12$ gives

$$\sqrt{25} = 5$$

a correct statement. However, $x = 4$ gives

$$\sqrt{9} = -3$$

which is false since we have agreed that we want only a positive square root. Thus, $x = 12$ is the only solution; $x = 4$ is extraneous.

In this fashion, irrational equations (equations with surds) may be changed into polynomial equations. It is then a question of solving these, which, as we have seen, may be easy or difficult depending on the degree and form.

The last types of equation that we consider are the *transcendental* equations. These are equations that include the exponential, logarithmic, and trigonometric functions. Unless of especially simple form, they cannot be solved analytically. The equation

$$e^x = 3$$

has the obvious solution $x = \ln 3$. But the slightly more complicated

$$xe^x = 3$$

has no such simple solution; it must be solved numerically. In general, transcendental equations must be solved (approximately) by numerical methods.

2. THE PRINCIPLE OF INDUCTION

It is very often necessary to prove that some statement is true for all the natural numbers: 1, 2, 3, ... Since it is not always possible to prove the statement for all of them simultaneously, and since it is definitely impossible to prove it for all the numbers, one at a time, recourse is often had to the *principle of mathematical induction*. We give the principle in one form (there are several other possible forms):

A.2.1   Let T be a subset of the natural numbers, and suppose that T satisfies conditions (a) and (b):

(a) $1 \in T$

(b) If $n \in T$, then $n + 1 \in T$

Then T contains all the natural numbers.

The principle (A.2.1) is a basic part of the foundations of all mathematics. It is, in fact, generally used as part of the definition of our number system. Heuristically, we can see the justification for this principle. Suppose T satisfies the two conditions (a) and (b). Then by (a), it must contain the number 1. Since T contains 1, 2 must, by (b), also belong to T. But then 3 must also belong to T, and so on. It will follow that all the natural numbers (i.e., positive integers) belong to T.

Let us suppose, then, that we wish to prove the truth of a statement for all natural numbers. We can let T be the set of all natural numbers for which this statement is true. We must then prove that T has properties (a) and (b) of the principle (A.2.1). To prove (a), we must show that the statement is true for the natural number 1. To prove (b), we must show that whenever the statement is true for the natural number n, it is also true for the next natural number, $n + 1$.

Thus, a *proof by induction* will consist of two steps. First, prove for 1. Second, prove that truth for n implies truth for $n + 1$. This effectively proves it for all.

Let us suppose that we have an endless succession of light bulbs (having a first bulb, but no last bulb) which we wish to light. It is obviously impossible to light them all one at a time, since we simply do not have the time to do this. Suppose, however, that there is a mechanism connecting each one to the next, in such a way that whenever a bulb is lit, the next bulb is necessarily lit. Then lighting the first bulb will cause the second one to light, the second will light the third, and so on. To guarantee that all the bulbs are lit, it is sufficient to check: (a) that the first bulb is lit; (b) that the connections between bulbs have the desired property—that, if a bulb is lit, the next is also lit.

To see how this principle is used in practice, consider the problem of finding the sum of the first n squares. We wish to prove that:

A.2.2 $$1^2 + 2^2 + \ldots + n^2 = \frac{n(n+1)(2n+1)}{6}$$

for all values of n.

We prove this by the induction method. Consider, then, the case $n = 1$. In this case, the left side of (A.2.2) is simply 1^2, or 1. The right side is

$$\frac{1(1+1)(2+1)}{6}$$

which is also equal to 1. Thus, (A.2.2) is true when $n = 1$.

We must now prove that, if (A.2.2) is true for one value of n, it is also true for the next larger value. If we replace n by $n + 1$, (A.2.2) will take the form

A.2.3 $$1^2 + 2^2 + \ldots + n^2 + (n+1)^2 = \frac{(n+1)[(n+1)+1][2(n+1)+1]}{6}$$

We must prove that, if (A.2.2) is true for a value of n, then (A.2.3) will also be true for that same value of n.

To do this, we add the term $(n+1)^2$ to both sides of (A.2.2). This gives us

A.2.4 $$1^2 + 2^2 + \ldots + n^2 + (n+1)^2 = \frac{n(n+1)(2n+1)}{6} + (n+1)^2$$

It is now easy to check that (A.2.3) and (A.2.4) are the same equation. In fact, their left sides are clearly equal; it is simply a matter of expanding their right sides to show that these, also, are equal.

We have now shown that (A.2.2) is true when $n = 1$, and moreover, that if it is true for one value of n, then it is also true for the next value, $n + 1$. By the principle of induction, this means that the statement is true for all natural numbers. The proof is complete.

408   3.   EXPONENTS AND LOGARITHMS

Throughout much of this book, great use has been made of the *exponential* and *logarithmic* functions. We give here some of the elements of exponents and logarithms.

If a is any number, we shall write

$$a^2 = a \cdot a$$
$$a^3 = a \cdot a \cdot a$$

and so on; in general, if m is a positive integer, a^m is the product of m terms, each equal to a:

$$a^m = \underbrace{a \cdot a \cdot \ldots \cdot a}_{m \text{ terms}}$$

A stricter mathematical definition would be the *inductive* definition

A.3.1 $$a^1 = a$$

A.3.2 $$a^{m+1} = a^m \cdot a$$

Note how this inductive definition works: equation (A.3.1) defines a^m for $m = 1$; (A.3.2) defines a^{m+1} when a^m is known. According to the principle of induction this defines the power a^m for all positive integers m.

There are three important rules for working with exponents. The first deals with multiplying two powers of the same number. We have

$$a^m \cdot a^n = \underbrace{a \cdot a \cdot \ldots \cdot a}_{m \text{ terms}} \cdot \underbrace{a \cdot a \cdot \ldots \cdot a}_{n \text{ terms}}$$

It may be seen that the right side of this equation has $m + n$ terms, and so we conclude that

A.3.3 $$a^m \cdot a^n = a^{m+n}$$

The second rule deals with "a power of a power." We have

$$(a^m)^n = \underbrace{a^m \cdot a^m \cdot \ldots \cdot a^m}_{n \text{ terms}}$$

Each of the n terms on the right side of this equation can itself be expanded; doing this, we will have

$$(a^m)^n = \left.\begin{matrix} (a \cdot a \cdot \ldots \cdot a) \\ \cdot (a \cdot a \cdot \ldots \cdot a) \\ \cdots \cdots \cdots \cdots \\ \cdot (a \cdot a \cdot \ldots \cdot a) \end{matrix}\right\} \begin{matrix} n \\ \text{terms} \end{matrix}$$

$$m \text{ terms}$$

Now, the right side of this has m terms in each of n rows. There are mn terms all told, and so

A.3.4
$$(a^m)^n = a^{mn}$$

The third rule deals with obtaining a power of the product of two numbers, a and b.

$$(ab)^m = \underbrace{ab \cdot ab \cdot \ldots \cdot ab}$$

$$m \text{ times}$$

The right side of this equation may be rearranged to give us

$$(ab)^m = \underbrace{a \cdot a \ldots a}_{m \text{ terms}} \cdot \underbrace{b \cdot b \ldots b}_{m \text{ terms}}$$

and so

A.3.5
$$(ab)^m = a^m b^m$$

Note that this last can only be obtained because *multiplication of numbers is commutative*, i.e., $ab = ba$ for any numbers. If this were not so, the rearrangement of terms would not be possible. If we deal, say, with matrices (Chapter II) we will find that laws analogous to (A.3.3) and (A.3.4) hold, but the analogue of (A.3.5) does not hold.

For the reader who is not entirely convinced by the proofs of (A.3.3), (A.3.4), and (A.3.5), a more formal proof, depending on the definitions (A.3.1) and (A.3.2) and on the principle of induction (A.2.1), is available. Suppose that we wish to prove (A.3.3). We do this by induction on n. Letting $n = 1$, we have

$$a^m \cdot a^1 = a^m \cdot a$$

since $a^1 = a$ by definition. But also by definition, the right side of this is equal to a^{m+1}. Thus,

$$a^m \cdot a^1 = a^{m+1}$$

Note that this is true for *all* values of m.

Proceeding by induction, let us assume that (A.3.3) is true for *all* values of m and for a *particular* value of n. Now we will have

$$a^m \cdot a^{n+1} = a^m(a^n \cdot a)$$

by the definition of a^{n+1}. The associative law for multiplication gives us

$$a^m(a^n \cdot a) = (a^m \cdot a^n) \cdot a$$

By the induction hypothesis, $a^m a^n = a^{m+n}$, and so

$$(a^m \cdot a^n)a = a^{m+n} \cdot a$$

and, by definition, the right side here is a^{m+n+1}. Thus,

$$a^m \cdot a^{n+1} = a^{m+n+1}$$

By the principle of induction, this proves (A.3.3) for all values of m and n (among the positive integers). Similar proofs may be obtained for both (A.3.4) and (A.3.5).

We have, thus, defined the meaning of the expression a^m, when m is a positive integer. We would like, now, to define the analogous expressions a^r, when r is negative, fractional, or even, for that matter, irrational. We define these by *extension,* i.e., in such a way as to preserve the properties (A.3.3) and (A.3.4).

Consider, for example, the expression a^0. It is, clearly, meaningless to talk about the product of 0 terms. Let us however, use rule (A.3.3), with $n = 0$. We have

$$a^m \cdot a^0 = a^{m+0} = a^m$$

and so (if a is not zero) we see that a natural definition is

A.3.6
$$a^0 = 1$$

(Note, however, that the expression 0^0 is meaningless.)

Similarly, we can find a meaning for negative integer exponents. We have, indeed,

$$a^m \cdot a^{-m} = a^{m-m} = a^0 = 1$$

and it follows that we can define

A.3.7
$$a^{-m} = \frac{1}{a^m}$$

Let us now consider fractional exponents. If p and q are integers, we see by rule (A.3.4) that

$$(a^{p/q})^q = a^{pq/q} = a^p$$

Thus, $a^{p/q}$ should be defined so that its qth power is equal to a^p. But this is precisely what is known as a qth root. We define

A.3.8
$$a^{p/q} = \sqrt[q]{a^p}$$

The definitions (A.3.7) and (A.3.8) allow us to deal with all rational exponents. We will have

$$a^{-1/2} = \frac{1}{\sqrt{a}}$$

$$a^{1.3} = \sqrt[10]{a^{13}}$$

and so on. There remains only the case of irrational exponents to be disposed of.

For irrational exponents, the method of definition depends on continuity. In effect, we know that every irrational number can be approximated as closely as desired by rational numbers. If r is any real number, we know that there is a sequence $q_1, q_2, \ldots$ of rational numbers that *converges* to r. Then the power a^r is defined as the *limit* of the sequence:

$$a^{q_1}, a^{q_2}, \ldots \to a^r$$

To give an example, we know that the irrational $\sqrt{2}$ is the limit of the sequence

$$1, 1.4, 1.41, 1.414, 1.4142, \ldots$$

(represented, of course, by the decimal expansion of $\sqrt{2}$). This will mean that the power $a^{\sqrt{2}}$ can be defined as the limit of the sequence

$$a, a^{1.4}, a^{1.41}, a^{1.414}, a^{1.4142}, \ldots$$

In other words, a term of this sequence, such as $a^{1.414}$, is a good approximation to the desired $a^{\sqrt{2}}$ —just as 1.414 is a good approximation to $\sqrt{2}$.

Next we consider *logarithms*. Briefly speaking, a logarithm is the same as an exponent. If

A.3.9
$$a^y = x$$

we say that y is the logarithm of x, with base a. The notation for this is

A.3.10 $$y = \log_a x$$

Statements (A.3.9) and (A.3.10) are equivalent. From (A.3.9), we say that x is the *exponential* of y (with base a); we see that the exponential and logarithmic functions are (by definition) mutually inverse. We shall now develop rules for logarithms. These shall all be natural consequences of the rules (A.3.3) and (A.3.4) for exponents.

Let us suppose we have

$$x = a^y; \; z = a^w$$

or, equivalently,

$$y = \log_a x; \; w = \log_a z$$

Now, by (A.3.3),

$$xz = a^y \cdot a^w = a^{y+w}$$

and so

$$\log_a(xz) = y + w$$

or

A.3.11 $$\log_a(xz) = \log_a x + \log_a z$$

Similarly, we have, by (A.3.7)

$$\frac{1}{x} = \frac{1}{a^y} = a^{-y}$$

or

A.3.12 $$\log_a\left(\frac{1}{x}\right) = -\log_a x$$

Properties (A.3.11) and (A.3.12) can be combined to give us

A.3.13 $$\log_a\left(\frac{x}{z}\right) = \log_a x - \log_a z$$

Once again, assume $y = \log_a x$. We have, by (A.3.4),

$$x^r = (a^y)^r = a^{ry}$$

and so

A.3.14 $$\log_a x^r = r \log_a x$$

Rules (A.3.11) to (A.3.14) explain the usefulness of logarithms. In fact, we see that multiplications and divisions can be replaced by the simpler operations of addition and subtraction while the very difficult operation of raising to a power reduces to a multiplication.

One last property of logarithms interests us now, and this deals with a *change in base*. Let us suppose once again that

$$x = a^y$$

and, moreover,

$$a = b^c$$

We have, then,

$$x = (b^c)^y = b^{cy}$$

and therefore

A.3.15 $$\log_b x = (\log_b a)(\log_a x)$$

For a change in base, it is sufficient to multiply the logarithm using the old base, times the logarithm of the old base with respect to the new base. Equivalently, we have

A.3.16 $$\log_a x = \frac{\log_b x}{\log_b a}$$

so that it is sufficient to divide the logarithm of the variable by the logarithm of the new base.

As base, any positive number other than 1 may be used. Negative numbers are not used as base, since many of their powers are imaginary. The numbers 0 and 1 are useless, since all their powers are equal. Numbers smaller than 1 may be used as base, but they have the disadvantage that, with such bases, the logarithm of a number decreases as the number increases. With base 1/2, for example,

$$\log_{1/2} 2 = -1$$
$$\log_{1/2} 8 = -3$$

and so on. As a practical matter, the base should be a number greater than 1. In actual practice, only three bases have any use at all (and one of those is only used by computers). These three bases are the numbers 10, e, and 2.

The most commonly used logarithms are logarithms to the base 10. Their main advantage is the well-known fact that, with our numerical system, multiplication by 10 or any integral power of 10 can be accomplished by shifting a decimal point, with no change of digits. The importance of this can best be seen if we consider that, for example,

$$\log 2352 = \log 1000 + \log 2.352$$

But $\log_{10} 1000 = 3$, and so

$$\log_{10} 2352 = \log_{10} 2.352 + 3$$

Similarly,

$$\log_{10} 0.0158 = \log_{10} 1.58 - 2$$

and so on. It is sufficient to tabulate logarithms for numbers from 1 to 10; logarithms for smaller or larger numbers can easily be obtained from them.

The advantages of the base e, while not inconsiderable, are of a more theoretical nature. Logarithms to the base e are known as *natural* logarithms. They are also known as *Napierian* logarithms, after Napier, who was one of the first men to study logarithms. We use the notations

$$\log_e x, \ \ln x$$

interchangeably, to denote these logarithms. Such logarithms are also commonly tabulated. They may be obtained from the common (base 10) logarithms, however, by using (A.3.15). This gives us,

$$\ln x = (\ln 10)(\log_{10} x)$$

Since the natural logarithm of 10 is approximately 2.3026, we have

A.3.17 $$\ln x = 2.3026 \log_{10} x$$

One last number, 2, is in current use as base for logarithms. Its use is mainly confined to computers, which can work best with 2 as basis for all their calculations; there is also some use for these logarithms in the computer-related science of information theory. Should the reader ever need these logarithms, they can be obtained from (A.3.15) with the formula

A.3.18 $$\log_2 x = 3.3219 \log_{10} x$$

4. THE SUMMATION SYMBOL

Throughout this book, we have used the summation symbol, Σ. We will explain its use simply in this appendix.

Let us suppose we have an *indexed* set of numbers, with indices among the integers:

$$a_1, a_2, a_3, \ldots$$

Let k and l be integers such that $k < l$. Then the symbol

$$\sum_{j=k}^{l} a_j$$

will be taken to mean the sum of all the numbers a_j, as j takes on all integral values from k to l, inclusive:

A.4.1
$$\sum_{j=k}^{l} a_j = a_k + a_{k+1} + \ldots + a_{l-1} + a_l$$

For example,

$$\sum_{j=5}^{9} b_j = b_5 + b_6 + b_7 + b_8 + b_9$$

$$\sum_{i=0}^{5} a_i = a_0 + a_1 + a_2 + a_3 + a_4 + a_5$$

Sometimes, of course, the indexed number, a_i, may be defined as a function, $f(i)$, of the index. Technically, this is no difference at all. We have, then, for example,

$$\sum_{j=1}^{4} \frac{1}{j} = \frac{1}{1} + \frac{1}{2} + \frac{1}{3} + \frac{1}{4}$$

$$\sum_{i=11}^{13} i^2 = 11^2 + 12^2 + 13^2$$

Strictly speaking, it is not even necessary that the indices be among the integers; they can come from any set at all. If Ω is such a set, the notation

$$\sum_{\omega \in \Omega} a_\omega$$

represents the sum of the terms a_ω, for all ω in the index set Ω.

As an example, we can let Ω be the set of people inside an elevator, and a_ω represent the weight, in pounds, of the individual ω. The inequality

$$\sum_{\omega \in \Omega} a_\omega \leq 1600$$

is then symbolic notation for a sign that may appear inside this elevator.

There are certain rules for working with summations, following from the well-known rules for sums and products of numbers. Thus

$$\sum_{j=k}^{l} c\,a_j = c\,a_k + c\,a_{k+1} + \ldots + c\,a_l$$

$$= c(a_k + a_{k+1} + \ldots + a_l)$$

so

A.4.2
$$\sum_{j=k}^{l} ca_j = c \sum_{j=k}^{l} a_j$$

i.e., a constant factor (one not dependent on the index j) can be "factored out" of a sum.

Similarly,

$$\sum_{j=k}^{l} (a_j + b_j) = (a_k + b_k) + \ldots + (a_l + b_l)$$

$$= (a_k + a_{k+1} + \ldots + a_l) + (b_k + b_{k+1} + \ldots + b_l)$$

and so

A.4.3
$$\sum_{j=k}^{l} (a_j + b_j) = \sum_{j=k}^{l} a_j + \sum_{j=k}^{l} b_j$$

i.e., a sum can be "split" into two sums in this manner.

Double sums are also encountered on occasion. Given a doubly indexed set of numbers a_{ij}, we have the notation

A.4.4
$$\sum_{i=m}^{n} \sum_{j=k}^{l} a_{ij}$$

to represent the sum of the terms a_{ij}, as i and j are allowed, independently, to take all values between m and n, and between k and l, respectively. These double sums can also be rewritten as *iterated sums*:

A.4.5
$$\sum_{i=m}^{n} \left\{ \sum_{j=k}^{l} a_{ij} \right\}$$

To evaluate this expression, we must first compute the sum

$$b_i = \sum_{j=k}^{l} a_{ij}$$

for each i, and then the sum

$$\sum_{i=m}^{n} b_i$$

It is not difficult to see that (A.4.4) and (A.4.5) are equal to each other, and also to the other iterated sum

$$\sum_{j=k}^{l} \left\{ \sum_{i=m}^{n} a_{ij} \right\}$$

in which the sum is taken first over i and then over j.

Similarly, triple and, more generally, n-tuple sums may be defined. These can always be evaluated as iterated sums. Thus

$$\sum_{i \in I} \sum_{j \in J} \sum_{k \in K} a_{ijk} = \sum_{i \in I} \left\{ \sum_{j \in J} \left[\sum_{k \in K} a_{ijk} \right] \right\}$$

and so on.

INDEX

419

SOLUTIONS TO PROBLEMS

1.   (a) $2x - y = -1$ (m) $x - 2y = -7$

    (b) $x - y = -2$ (n) $x - y = 4$

    (c) $2x + y = 0$ (o) $x - y = 3$

    (d) $x = -1$ (p) $x + 3y = 13$

    (e) $y = 4$ (q) $2x - y = 10$

    (f) $-x + 2y = 8$ (r) $x + 2y = 4$

    (g) $2x - y = 0$ (s) $-2x + 3y = -6$

    (h) $y = 1$ (t) $-4x - y = 4$

    (i) $x - y = 5$ (u) $y = 3x + 6$

    (j) $x - 4y = 1$ (v) $y + x = 5$

    (k) $2x + 4y = 14$ (w) $y = 2$

    (l) $x = 0$ (x) $2x + y = -6$

                                 (y) $2x + y = 4$

2.   (a) $m = -1/3$, $x_0 = 6$, $y_0 = 2$

    (b) $m = 3/2$, $x_0 = 5/3$, $y_0 = -5/2$

    (c) $m = -1/4$, $x_0 = 12$, $y_0 = 3$

    (d) $m = 1/2$, $x_0 = 3$, $y_0 = -3/2$

    (e) $m = 1/4$, $x_0 = 16$, $y_0 = -4$

3.   (a) 1 (b) $\sqrt{10}$ (c) $\sqrt{13}$ (d) $\sqrt{26}$ (e) $\sqrt{73}$

4.   (a) $(7/2, 5/2)$, $(3, 1/2)$, and $(3/2, 2)$.

    (b) $2x + 7y = 17$ (c) $x + 4y = 9$
         $7x + 2y = 22$ $4x + y = 12$
           $x - y = 1$ $x - y = 1$

1.   (a) $x = 1$, $y = 2$.   (j) $x = 6$, $y = 3$.

    (b) $x = 10$, $y = 1$.   (k) x arbitrary, $y = 5/2 - 3/2\,x$.

    (c) $x = 11/7$, $y = 9/7$.   (l) no solution.

    (d) $x = 3$, $y = 2$.   (m) $x = 1$, $y = 2$.

    (e) x arbitrary, $y = 2 - 1/2x$.   (n) $x = -1$, $y = 1$.

    (f) $x = 3$, $y = 1$.   (o) no solution.

    (g) $x = -1$, $y = -1$.   (p) no solution.

    (h) no solution.   (q) $x = 2$, $y = 1$.

    (i) $x = 2$, $y = 3$.   (r) $x = 0$, $y = 3$.

425

2. The points are (a), (8/3, 5/3), and (b), (13/5, 8/5).

PAGES 43–44

1. The extreme points are:

(a) (6/5, 27/5), (0,6), (3,0), and (0,0).

(b) (9, 2), (39/5, 14/5), (5,0), (0,0), and (10, 0).

(c) (12/5, 44/5), (16/3, 0), and (20, 0).

(d) (9, 2), (25/3, 10/3), and (12, 0).

(e) (8/5, 16/5), (0, 4), (10/3, 4/3), (2, 0), and (0, 0).

2. Let x = quantity of expensive mixture.

y = quantity of cheaper mixture.

z = quantity of unmixed peanuts.

Then

$$\begin{array}{rl} 5x + y & \leqq 6000 \\ x + y + z & \leqq 4000 \\ x & \geqq 0 \\ y & \geqq 0 \\ z & \geqq 0. \end{array}$$

PAGE 55

1. (a) $x = 1$, $y = 1$, $z = -1$.

(b) $x = 1$, $y = -1$, $z = 2$.

(c) no solution.

(d) x arbitrary, $y = 3x - 11$, $z = 5/2x - 17/2$.

(e) $x = 3$, $y = 1$, $z = 1$.

(f) $x = 1$, $y = -4$, $z = 1$.

PAGES 62–63

1. (a) $(5, 13, -3)$.

(b) $\begin{pmatrix} 13 \\ -6 \end{pmatrix}$.

(c) $(-6, -14, 20, 6)$.

(d) $\begin{pmatrix} 1 \\ 13 \\ 10 \end{pmatrix}$

(e) $(-9, +28)$

(f) $(-5, 29, -15)$.

(g) $(8, 12, 8)$

(h) $\begin{pmatrix} 6 \\ 14 \end{pmatrix}$

(i) $\begin{pmatrix} 12 \\ 13 \\ -5 \\ -8 \end{pmatrix}$

(j) $(-6, -1, 31)$.

2. (a) $x = 3$, $y = 5$.   (f) $x = 1$, $y = 5$.

(b) $x = -2$, $y = 1$.   (g) $x = 3$, $y = 1$.

(c) $x = 11/5$, $y = 6/5$.   (h) no solution.

(d) $x = -2$, $y = 3$.   (i) $x = -1$, $y = 2$.

(e) $x = -1$, $y = 2$.   (j) $x = 102/43$, $y = 25/43$, $z = 115/43$.

(k) $x = 1$, $y = 1$, $z = -3$.

3. (a) yes   (b) yes   (c) no   (d) no   (e) yes.

4. (a) $\begin{pmatrix} 1 & 3 \\ 11 & 6 \end{pmatrix}$ (c) $\begin{pmatrix} 17 & 14 & -5 & 24 \\ 34 & 17 & -8 & -4 \\ -27 & 23 & 0 & 7 \end{pmatrix}$ (e) $\begin{pmatrix} 2 & -1 & -1 \\ 12 & -15 & 19 \\ -15 & 11 & 13 \end{pmatrix}$

(b) $\begin{pmatrix} 6 & 43 \\ 32 & -9 \\ 29 & -29 \end{pmatrix}$ (d) $\begin{pmatrix} 2 & -21 \\ 12 & -8 \\ 15 & -10 \\ 13 & 15 \end{pmatrix}$

PAGES 77–79

1. (a) 40

(b) -6

(c) -10

(d) 19

(e) $\begin{pmatrix} -16 & -2 & 29 \\ 6 & -34 & 34 \end{pmatrix}$

(f) $\begin{pmatrix} 19 & -4 \\ 2 & 4 \\ 6 & -3 \end{pmatrix}$

(g) $\begin{pmatrix} 14 & -7 & -3 \\ 36 & 26 & -2 \\ 29 & 5 & 7 \end{pmatrix}$

(h) $\begin{pmatrix} 8 & 19 & -11 \\ 17 & 9 & 21 \\ -16 & -28 & 16 \end{pmatrix}$

(i) $\begin{pmatrix} 14 & 36 & 29 \\ -7 & 26 & 5 \\ -3 & -2 & -7 \end{pmatrix}$

(j) $\begin{pmatrix} 0 & 0 & 0 \\ 0 & 0 & 0 \\ 0 & 0 & 0 \end{pmatrix}$

(k) $\begin{pmatrix} 7 \\ 8 \\ -1 \end{pmatrix}$

(l) $\begin{pmatrix} 25 & 32 \\ 11 & -4 \end{pmatrix}$

(m) $\begin{pmatrix} -10 & 30 & 14 \\ 3 & 19 & 7 \\ 4 & 32 & 12 \end{pmatrix}$

(n) $\begin{pmatrix} 7 & 46 & 54 & 55 \\ -7 & 37 & 0 & 4 \end{pmatrix}$

(o) $\begin{pmatrix} 19 & 2 \\ 26 & 6 \end{pmatrix}$

(p) $\begin{pmatrix} 26 & 54 \\ 7 & 17 \end{pmatrix}$

(q) $\begin{pmatrix} 29 & 6 \\ 38 & 20 \end{pmatrix}$

(r) $\begin{pmatrix} 6 & 34 \\ 2 & 16 \\ -5 & -32 \end{pmatrix}$

(s) 6

(t) 19.

2. (a) $\begin{pmatrix} -7 & -21 & -7 \\ 14 & 42 & 14 \end{pmatrix}$

(b) $\begin{pmatrix} 48 & 82 \\ -16 & -4 \end{pmatrix}$

(c) $\begin{pmatrix} -55 & -36 & 95 \\ -46 & -46 & 108 \\ 233 & 81 & -229 \end{pmatrix}$

(d) $\begin{pmatrix} -6 & -164 & -200 \\ -23 & -266 & -336 \end{pmatrix}$

(e) 99.

3. (a) $x = 2$, $y = 1$. (d) $x = 5$, $y = -1$, $z = -3$, $w = 2$.

(b) $x = 1/2$, $y = 0$. (e) $x = -7$, $y = -7$, $z = 5$, $w = 6$.

(c) $x = -2$, $y = 1$. (f) $x = -1$, $y = -17$, $z = 1$, $w = 7$.

PAGES 93–94

1. (a) $x = 2$, $y = 0$.

(b) $x = -27$, $y = 17$.

(c) $x = 31$, $y = 20$, $z = -47$.

(d) $x = 1$, $y = 3$, $z = 5$.

(e) $x = -7$, $y = -4$, $z = 2$, $w = -5$.

(f) $x = -1$, $y = 1$, $z = 4$.

(g) $x = -5/2\,z + 5/2$, $y = 9/4\,z + 1/4$, $w = -6z + 3$.

(h) $x = -137/14$, $y = 39/14$, $z = 25/2$, $w = 9/14$.

(i) $x = 1$, $y = 1$, $z = 1$, $w = 1$.

(j) $x = -6$, $y = 7$, $z = 3$, $w = 4$.

(k) $x = -25/44$, $y = -127/44$, $z = 61/44$, $w = -133/44$.

(l) $x = 1$, $y = 2$, $z = 1$, $w = 3$.

(m) no solution.

(n) no solution.

(o) w arbitrary, $x = -1/20w + 21/20$, $y = -2/5w - 8/5$, $z = 31/40w + 89/40$.

PAGES 98–99

1. (a) $\begin{pmatrix} -1/28 & 5/28 \\ 6/28 & -2/28 \end{pmatrix}$

(b) $\begin{pmatrix} -2 & 1 \\ 7/4 & -3/4 \end{pmatrix}$

(c) $\begin{pmatrix} 5/11 & 3/11 \\ -2/11 & 1/11 \end{pmatrix}$

(d) $\begin{pmatrix} -5 & 6 \\ 1 & -1 \end{pmatrix}$

(e) $\begin{pmatrix} 2 & -3/2 \\ 1 & -1/2 \end{pmatrix}$

(f) $\begin{pmatrix} -23 & -13 & 12 \\ 60 & 34 & -31 \\ -2 & -1 & 1 \end{pmatrix}$

(g) $\begin{pmatrix} 14/99 & -53/198 & 2/9 \\ 1/9 & 1/9 & -1/9 \\ -1/11 & 3/22 & 0 \end{pmatrix}$

(h) not invertible.

(i) $\begin{pmatrix} 389/2268 & -34/2268 & -344/2268 & -13/2268 \\ -412/2268 & 380/2268 & 376/2268 & 212/2268 \\ 455/2268 & -238/2268 & -140/2268 & -91/2268 \\ -47/2268 & 214/2268 & 164/2268 & -185/2268 \end{pmatrix}$

(j) not invertible.

2. (b) $x = y(I-A)^{-1}$

(c) $A = \begin{pmatrix} 0 & 100 \\ 1/1000 & 0 \end{pmatrix}$, $x = (1139,\ 138889)$.

3. (b) $55/9$ ¢ per lb., $200/9$ ¢ per T-mi.

PAGES 109–110

1. (a) $x=0$, $y=4$, value 20.   (c) $x=3/5$, $y=9/5$, value $27/5$.

(b) $x=4$, $y=0$, value 8.   (d) $x=12$, $y=0$, value 36.

2. $70/3$ of A, $40/3$ of B, cost \$6.17.

3. 0, 640 lb., 360 lb. Profit \$404.

4. 2000 lb., 4000/3 lb. Profit \$2333.33.

PAGE 119

1. (a) $-1/8\,s -1/8\,t -3/4\,u +7/2 = -x$
$9/4\,s +1/4\,t +1/2\,u \qquad = -y$
$11/8\,s +3/8\,t + 1/4\,u -1/2 = -z.$

(b) $7s -2t -23 = -x$
$-3s + t + 10 = -y$
$-5s + t +11 = -z.$

(c) $5r +22s +2t +11 = -x$
$-20r -89s -7t -42 = -y$
$3r +13s + t + 6 = -z.$

(d) $-0.8s -0.5t +0.7r - 8.9 = -x$
$1.2s + t -0.8r +13.6 = -y$
$0.2s +0.5t -0.3r + 5.1 = -z.$

(e) $-r -2t -12s + 9 = -x$
$2r +3t +21s -13 = -y$
$-r - t - 7s + 8 = -z$

(f) $5/7r +2/7s -1/7t -19/7 = -x$
$16/7r +5/7s +1/7t -23/7 = -y$
$62/7r +8/7s +3/7t + 1/7 = -z.$

(g) $-\frac{3}{8}s -\frac{7}{8}t +\frac{5}{8}u +1 = -x$
$s +2t - u \qquad = -y$
$\frac{1}{8}s -\frac{3}{8}t +\frac{1}{8}u -1 = -z.$

PAGES 142–144

1. $x=40/7$, $y=64/35$, $z=53/35$, $w=129/7$.

2. $x=4/5$, $y=0$, $z=44/5$, $w=48/5$.

3. $x=0$, $y=10/3$, $z=25/3$, $w=80/3$.

4. $x=5$, $y=4$, $z=3$, $w=42$.

SOLUTIONS TO PROBLEMS

5. 135/34 of A, 175/34 of C. Cost $5.52.

6. 5 of A, 15 of C. Revenues $155.

7. 25/2 chairs, 5/2 tables, profit $187.50.

8. 4273 lb. of A, 21,818 lb. of C, 2909 lb. of D. Revenues, $18,391.

9. 50 of A, 175/2 of B. Revenue $1025.

10. 485 $T.$ of A, 50 $T.$ of C. Cost $2725.

PAGES 156–157

1. (a) Primal solution, $x = 0$, $y = 5$, $z = 10$.
Dual solution, $r = 24/7$, $s = 0$, $t = 1/7$.
Value 90.

(b) Primal solution, $x = 0$, $y = 25/3$, $z = 0$.
Dual solution, $r = 0$, $s = 2/3$, $t = 0$.
Value 50/3.

(c) Primal solution, $x = 45/4$, $y = 25/4$, $z = 0$.
Dual solution, $r = 1/8$, $s = 0$, $t = 5/8$.
Value 115/4.

(d) Primal solution, $x = 5$, $y = 15$, $z = 0$.
Dual solution, $r = 0$, $s = 3$, $t = 5$.
Value 45.

2. (5) $r = 31/4$, $s = 0$, $t = 5/34$.
(6) $r = 7/2$, $s = 3/4$, $t = 0$.
(7) $r = 25/4$, $s = 5/4$. $t = 0$.
(8) $r = 215/88$, $s = 375/88$, $t = 45/8$.
(9) $r = 3/2$, $s = 0$, $t = 11/2$.
(10) $r = 50$, $s = 0$, $t = 225$.

PAGES 174–176

1. (a) $\begin{pmatrix} 0 & 80 & 20 \\ 140 & 10 & 0 \\ 0 & 0 & 120 \end{pmatrix}$ cost 950

(b) $\begin{pmatrix} 5 & 45 & 0 & 0 \\ 0 & 0 & 15 & 55 \\ 0 & 25 & 15 & 0 \end{pmatrix}$ cost 530

(c) $\begin{pmatrix} 0 & 50 & 0 & 25 & 5 \\ 50 & 0 & 20 & 10 & 0 \\ 0 & 0 & 0 & 0 & 40 \end{pmatrix}$ cost 425

(d) $\begin{pmatrix} 5 & 60 & 25 & 0 & 0 \\ 35 & 0 & 0 & 15 & 40 \\ 0 & 0 & 0 & 20 & 0 \end{pmatrix}$ cost 560

(e) $\begin{pmatrix} 40 & 10 & 0 & 0 & 0 \\ 0 & 0 & 20 & 0 & 30 \\ 0 & 30 & 20 & 0 & 0 \\ 0 & 0 & 0 & 40 & 10 \end{pmatrix}$ cost 470

3. (a) (1,1), (2,3), (3,5), (4,2), (5,4). Value 23.
Salaries (dual variables) 4, 2, 3, 2, 1.

(b) (1,5), (2,7), (3,6), (4,3), (5,1), (6,4), (7,2).
Salaries 4, 6, 3, 3, 4, 4, 3.

(c) (1,3), (2,4), (3,1), (4,5), (6,2). Value 30.
Salaries 2, 1, 3, 2, 0, 2.

(d) (1,2), (2,5), (3,4), (4,3), (5,6), (6,1).
Value 41. Salaries 1, 6, 1, 3, 1, 2, 0.

PAGES 189–190

1. 84.

2. $10!/3! = 604, 800.$

3. 1764.

4. 1596, 1617.

5. 64.

6. 120.

7. 15.

8. 1,048,576.

10. Hint: note that both sides of the equation give the total number of subsets in a set with n elements.

PAGES 191–192

1. (a) The grass is green. The weather is hot. It has rained.

(b) The people are happy. The weather is bad.

(c) The nation is prosperous. There is war.

(d) We are together. All are good friends.

(e) We are arming. The enemy will attack us.

2. (a) We are singing or he is playing the piano.

(b) We are singing, and either he is not playing the piano, or they are dancing.

(c) Either it is not true that we are singing and he is playing the piano, or they are not dancing.

(d) Either we are singing, or we are not singing or he is playing the piano, and they are dancing.

(e) We are singing or they are dancing, and we are singing and he is not playing the piano, or, we are not singing and they are dancing.

PAGES 195–196

1. (a)

p	T	T	F	F
q	T	F	T	F
compound	T	F	F	T

(b)

p	T	T	F	F
q	T	F	T	F
compound	F	T	F	T

(c)

p	T	T	T	T	F	F	F	F
q	T	T	F	F	T	T	F	F
r	T	F	T	F	T	F	T	F
compound	T	T	T	T	T	T	F	F

(d)

p	T	T	T	T	F	F	F	F
q	T	T	F	F	T	T	F	F
r	T	F	T	F	T	F	T	F
compound	T	F	T	T	F	F	T	T

(e)

p	T	T	F	F
q	T	F	T	F
compound	F	F	F	F

2. (a) true whenever p and q are both F.

(b) true unless p and q are both T.

(c) always true.

(d) always false.

PAGE 199

1. (a)

p	T	T	F	F
q	T	F	T	F
compound	T	F	F	T

(b)

p	T	T	F	F
q	T	F	T	F
compound	T	F	F	T

(c)

p	T	T	F	F
q	T	F	T	F
compound	T	F	T	T

(d)

p	T	T	F	F
q	T	F	T	F
compound	F	F	T	T

(e)

p	T	T	T	T	F	F	F	F
q	T	T	F	F	T	T	F	F
r	T	F	T	F	T	F	T	F
compound	T	F	T	T	T	T	T	T

(f)

p	T	T	F	F
q	T	F	T	F
compound	T	F	F	T

(g)

p	T	T	F	F
q	T	F	T	F
compound	T	T	T	F

(h)

p	T	T	F	F
q	T	F	T	F
compound	T	T	T	T

(i)

p	T	T	T	T	F	F	F	F
q	T	T	F	F	T	T	F	F
r	T	F	T	F	T	F	T	F
compound	T	T	T	T	F	T	F	T

(j)

p	T	T	F	F
q	T	F	T	F
compound	T	T	T	T

PAGES 203–204

1. (a) yes   (f) no

    (b) no   (g) yes

    (c) no   (h) yes

    (d) yes   (i) yes

    (e) yes   (j) yes

2. (a) yes   (d) yes

    (b) yes   (e) no

    (c) no   (f) no

PAGES 213–214

1. $P(E) = 1/2, P(S) = 1/3, P(E\cap S) = 1/6, P(E-S) = 1/3, P(E\cup S) = 2/3,$
$P(S-E) = 1/6.$

2. (a) 1/2   **3.** (a) 1/13

    (b) 3/4   (b) 1/52

    (c) 1/4   (c) 4/13

    (d) 1/8   **4.** (a) 5/36

    (e) 7/8   (b) 15/36

    (f) 1/4   (c) 6/36

    (g) 1/2

PAGES 221–222

1. 1/2   **6.** 0.777

2. 3/4   **7.** 244/495

3. 211/243, 1/243   **8.** $(0.95)^{10}$, or .599

4. 203/396   **9.** 4/9

5. $29!/(20!\ 30^9)$, or .185

PAGE 227

1. 26/33   **4.** 1/26

2. 3/14   **5.** 24/41

3. (a) 7/178
    (b) 9/22

PAGES 234–235

1. (a) 924/4096   (b) 794/4096   (c) 2508/4096

2. (a) 3125/15,552

(b) 1453/23,328

3. 75%

4. .085

5. 11097/15625

6. 938/969

7. 21/46

8. 27/28

9. (a) 3679/3876

(b) 1144/1615

(c) 284/285

PAGES 255–256

1. mean 50/3, variance 125/9

2. 0

3. 3 units

4. .01

7. $\sigma_U^2 = \sigma_X^2 + \sigma_Z^2$, $\sigma_V^2 = \sigma_Y^2 + \sigma_Z^2$, $\sigma_{UV} = \sigma_Z^2$

$$\rho_{UV} = \frac{\sigma_Z^2}{\sqrt{(\sigma_X^2 + \sigma_Z^2)(\sigma_Y^2 + \sigma_Z^2)}}$$

8. .978

9. .398

10. .303

PAGES 264–266

1. $A = \begin{pmatrix} .5 & .2 & .3 \\ .3 & .6 & .1 \\ 0 & 0 & 1.0 \end{pmatrix}$, .391

2. .424

3. (.396, .372, .232)

4. (3/8, 3/8, 1/4)

5. (a) $\begin{pmatrix} 0.7 & 0.3 & 0 & 0 & 0 \\ 0.14 & 0.62 & 0.24 & 0 & 0 \\ 0 & 0.14 & 0.62 & 0.24 & 0 \\ 0 & 0 & 0 & 0.2 & 0.8 \end{pmatrix}$

(b) 160/301

(c) 245/3483

8. Note that $(I - B)(I + B + B^2 + \ldots + B^K) = I - B^{K+1}$. Then note that, for large values of K, all entries of B^{K+1} will be small.

10.

$$\begin{pmatrix} 1 & 0 & 0 & 0 & 0 & 0 \\ 1/2 & 0 & 1/2 & 0 & 0 & 0 \\ 0 & 1/2 & 0 & 1/2 & 0 & 0 \\ 0 & 0 & 1/2 & 0 & 1/2 & 0 \\ 0 & 0 & 0 & 1/2 & 0 & 1/2 \\ 0 & 0 & 0 & 0 & 0 & 1 \end{pmatrix}$$

(a) 8/5

(b) 5 plays

(c) 3/5

PAGES 276–277

1. (a) row 2, column 1.

(b) row 1, column 1.

(c) row 2, column 2.

(d) row 3, column 3.

(e) row 2, column 3.

2. Both build in B.

3. Since the matrix is singular, the first row must be k times the second. If $k > 1$, row 1 dominates row 2; otherwise row 2 dominates row 1 (unless they are equal). The same holds for the columns.

4. Say $a_{11} = a_{12}$. If $a_{11} \geqq a_{21}$, then a_{11} is a saddle point. Similarly, if $a_{12} \geqq a_{22}$, then a_{12} is a saddle point. If $a_{11} < a_{21}$ and $a_{12} < a_{22}$, then the smaller of a_{21} and a_{22} is a saddle point.

5. Problem 1(c), on page 276, is a counter-example.

PAGE 289

1. (a) $x = (7/9, 2/9)$, $y = (4/9, 5/9)$, $v = 37/9$.

(b) $x = (4/5, 1/5)$, $y = (2/5, 0, 3/5)$, $v = 17/5$.

(c) $x - (1, 0)$, $y - (1, 0, 0)$, $v = 2$.

(d) $x = (1/2, 1/2)$, $y = (3/8, 5/8, 0)$, $v = 7/2$.

(e) $x = (1/5, 4/5)$, $y = (2/5, 3/5, 0)$, $v = 18/5$.

2. For P_1: 1/4 Ace of Spades, 3/4 Ace of diamonds.
For P_2: 3/4 Ace of Hearts, 1/4 Deuce of Spades.
Value is $-1/4$.

3. Blotto's optimal strategy is to send out one division with probability 2/3, and two divisions with probability 1/3.

1. (a) $x = (1/2, 0, 1/2)$, $y = (1/2, 0, 0, 1/2)$, $v = 5/2$.

   (b) $x = (1/2, 1/2, 0)$, $y = (3/4, 1/4, 0)$, $v = 7/2$.

   (c) $x = (11/27, 2/27, 14/27)$, $y = (0, 11/27, 13/27, 1/9)$, $v = 61/27$.

   (d) $x = (1/3, 2/3, 0)$, $y = (1/3, 0, 2/3)$, $v = 5/3$.

   (e) $x = y = (1/2, 1/3, 1/6)$, $v = 0$.

2. Note that $xAx^t = (xAx^t)^t = xA^t x^t$ as this is a number. But $xAx^t = x(-A^t)x^t$, so we see that $xAx^t = -xAx^t$ and thus $xAx^t = 0$. This means that $v \le 0$. Similar reasoning proves that $v \ge 0$. Therefore $v = 0$.

3. (a) The game matrix is $A = (a_{ij})$, given by

   $$a_{ij} = \begin{cases} (i-j)/n + ij/n^2 & \text{if } i < j. \\ (i-j)/n - ij/n^2 & \text{if } i > j. \\ 0 & \text{if } i = j. \end{cases}$$

   for $i,j = 1, 2, \ldots, n$. (It is clearly wrong to shoot at 0.)

   (b) For $n = 2$, any strategy is optimal.
   For $n = 3$, $x = y = (0, 1, 0)$.
   For $n = 4$, $x = y = (0, 1, 0, 0)$.
   For $n = 5$, $x = y = (0, 5/11, 5/11, 0, 1/11)$.

4. There are six strategies for this game, corresponding to the six permutations of three cards. The game has a saddle point, corresponding to the strategy of placing ace on ace, deuce on deuce, trey on trey.

5. The optimal strategy is: 1/3 (deuce on deuce, trey on trey, four on four), 1/3 (trey on deuce, deuce on trey, four on four), 1/3 (trey on deuce, four on trey, deuce on four).

6. Blotto's optimal strategy is: 4/7 (one division), 2/7 (two divisions), 1/7 (three divisions). His opponent's optimal strategy is (4/7) (no divisions), 2/7 (one division), 1/7 (two divisions). Value 1/7.

PAGE 299

1. $x = (0, 2/3, 0, 1/3)$, $y = (1/2, 0, 0, 1/2)$, $v = 3$.

2. $x = (14/27, 0, 10/27, 1/9)$, $y = (0, 0, 1/9, 11/27, 13/27)$, $v = 103/27$.

3. $x = (2/5, 3/5, 0, 0)$, $y = (3/5, 0, 0, 2/5, 0)$, $v = 11/5$.

5. The actual solution is $(16/11, 0, 0, 18/11, 0)$, with value $152/11$.

PAGES 310–311

1. The intensity and price vectors will be any vectors proportional to the x and y given.

   (a) $x = (4.27, 7.27)$, $y = (6.13, 5.41)$, $\alpha = 1.14$.

   (b) $x = (4.53, 7.59)$, $y = (6.59, 5.53)$, $\alpha = 1.53$.

(c) $x = (7.60, 10.24)$, $y = (11.60, 6.24)$, $\alpha = 1.12$.

(d) $x = (3,4)$, $y = (2,1)$, $\alpha = 2$.

2. At equilibrium, the production should be 8.944 lb. of fuel for each T.-mi. of transportation. The expansion rate is 5 or 2.236. Each T-mi. of transportation should cost as much as 8.944 lb. of fuel.

3. The new expansion rate is 1.90. Each T-mi. of transportation should cost as much as 8.62 lb. of fuel.

4. Each T-mi. should cost as much as 8.43 lb. of fuel.

PAGES 315–316

1. (a) $x = (1, 0)$, $y = (0, 1)$.

   (b) $x = (1, 0)$, $y = (1, 0)$. Also $x = (0, 1)$, $y = (0, 1)$, and finally, $x = (1/2, 1/2)$, $y = (1/2, 1/2)$.

   (c) $x = (4/5, 1/5)$, $y = (2/5, 3/5)$.

   (d) $x = (0, 1)$, $y = (1, 0)$. Also $x = (1, 0)$, $y = (0, 1)$, and, finally, $x = (3/4, 1/4)$, $y = (3/5, 2/5)$.

2. (b) the expected payoffs are $-1/3$, $1/3$.

   (c) The expected payoffs are then $+0.32$, 0.34.

PAGES 327–328

1. The value is 13/42 for the chairman, 1/6 for each senior member, and 1/14 for each junior member.

2. With power to break ties, 1/2. With veto power, 3/4.

3. This strengthens them (their value goes from 1/5 each to 1/2 total.)

4. This weakens them (value goes from 1/5 each to 1/4 total).

PAGES 340–341

1. Assign one man to job 1, none to job 2, one to job 3, eight to job 4. Revenue: 106.

2. Send one gadget, two widgets, three beepers. The expected revenue is then $3895.

3. (a) $\begin{pmatrix} 0 & 0 & 0 & 4 & 4 \\ 3 & 3 & 5 & 0 & 2 \end{pmatrix}$ cost 74

   (b) $\begin{pmatrix} 0 & 7 & 3 & 3 & 0 & 0 \\ 5 & 0 & 5 & 0 & 2 & 5 \end{pmatrix}$ cost 92

   (c) $\begin{pmatrix} 3 & 0 & 3 & 3 \\ 0 & 4 & 2 & 0 \end{pmatrix}$ cost 57

PAGES 347–348

1. Produce 640 T at the beginning of the first and fifth months.

2. Produce 5000 cans each month.

3. Produce 1200 lb. the first month, 800 lb. the fourth month.

4. Produce 1000 in August, 600 in December.

PAGES 352–353

1. Keep 5 cars in stock at all times (i.e., at the beginning of each month, raise stock to 5).

2. Order 5 in July. In August, if level falls to 1 or less, raise the level to 3; otherwise, no order. In September, if level falls to 0, order 1. Otherwise no order.

3. Assume model year starts in January. For January, order 4, and, until September, raise stock to 4 whenever it falls to 0. For October, raise to 3 if it falls to 0. November, raise to 2 if it falls to 0 or 1. For December, raise to 1 if it falls to 0.

PAGES 363–364

1. (a) Early start time: $a,0$; $b,0$; $c,4$; $d,4$; $e,7$; $f,4$; $g,7$; $h,8$; $i,10$; $j,13$.

    (b) Late start times: $a,0$; $b,0$; $c,4$; $d,6$; $e,7$; $f,5$; $g,7$; $h,8$; $i,10$; $j,13$.

    (b) There are two critical paths, namely, a,b,c,e,i,j, and a,b,c,g,h,i,j.

    (c) 13.

2. (a) 13 weeks.

    (b) a,b,d,f,i,k.

3. All are zero, except $c,5$; $h,8$; and $j,5$.

4. All are zero except $c,5$; $h,6$; $j,3$.

PAGE 368

1. The minimal route is 0, 1, 5, 10, 13. Its length is 2021.

2. The minimal route is 0, 1, 3, 6, 9, 11, with length 12.

PAGES 372–373

1. Arcs (2,9), (1,2), (1,10), (8,10), (6,8), (6,7), (3,7), (4,5) and (2,5).

2. (Fig. VIII.4.7) Arcs (A,B), (A,G), (B,F), (B,C), (C,E), (D,G).
    (Fig. VIII.4.8) Arcs (A,B), (A,C), (C,E), (B,G), (D,G), (D,F), (D,H).
    (Fig. VIII.4.9) Arcs (A,D), (A,H), (B,C), (C,H), (C,F), (D,E), (G,H).
    (Fig. VIII.4.10) Arcs (A,C), (B,C), (C,E), (C,I), (D,F), (F,G), (G,H), (H,L), (H,M), (I,J), (I,K), (K,R), (I,L), (M,Q), (N,P), (R,S), (M,N).

1.

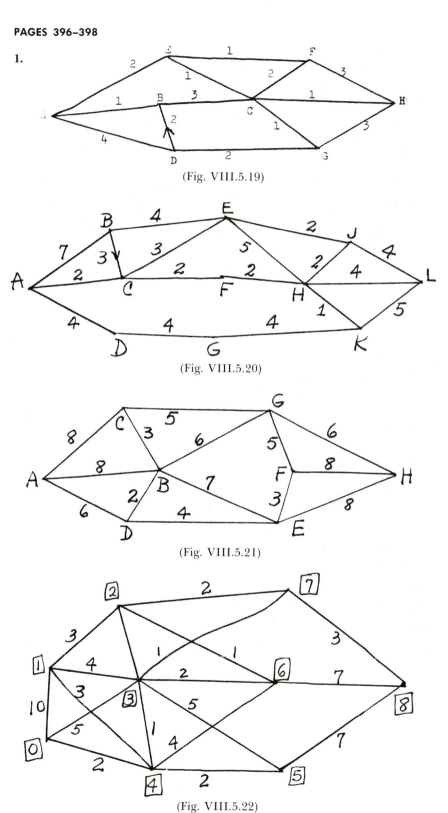

(Fig. VIII.5.19)

(Fig. VIII.5.20)

(Fig. VIII.5.21)

(Fig. VIII.5.22)

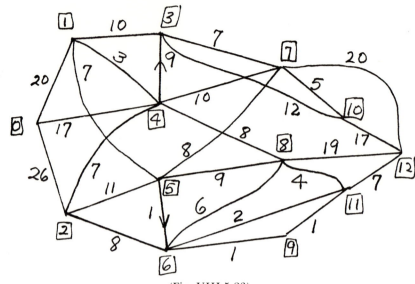

(Fig. VIII.5.23)

2.	To	1	2	3	4	5	6	7	8	9	10	11	12	13
	From													
	0	29	29	3	0	0	0	0	0	0	0	0	0	0
	1	–	0	0	12	17	0	0	0	0	0	0	0	0
	2	0	–	0	20	1	8	0	0	0	0	0	0	0
	3	0	0	–	0	0	3	0	0	0	0	0	0	0
	4	0	0	0	–	0	0	7	25	0	0	0	0	0
	5	0	0	0	0	–	0	0	0	10	8	0	0	0
	6	0	0	0	0	0	–	0	0	4	7	0	0	0
	7	0	0	0	0	0	0	–	0	5	0	0	0	2
	8	0	0	0	0	0	0	0	–	0	0	25	0	0
	9	0	0	0	0	0	0	0	0	–	0	5	14	0
	10	0	0	0	0	0	0	0	0	0	–	0	15	0
	11	0	0	0	0	0	0	0	0	0	0	–	0	30
	12	0	0	0	0	0	0	0	0	0	0	0	–	29

Total flow 61.

DATE DUE